T0317662

Theory and Practice of Additive Manufacturing

Theory and Practice of Additive Manufacturing

Tuhin Mukherjee
Pennsylvania State University

Tarasankar DebRoy
Pennsylvania State University

Library of Congress Cataloging-in-Publication Data
Names: Mukherjee, Tuhin, author. | DebRoy, Tarasankar, author.
Title: Theory and practice of additive manufacturing / Tuhin Mukherjee, Tarasankar DebRoy.
Description: Hoboken, New Jersey : John Wiley & Sons Inc., [2024] | Includes bibliographical references and index.
Identifiers: LCCN 2023018557 (print) | LCCN 2023018558 (ebook) | ISBN 9781394202263 (hardback) |
    ISBN 9781394202324 (adobe pdf) | ISBN 9781394202270 (epub)
Subjects: LCSH: Additive manufacturing.
Classification: LCC TS183.25 .M85 2024 (print) | LCC TS183.25 (ebook) | DDC 621.9/88--dc23/eng/20230506
LC record available at https://lccn.loc.gov/2023018557
LC ebook record available at https://lccn.loc.gov/2023018558

Cover Image: Courtesy of 3D Systems
Cover Design: Wiley

Set in 9.5/12.5pt STIXTwoText by Integra Software Services Pvt. Ltd, Pondicherry, India

Contents

Preface

In the past few decades, additive manufacturing, also known as 3D printing, has gained wide acceptance in various industries because of its ability to make important components that cannot be made by other processes easily and economically. Today, additive manufacturing technologies are advancing so rapidly that it is challenging to keep up with the details of the processes in a textbook. This textbook on the *Theory and Practice of Additive Manufacturing* addresses this issue by combining the key features of the existing technologies with the underlying scientific theories. Connecting practice with the theory can explain not just how things work but why allowing readers to critically examine potential improvements based on solid scientific principles.

The theoretical basis of additive manufacturing has its origin in fusion welding, cladding, and prototyping. For metallic materials, the theories of heat and fluid flow, the evolution of microstructure and mechanical properties, defect formation, and the accumulation of residual stresses and distortion are important to understand both the process and the product. Besides metallic materials, a basic scientific understanding of additive manufacturing of ceramics, polymers, and composite materials is also included in the book. Common practices related to sensing and control, design, qualification, safety, sustainability, and economic issues are also covered.

The engineering and scientific backgrounds of educators, working professionals, and graduate and undergraduate students interested in additive manufacturing are diverse. This book provides a good range of depth of discussions, many worked-out examples, practice problems, and aids to promote visual learning to accommodate the diversity of learners' backgrounds. Some engineering problems have specific solutions and others have a variety of open-ended solutions. The worked-out examples demonstrate a variety of solutions in important applications of additive manufacturing, considering fundamental scientific principles.

In the past, the progress of manufacturing technologies such as casting, welding, machining, and metal forming have followed paths of incremental advancements, in many cases by time-consuming trial–and-error testing. Today's manufacturing culture is benefiting from unrelated developments in emerging digital tools such as mechanistic modeling, artificial intelligence, and information technology. By including these emerging technologies, this textbook will help learners to appreciate this new path for advancement of additive manufacturing that incorporates digital tools.

We are grateful to all those who have encouraged and supported the writing of this book. Many colleagues, friends, and students have generously provided comments and contributed valuable suggestions. Special thanks go to Professors Harry Bhadeshia, Kwadwo Osseo Asare, Amitava De, Bikramjit Basu, Dr. Qianru Wu, and John Milewski. We are grateful to our families for their encouragement, understanding, and the warmth of smiles.

We welcome reader feedback to help us enhance this book in a future edition.

Tuhin Mukherjee and Tarasankar DebRoy,
University Park, Pennsylvania.

1

Introduction

Learning objectives

After reading this chapter the reader should be able to do the following:

1) Understand the principles of additive manufacturing and the commonly used terms in this field.
2) Appreciate the unique capabilities of additive manufacturing in overcoming limitations of conventional manufacturing.
3) Understand the common uses of metallic, ceramics, polymeric, and composite printed parts in different industries.
4) Comprehend the scientific synergies and differences between additive manufacturing and multi-pass fusion welding.
5) Understand the roles of computers and the emerging digital tools in additive manufacturing.
6) Recognize the important scientific, technological, and commercial issues in additive manufacturing that need to be addressed for its continued growth.

CONTENTS

Theory and Practice of Additive Manufacturing, First Edition. Tuhin Mukherjee and Tarasankar DebRoy.
© 2024 John Wiley & Sons, Inc. Published 2024 by John Wiley & Sons, Inc.

1.1   What is additive manufacturing?

Additive manufacturing (AM) is a process of building parts by progressively adding thin layers of materials, sometimes layers thinner than a human hair. Computers play a central role in AM because the printing process is guided by a digital model. Imagine a computer slicing a three-dimensional object into many parallel thin slices, figuring out how to print each slice one after the other, and then having a mechanism to combine each layer with those previously deposited. Parts are made with metals, ceramics, polymers, and composite materials. There are many types of additive manufacturing. The type of material printed, its size, cost competitiveness, and other part attributes all influence the choice.

No specialized cutting tools, dies, molds, or other equipment are needed although depending on the part, some post-processing may be required. The process can produce unique parts such as the ones that are hard on the surface and soft in the core made with site-specific chemical composition; or parts that have complex internal cooling channels that are hard to make by other means. Complicated shapes that cannot be made by conventional methods in one go can be produced by depositing materials layer by layer. For example, the complex shape in Figure 1.1 cannot be made by machining a solid block. It could be made in multiple parts and then joined together to create the overall shape followed by intricate machining to get the required dimensions and surface topology. However, this process will be expensive and time-consuming.

Figure 1.1   An abstract-shaped bronze-tinted, stainless steel bottle opener produced by depositing thin metal layers layer-by-layer [1]. Image courtesy: Bathsheba Grossman.

In additive manufacturing, the same equipment can produce diverse components for different industries. This flexibility to select products is important because it enables the equipment to make critical parts for which the supply chain no longer exists. For example, an AM facility can start the production of any part as soon as a digital design is available. This can have life-changing consequences. The COVID-19 pandemic led to a massive shortage of components for ventilators and equipment for the protection of medical staff. AM facilities in many parts of the world switched to the production of these vital products, sometimes with innovative design features.

Since a variety of parts can be produced on demand, the cost of maintaining a sizeable inventory can be reduced. If the supply chain for a part did not exist, it was considered obsolete and not usually obtainable. That is no longer so because AM allows the printing of just a few parts without any new equipment and tooling. A case in point is an impeller additively manufactured by Siemens for a nuclear power station for which spare parts were no longer available. Patient-specific implants, customized parts, and repair of existing equipment have also become possible in many cases.

In summary, there are many advantages of additive manufacturing over conventional manufacturing [2-4]. They include parts that previously required assembly of multiple individual parts that can be printed in one step; printing of customized parts as soon as the digital drawing is available thus reducing inventory; parts with site-specific composition and properties can be manufactured; printing of intricate parts containing internal features such as cooling channels can be made. Several selected examples of the unique capabilities of AM are presented in Worked out example 1.1. Because of these advantages, AM is now widely used to make parts for various industries. The market revenue, the growing sales of metal printing machines in recent years, and the number of patents granted globally show the growing acceptance of AM in the industry.

Worked out example 1.1

Do a literature search and give three examples where the unique capabilities of additive manufacturing were shown to have significant advantages over conventional manufacturing.

Solution:

a) A ratcheting socket wrench was requested by the crew of the International Space Station when no supply mission was planned. A digital file was sent electronically by NASA to an AM machine located in the Space Station. The machine then built a wrench including the movable parts in 104 layers from the plastic feedstock. After printing, an astronaut could use it in the space station (see Figure E1.1 below). This is an example of manufacturing a part on demand that could not have been imagined before the advent of additive manufacturing.

Figure E1.1 Bruce Wilmore, commander of the international space station showing a ratchet wrench made using a 3D printer. Image credit: NASA.

b) General Electric, a leading producer of jet engines, now prints fuel injector nozzles for jet engines (see Figure E1.2 below [3]) in Alabama, USA, in one piece that previously required assembly of about 20 parts. Manufacturing of these fuel nozzles by AM has avoided the assembly of multiple parts and increased the reliability of jet engines.

c) A compositionally graded material (see Figure E1.3 below) produced by AM can replace a dissimilar metal joint (for example, joints of steel and nickel-iron-chromium alloy 800H) often used in power plants. The compositionally graded alloy is thought to increase the creep life (see Chapter 9) by hundreds of years over the dissimilar metal joints [4]. In this material, the alloy composition varies across the length. The compositionally graded parts are difficult if not impossible to produce by conventional manufacturing such as casting.

(Continued)

Worked out example 1.1 (Continued)

Figure E1.2 Additively manufactured fuel injector nozzles for jet engines produced by General Electric [3]. The figure is taken from open-access articles under the terms and conditions of the Creative Commons Attribution (CC BY) license.

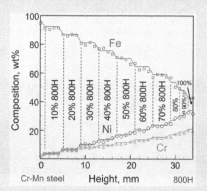

Figure E1.3 Variations in composition across a compositionally graded joint between a Cr-Mn steel and a nickel alloy 800H. Image source: T. DebRoy.

1.2 Terminology

The name "additive manufacturing" contrasts with most other manufacturing processes such as machining or drilling used to remove materials from a larger piece to make a smaller, value-added part. Here, machining or drilling may be viewed as a subtractive process. Additive manufacturing is also known by many other names. For example, additive manufacturing is also called 3D printing. Although rapid prototyping is sometimes used interchangeably with additive manufacturing, it is not a manufacturing process because the goal is to make a single part for display, analysis, and conceptualization. However, additive manufacturing can be used for rapidly making a prototype.

Printing, i.e., placing ink on the surface of a paper is essentially a two-dimensional process. 3D printing extends the printing process to a third dimension, i.e., along the build direction, up above a flat substrate on which the part is commonly made. Thus, the keyword "3D" adds a new third dimension. Polymer, ceramic, and composite materials are printed but metal printing is the fastest-growing area of AM because of its widespread practical applications. The term "metal printing" is widely used, particularly in scientific literature. Lithography is a process of printing text or pictures using a metal or polymer plate on which an image is made using a polymer coating. The printing can be done directly from the plate or can be offset to a flexible material such as a sheet of rubber from which the image can be printed. While lithography is a printing process, the term "stereo" adds a third dimension, and stereolithography is often used to describe additive manufacturing.

The first step of manufacturing a product is building a prototype or a trial part to check out that the design of a part represented in a drawing is correctly replicated in the product. The traditional way of building a prototype is slow and expensive because anything made for the first time from a drawing may not correctly reflect all the functionality intended in the design. The use of additive manufacturing (AM) also expedites the production of a prototype.

AM is often called freeform fabrication to emphasize the ease with which various geometry or "form" of a part can be manufactured using the same machine. For example, making a complex part with internal channels can be often made as easy as making a simple cube. In other words, the geometrical complexity of a part is easier to accommodate in AM than in conventional manufacturing and complex parts that cannot be easily made in conventional manufacturing can be made in AM. The ability to handle complex part geometry is an important advantage of AM but the geometrical freedom is not the only factor for the selection of AM. The following Worked out example 1.2 illustrates the similarities and differences between additive manufacturing, rapid prototyping, freeform fabrication, and stereolithography.

Worked out example 1.2

Briefly discuss the similarities and differences between additive manufacturing, rapid prototyping, freeform fabrication, and stereolithography.

Solution:

The process of adding materials layer by layer is adequately described by the term additive manufacturing or in short, AM. This term is widely used in contemporary literature and this book. However, this technology is also known by several other names such as rapid prototyping, freeform fabrication, and stereolithography to describe the context and emphasize one or more aspects of the same manufacturing technology.

Features	Additive manufacturing/3D printing	Rapid prototyping	Freeform fabrication	Stereolithography
Number of parts	One or more parts on demand; not limited by the economy of scale	Made to test one part, not always a finished part	One or more parts; a complex shape can be produced as easily as a cube	One or more parts
Material(s)	Metals, ceramics, polymers, composites	Polymers, glass fiber, carbon fiber, metals, wood	Metals, ceramics, polymers, composites	Thermoset plastic, photopolymers

(Continued)

	Worked out example 1.2 (Continued)			
Features	**Additive manufacturing/3D printing**	**Rapid prototyping**	**Freeform fabrication**	**Stereolithography**
Shapes	Curved surfaces may be approximated by small steps and triangles	Building a prototype of any shape quickly, often by AM	Emphasizes independence of shapes and complexity from process	The emphasis is on extending 2D printing to 3D
Design files	Most machines make parts from .STL files prepared from CAD files	Many machines can use CAD files directly but the AM machines use larger .STL files	.STL format is used but new formats are developed to avoid redundancies and improve efficiency	.STL stands for StereoLithography or Standard Tessellation Language
Hardware	Many variants of the process use different hardware, see Chapters 2, 3, and 4.	AM machines, subtractive processes such as CNC machines and stack and cut thin sheets of solid feedstock	AM machines, laminated stacks of shaped thin sheets, Stereolithography equipment	Equipment that use photopolymer as a binder of ceramic and metal slurries and sintering furnaces.

1.3 Uses

AM of metallic materials is the fastest growing sector because of the significant demand for these parts, although ceramic, composite, and polymer parts are also used in the industry. This section provides examples of different applications of AM for the printing of different types of materials.

1.3.1 Metals and alloys

AM is now widely used to make metallic parts [5] for consumer products, healthcare, aerospace, energy, marine, automotive, and other industries (Table 1.1). Figure 1.2 shows the distribution of revenues from these industries. These products manufactured by AM have shown significant advantages over conventional methods. AM is used to manufacture heat exchangers, compressors, and turbine parts of aero engines (Figure 1.3 (a-c)). The aerospace industry now uses many lightweight parts that result in significant cost savings (Figure 1.3 (d-f)). AM can now be used to fabricate large parts. For example, NASA printed a rocket nozzle with 60″ (1.52 m) diameter and 70″ (1.78 m) height in 90 days using a laser-based directed energy deposition process of Inconel 718 (Figure 1.4 (a-b)). NASA has also printed large-scale rocket nozzle parts (Figure 1.4 (c-e)) using a laser-based powder bed fusion process made of copper-chrome-niobium and iron-nickel alloys. AM parts are also widely used in automotive (Figure 1.5 (a)), healthcare (Figure 1.5 (b)), and energy industries (Figure 1.5 (c)). Patient-specific medical and dental implants can be made with biocompatible alloys using data from the patient's medical imaging. Additively manufactured metallic gauges are used in the automotive industry for precise measurements of complex parts. Worked out examples 1.3 and 1.5 show how alloys are selected for specific additive manufacturing applications. Worked out example 1.4 discusses why the market penetration of AM is limited in spite of its success over conventional manufacturing.

Table 1.1 Common additive manufacturing alloys and their applications [9].

Applications	Aluminum	Maraging steel	Stainless steel	Titanium	Cobalt-chrome	Nickel alloys	Precious metals
Aerospace	X		X	X	X	X	
Medical implant			X	X	X		X
Energy, oil, and gas			X				
Automotive	X		X	X			
Marine			X	X		X	
Machinability and weldability	X		X	X		X	
Corrosion resistance			X	X	X	X	
High temperature			X	X		X	
Tools and molds		X	X				
Consumer products	X		X				X

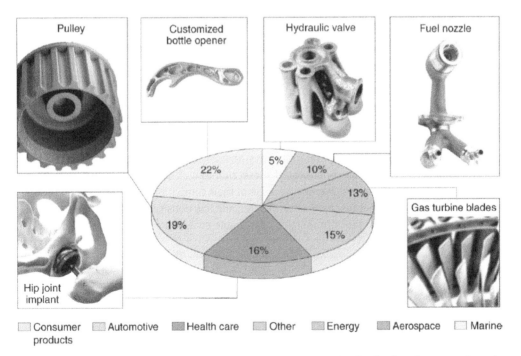

☐ Consumer products   ☐ Automotive   ▨ Health care   ☐ Other   ▨ Energy   ▨ Aerospace   ☐ Marine

Figure 1.2 Applications of metal printing in various industries and the distribution of revenues from the printed parts among various industries. *Source:* T. Mukherjee and T. DebRoy.

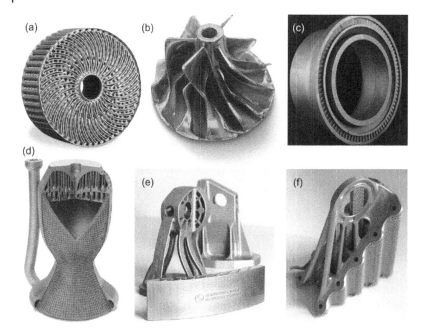

Figure 1.3 Additively manufactured alloy parts in aerospace applications. (a) Component for an aircraft heat exchanger made of an aluminum-silicon alloy by 3D Systems [image courtesy: 3D Systems]. (b) An impeller for a gas turbine engine made of stainless steel 316 by 3D Systems [image courtesy: 3D Systems]. (c) An engine liquid oxygen turbopump stator of a spacecraft printed by NASA [3]. (d) A prototype rocket nozzle of Inconel 718 featuring internal cooling channels [3]. (e) A stainless-steel bracket optimized for weight reduction (front) and the traditional cast bracket in the back [6]. (f) A titanium alloy bracket before support removal oriented at a 45° angle for Airbus A350 manufactured by Concept Laser [7]. The figures (c-f) are taken from open-access articles under the terms and conditions of the Creative Commons Attribution (CC BY) license.

Worked out example 1.3

Recommend an alloy for a lightweight dental implant.

Solution:

Table 1.1 shows that stainless steel, titanium alloys, and cobalt-chrome alloys are used to print medical implants. They are biocompatible. Of these, titanium alloys, particularly Ti-6Al-4V is the material of choice for a dental implant because of their compatibility with bone and tissue, corrosion resistance, excellent mechanical properties, and low density. Ti-6Al-4V is widely used as a dental implant for its superior properties, reasonable cost, and excellent durability.

Worked out example 1.4

AM overcomes many of the difficulties of conventional manufacturing and yet only 0.1% or so of the manufactured products are now made by AM. Discuss the main reasons for the limited market penetration by AM so far.

Solution:

- Applications of the commonly used manufacturing technologies have developed over many decades. It is time consuming to develop cost-competitive applications of a new technology such as additive manufacturing. The data on the sales of AM equipment and the number of patents indicate promise for the growth of AM.

Worked out example 1.4 (Continued)

- AM needs appropriate feedstock materials, mostly in powder and wire forms. The availability of feedstock materials for most commercial alloys is limited by the market demand for these materials. The feedstock materials are made by conventional processes and a certain minimum demand for these materials is a prerequisite for their manufacture.
- Control of microstructure, properties, and defects of printed metallic parts are still evolving and not well understood. Solutions of these difficulties are time consuming and expensive.
- AM has its share of scientific, technological, and commercial issues that need to be solved to produce and qualify commercial parts. Susceptibility to defect formation and serviceability of components affect cost competitiveness.
- High cost of machines, particularly for metal printing, the time consuming and expensive qualification process, the need for quality consistency, post processing, and defect minimization favors large corporations over small and medium sized companies. The limited participation of small and medium size companies limits the development of AM processes.

Worked out example 1.5

Corrosion resistance is an important criterion to select an alloy for printing both aeroengine parts and medical implants. Both cobalt-chrome alloys and nickel alloys have excellent corrosion resistance (Table 1.1). Why are cobalt-chrome alloys used in the medical industry, but nickel alloys are popular in the aerospace industry?

Solution:

Cobalt-chrome alloys are biocompatible which is an essential criterion for medical implants. In addition, they have good wear resistance and their Young's modulus is higher than those of titanium alloys. Although Co-Cr alloys are commonly used in hip replacement, cobalt poisoning has been reported in rare instances. Nickel base superalloys can retain good mechanical properties at high temperatures. Therefore, they are excellent choices for high temperature applications in the aerospace industry such as gas turbine blades (Table 1.1). They have high corrosion and wear resistance and excellent toughness.

1.3.2   Polymers, composites, and ceramics

Several parts such as medical implants require high strength, thermal and wear resistance as well as biocompatibility. Ceramics are often used to fulfill these requirements. Limitations on the geometry and shrinking during the conventional manufacturing processes of ceramics hinder their use in various applications. In addition, high costs are associated with some conventional production methods such as milling because of high tool wear and low material yield. These limitations can often be overcome by additively manufactured ceramic materials (see Chapter 4). AM allows layer-wise fabrication of complex components with high densities and good strengths. Because of these advantages, additively manufactured ceramic parts are used in various applications. For example, Figure 1.6 (a-b) shows a ceramic chess piece, the castle, made with internal features with re-entrant angles. Ceramic materials provide the rigidity that polymeric materials lack in most cases. Similarly, the chess piece does not need to withstand stresses and strains, so a metallic material is also not needed. Printed ceramic parts are widely used in medical industries. Biocompatible ceramic bone implants can now be made by AM (Figure 1.6 (c)). 3D-printed dental implants can fulfill the requirement of good surface quality, hardness, biocompatibility, and aesthetics.

Figure 1.4 (a) and (b) Example of large-scale additive manufacturing. NASA printed a rocket nozzle with 60" (1.52 m) diameter and 70" (1.78 m) height of Inconel 718 in 90 days using a laser-based directed energy deposition process [3]. (c) and (d) A bimetallic chamber for the rocket nozzle made of copper-chrome-niobium and iron-nickel alloys. Figure (c) shows the chamber with complete manifolds [3]. (e) A rocket nozzle with a covering jacket printed by NASA using a laser-based powder bed fusion process [3]. The figures are taken from an open-access article [3] under the terms and conditions of the Creative Commons Attribution (CC BY) license.

Figure 1.5 (a) A car frame made of AlSi10Mg by 3D Systems [image courtesy: 3D Systems]. (b) A titanium alloy orthopedic implant fabricated by 3D Systems [image courtesy: 3D Systems]. (c) Additively manufactured gas turbine blade assembly made by Inconel 718 [8]. The figure is taken from an open-access article [8] under the terms and conditions of the Creative Commons Attribution (CC BY) license.

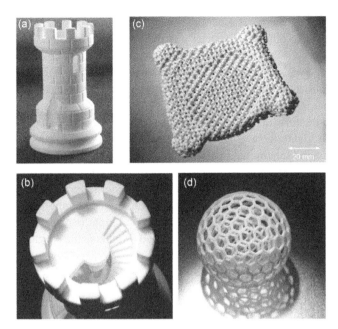

Figure 1.6 (a-b) A ceramic chess piece that is a castle, with a spiral staircase leading to the battlement, accompanied by a spiral handrail at the center. *Source:* T. DebRoy and H.K.D.H. Bhadeshia. (c) A biodegradable bone implant of β-tricalcium phosphate/polylactide ceramic [10]. The figure is reprinted with permission from Elsevier. (d) Conceptual artwork by printing of a polymer [11]. The figure is reprinted with permission from BMJ.

AM offers a means to fabricate metal matrix composites by adding small amounts of nonmetallic particles to the metallic feedstock. Examples include steel, aluminum, and titanium alloy-based metal matrix composites where Fe_2O_3, SiC, ZnO, Si_3N_4, and Mo_2C are added to form the composites. Additively manufactured metal matrix composites are uniquely made to achieve a high strength-to-weight ratio, good fatigue, wear, and corrosion resistance, low creep rate at elevated temperatures, and low coefficient of thermal expansion to reduce distortion. Because of these unique properties, additively manufactured composite parts are used in a variety of applications ranging from tools and molds to actual functional components such as helicopter blades, lightweight bicycle frames, and rocket parts. Depending on the shape, size, and type of the matrix and additive materials used to make the composite, a wide variety of AM processes are used to print parts of composite materials (see Chapter 3). In addition, the flexibility in adding materials during printing helps in rethinking the material choice for certain applications, allowing manufacturers to make more durable and cheaper composites.

Polymers and other soft materials are printed to produce sculptures, electronic materials, prototypes, and medical parts. For example, Figure 1.6 (d) shows a printed polymeric sculpture. Commonly used polymers in additive manufacturing are Acrylonitrile Butadiene Styrene (ABS), Polylactide (PLA), nylon, Polycarbonate (PC), and Polyvinyl Alcohol (PVA). Acrylonitrile Butadiene Styrene (ABS) has excellent heat resistance and good mechanical properties and is used to print parts for automobile and aerospace industries. Polycarbonates (PC) have excellent tensile strength, making them ideal for printing impact-resistant materials. They are often reinforced with carbon to make stronger and more heat-resistant components. Polyvinyl Alcohol (PVA) is used to print support structures because of its water-soluble properties. Once the part is printed, the polyvinyl alcohol is simply washed with warm water, leaving the main part intact. Polymer parts are printed using two main methods that are described in detail in Chapter 3. Heat processing methods involve melting the polymer partially or fully using a laser or other heat sources and then cooling the material to form the component. In contrast, light processing using liquid photopolymers is used to create a chemical reaction that forms new chemical bonds inside the polymers which solidify to fabricate the component. Printed polymeric components often need post-processing such as grinding, polishing, and painting to improve their appearance and aesthetics.

Additive manufacturing is also used to fabricate polymer composites, which are made up of a polymer matrix and reinforcement materials such as fibers or particles. AM of polymer composites allows for the creation of complex and customized parts. Additionally, it can create parts with improved properties such as increased strength-to-weight ratio, corrosion resistance, and stiffness, making it useful in various industries. Several major products include spacecraft components, such as wing and fuselage panels, and rocket motor casings, automotive components such as bumpers and exhaust systems, sporting goods such as golf clubs, tennis rackets, and fishing rods, and biomedical applications such as prosthetic limbs, dental implants, and surgical instruments.

1.4 Scientific synergy with welding

For metallic materials, fusion welding and the main variants of additive manufacturing share many similarities in their physical processes. Both processes use moving heat sources. The heat sources used in additive manufacturing, i.e., lasers, electron beams, electric arcs, and plasma arcs have been used in welding for a long time before the advent of AM. Both processes involve the melting and solidification of metallic materials. In many instances, the fusion welding processes use multiple passes to fabricate a joint which is similar to the layer-by-layer printing of parts in additive manufacturing. The heating, melting, solidification, and cooling of metallic materials affect the evolution of microstructure and properties of welds and AM parts. Defects such as porosity, lack of fusion, cracking, loss of volatile alloying elements, and the evolution of residual stresses and distortion are common in both welding and additive manufacturing.

Metal additions are an integral part of gas-metal-arc, electroslag, and other welding processes. Thick metal plates have been routinely joined by adding molten filler metals in a welding groove. Multiple tracks of molten filler metals are deposited next to each other with sufficient overlap to avoid any vacant space between the tracks. Successive layers are then deposited over the previously deposited layers and the process continues layer upon layer until the groove is filled with many layers of filler metals. Figure 1.7 shows the deposition of 42 robotically deposited welded tracks to fill a 1″ by 1″ square groove in a 9″ outer diameter metal pipe using gas metal arc welding [12]. Although metal additions have been important in the manufacturing of thick section joints, only the joints and not the entire pipes were made additively. What has changed in recent decades is that the entire parts are now made by adding metal layers, layer upon layer. As an emerging technology, AM faces many scientific problems, and the rich knowledge base of fusion welding is an important resource for their solution (see Worked out example 1.6 for a detailed discussion). In spite of many similarities between fusion welding and AM, there are many differences between these two processes as explained in Worked out example 1.7.

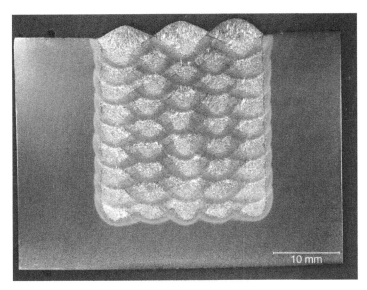

Figure 1.7 A pulsed gas-metal arc weld cross-section [12] showing a square groove 1" wide and 1" deep containing 42 deposited weld beads using pulsed gas metal arc welding. The groove is a part of a 9" outer diameter pipe welded at a heat input of 0.6 kJ/mm. Figure courtesy: Ms. Rebecca Gurk (Marketing Manager, EWI).

Worked out example 1.6

In Figure 1.7, a square groove 1" wide and 1" deep was filled up by multi-pass welding where 42 passes were required. What are the most important factors a welding engineer should consider to determine the required number of passes? How can the knowledge of multi-pass welding be useful to make a large part using wire-arc additive manufacturing (see Chapter 2)?

Solution:

The number of passes required to fill up a groove of a given size using multi-pass welding is controlled by the filler metal addition rate. For a given alloy, this rate depends on the heat input, which is defined as the ratio of the power of the heat source to the welding speed. The net addition of the filler metal can be estimated from the length of all the welding tracks times the cross-section of the molten zone after correcting for the remelting of both the previously deposited and the adjacent tracks and their mixing with the feedstock known as dilution. For a given alloy, heat input and the track location control both the fusion zone cross section and the extent of dilution. For a given alloy, the selection of power and welding speed determines the number of passes required for a multi-pass welding process.

In additive manufacturing where parts are made by depositing layer upon layer of materials, the knowledge of fusion zone cross-section and dilution will be useful to roughly estimate the number of tracks necessary to manufacture a part. However, the differences in the part geometry and the weld groove geometry will affect both the fusion zone cross-section and the dilution because of the differences in the heat transfer rates from the molten pool to its surroundings. As a result, the estimations of the number of tracks will be approximate. In addition, in many instances, the machine manufacturers will recommend the current, voltage, and scanning speed considering not only the deposition rate but also the need to avoid defects and minimize part distortion.

Worked out example 1.7

Much has been discussed in the literature about the synergy between welding and additive manufacturing. What are some of the main differences between the two processes?

Solution:

The main differences are itemized below.
a) Welding is used to join two semi-finished parts together to make a larger part, whereas AM fabricates a single component layer-by-layer that may or may not be later welded or joined to other components.
b) Weld metals are commonly deposited in wire form. However, both wire and powder feedstocks are used in AM.
c) Powder-based AM processes involve very rapid scanning which results in a higher cooling rate than welding.
d) Powder-based AM processes use a very focused beam to print intricate parts with a good surface finish. Welded parts in many cases need post-process machining to get good surfaces.
e) Keyhole mode welding is used to join thick plates. However, the formation of keyholes is generally avoided in AM to prevent the formation of pores due to keyhole instability.

1.5 The role of computers

Additive manufacturing largely relies on the advanced tools of the digital age in all stages of the product lifecycle (Figure 1.8). The roles of computers, software, and digital techniques in additive manufacturing are discussed below.

1.5.1 Computer-aided part design

After a product is conceived, CAD (computer-aided design) software is commonly used to design, modify, and visualize a design. It allows instant changes to the design thus allowing many alternative designs to be explored. There are three categories of CAD software to choose from depending on the intricacy of the product. The simplest is wireframe modeling which uses lines and curves, a more detailed design is solid modeling which includes mass and volume, and finally, surface modeling which uses connected surfaces. CAD software is complex and expensive, and its effective use requires specialized training.

1.5.2 Conversion of part design to a machine-readable file

Most of the AM machines make parts from a type of computer file that is called STL files, i.e., files with an extension of .STL. The file extension originates from the word stereolithography. However, it is also referred to as "Standard Triangle Language" or "Standard Tessellation Language." The STL files are widely used by various types of additive manufacturing machines. When a part or a prototype is available and the task is to replicate the part using AM, STL files can also be made from the product itself. Geometric data on the surface of a part can be captured using laser scanning or using a touch/contact probe. The STL file can then be made from the scanning or the touch probe data. The internal features are not always captured in these techniques and the files may have to be modified to represent the features of complex parts to include the internal features not

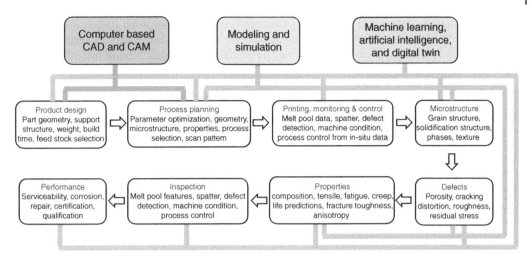

Figure 1.8 Use of computer-based digital tools such as computer-aided design (CAD), computer-aided manufacturing (CAM), modeling and simulation, machine learning, artificial intelligence, and digital twins in different stages of the product lifecycle. *Source:* T. Mukherjee and T. DebRoy.

detected by scanning or touch-probe data. Another approach is to use a high-energy x-ray to image the internal features using Computerized Tomography (CT) technology which can provide a detailed three-dimensional image of the part from which an STL file may be made. For example, a CT scan of a patient is useful to make an STL file for additively manufacturing a patient-specific implant. The CT scan can also be used to reverse engineer a part.

The CAD file containing the 3D design of the part is transformed into an STL file which is then used in the AM machine to print the part (Figure 1.9). The STL file also needs to include the support struc-

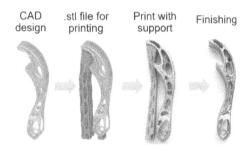

Figure 1.9 Step-by-step process to additively manufacture a part. *Source:* T. Mukherjee and T. DebRoy.

tures, if any are required. They are removed after the printing is done. The STL file format is now the industry standard that is recognized by all AM machines. However, both making CAD files from a design concept and converting these files to error-free STL files require considerable skills. Also, STL was a useful format for the initial phase of the development of AM when parts were made of single-phase homogeneous materials. Now functionally graded materials and multi-materials are used, and their texture and color are important features of the parts made by AM. Also, the STL files are larger than the CAD files and they are susceptible to the introduction of errors while converting from the CAD files. With improved demand for accuracy, approaching micrometer levels in some applications, the number of triangles needed to represent smooth surfaces requires a very large size of the STL files. Because of these difficulties, there is a need for improved file formats to translate CAD files to additive manufacturing machines.

Currently, work is continuing to adapt improved computer file formats with improved functionality and ease of use. For example, Microsoft has developed a file format .3MF for use with the Windows operating system. Unlike STL, it contains data on mesh, texture, and materials. The lack of any universally agreed-upon protocol prevented the adaptation of .3MF or any other file formats by machine manufacturers. ASTM and ISO have approved a plan for a file format .AMF to be managed jointly by ISO and ASTM. This file format can contain data for graded materials, multi-materials, texture, and color and has provisions for including parts that contain multiple objects that cannot be accommodated in standard STL files.

1.5.3 Machine learning and artificial intelligence for process control and quality improvement

Advanced digital tools such as artificial intelligence and machine learning using sophisticated computer programs are of considerable interest in AM. These tools enable computers to make reliable predictions by learning from existing data gathered from various sources. It extracts useful information and relations hidden in this data without any explicit programming. The availability of state-of-the-art, open-source computer programs facilitates the applications of machine learning and artificial intelligence for solving many complex problems in AM that may appear intractable at first. For example, machine learning is often used to monitor and control AM processes and mitigate defect formation. A camera is often used to obtain images of a part to detect flaws by comparing it with the CAD design. If a flaw is detected, the process variables can be adjusted guided by a machine learning algorithm. Machine learning and artificial intelligence are used to minimize AM defects such as porosity, lack of fusion, distortion, and surface roughness to improve part quality. Since the mechanisms of the formation of various types of defects are often not known, machine learning provides an excellent utility for their mitigation based on data. Details of machine learning and artificial intelligence and their applications in AM are discussed in Chapter 12 of this book.

1.5.4 Computer-based modeling and simulation of additive manufacturing

Modeling and simulation [13, 14] using computer programs are used to simulate temperature fields, deposit geometry, solidification, grain growth, microstructure evolution, formation of residual stresses, and defects such as lack of fusion, porosity, distortion, and surface roughness. These defects are often difficult to measure in real time because of the complexity of AM process. It is also time-consuming and expensive to experimentally evaluate processing conditions for a wide range of alloys. Therefore, well-tested computer models are often used for this purpose. For example, Figure 1.10 (a) shows the computed temperature and velocity fields and deposit geometry during AM. Half of the deposit is shown because of the symmetry. The color contours represent the isotherms whose values can be read from the color legend. Velocity vectors representing the flow of liquid metal are shown by black arrows. Figure 1.10 (b) shows the computed grain structure during AM which controls the microstructure and properties of the part. However, the mathematical representation of both the AM process and the product attributes is a challenging task. This complexity is addressed by modeling the most important physical processes and ignoring the less important ones. Recent advancements in computer programs and hardware help implement modeling of AM. Details of computer-based modeling and simulation and their applications in AM are discussed in Chapter 12 of this book.

1.5.5 Digital twins of additive manufacturing

A digital twin which is a digital replica of AM hardware may consist of modeling, sensing and control, machine learning, and big data and can reduce the parameter space for testing, shorten the time between the design and production, and accelerate part qualification. A digital twin is a synergistic combination of advanced digital tools. It can virtually test different combinations of

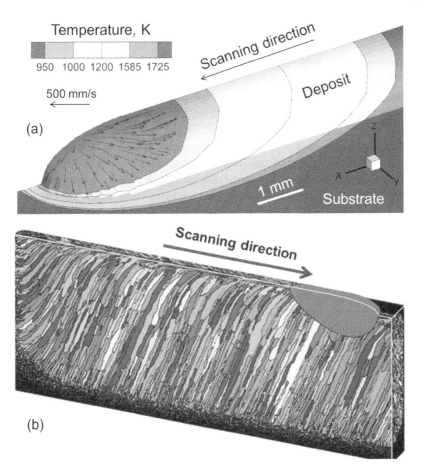

Figure 1.10 Use of computer-based modeling and simulation of AM to simulate (a) temperature and velocity fields and deposit geometry and (b) grain structure. *Source:* T. Mukherjee and T. DebRoy.

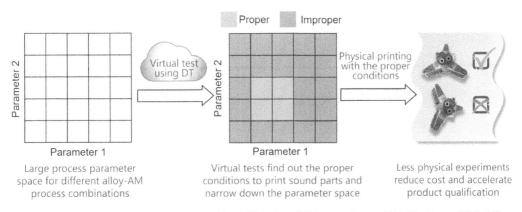

Figure 1.11 Use of digital twin (DT) for rapid qualification of AM parts. *Source:* T. Mukherjee and T. DebRoy.

process variables to find the proper set of conditions that can print structurally sound parts with minimum defects (Figure 1.11). Therefore, it significantly reduces the number of experiments to be performed. Thus, virtual tests using digital twins can accelerate the part qualification process. Details of digital twins and their applications in AM are discussed in Chapter 12 of this book.

1.6 The important scientific issues

Although AM is a rapidly progressing field, like all emerging technologies, AM has its fair share of scientific challenges (Figure 1.12). The AM process affects the microstructure, properties, and performance of metallic parts. The cooling rate, temperature gradients, and solidification rate determine the morphology of grains and their growth pattern, the solid-state phase transformations, the scale of the microstructure, and the nature of the defects that form [8]. The rich and mature knowledge base of metallurgy contains interrelation between processing, microstructure, properties, and performance. However, the relations are insufficient to tailor the microstructure of a part without resorting to complex mechanistic models that are out of reach of most practicing engineers. The main difficulty is that commercial alloys have a remarkable diversity of structure and properties depending on processing conditions including the selection of the AM process and the many process variables such as powder and speed of the heat source, the attributes of the feedstock, and the geometry of the part. The three common metal printing processes (Chapter 2) are laser-assisted powder bed fusion (PBF-L), laser-assisted directed energy deposition (DED-L), and wire-arc-based direct energy deposition (DED-GMA). In these processes, the scanning speed of the heat source and the heat source power vary widely (Figure 1.13 (a)). The scanning speed in PBF-L is about 100 times higher than those typically used in DED-GMA and DED-L. Also, the power

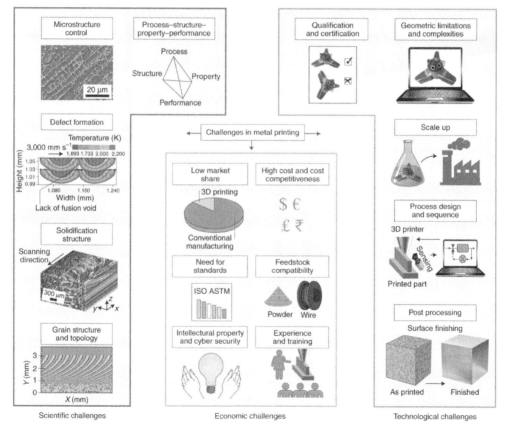

Figure 1.12 The main scientific, technological, and economic challenges in additive manufacturing. *Source:* T. Mukherjee and T. DebRoy.

used in DED-GMA is approximately 10 times higher than those in DED-L and PBF-L. Therefore, it is not surprising that the liquid alloys solidify at very different rates (Figure 1.13 (b)) in these processes. As a result, parts contain diverse shapes and sizes of grains, and different phases in the printed parts depending on the selection of diverse process variables of AM [15].

Grains may have planar, cellular, columnar dendritic, or equiaxed dendritic morphologies depending on the spatial gradient of temperature in the molten alloy, the velocity of the solidification front, and the alloy composition. The reported data of temperature gradients and growth rates vary significantly among the three processes (Figure 1.13 (c)). Therefore, diverse solidification morphologies are to be expected depending on the processing conditions. The importance of the processing conditions also becomes apparent when the hardness of the printed components is examined for a common AM alloy, stainless steel 316, printed using different techniques (Figure 1.13 (d)). Printing process parameters also affect the other mechanical properties such as the ductility of parts. Some trends can be appreciated by comparing different processes. For example, high heat source power and slow scanning speed reduce the cooling rate and more time increases grain size. Larger grains make the part softer and more ductile. Thinner layers in PBF-L facilitate cooling and make the parts harder and stronger. The faster rate of deposition in DED-L increases the cooling rate, hardness, and yield

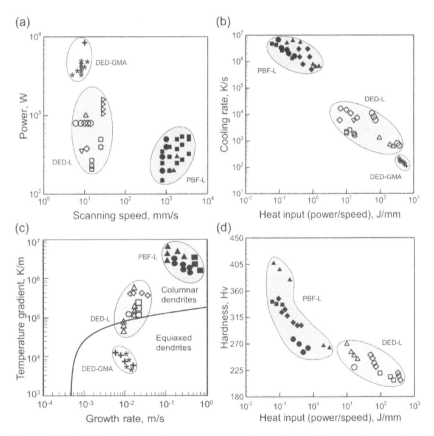

Figure 1.13 (a) Scanning speeds and powers used to make stainless steel parts using different AM processes. (b) Cooling rates during solidification as a function of linear heat input for different AM processes. (c) Temperature gradient versus solidification growth rate for different AM processes. (d) Effects of heat input on hardness for different AM processes. All data are for stainless steel 316. Values are either directly reported or calculated from the available data. *Source:* T. Mukherjee and T. DebRoy.

strength. As more layers are deposited away from the build plate, the heat accumulation progressively reduces cooling rates. As a result, the ductility of stainless steel builds exhibits anisotropy where the ductility varies owing to microstructural differences.

Columnar grains are detrimental to properties because they separate easily under transverse tensile loading. Controlling the temperature gradient and solidification growth rates are often ineffective to prevent their formation in certain alloys and other methods are needed. Well-known concepts in casting are showing promise in AM although more work is needed. For example, inoculants have been used during PBF-L of an aluminum alloy to increase the number of nucleation sites and form equiaxed grains. Changes in the scanning direction can also disrupt growth patterns in successive layers since the orientations of grains are affected by the heat flow direction which is affected by the scanning pattern. In short, it is now known that the formation of columnar grains can be prevented during AM of many alloys. However, there is no known methodology to obtain a desirable microstructure of most commercial alloys. The diversity of processing conditions offers ample opportunity to vary the processing condition to obtain acceptable structure and properties. However, selecting an appropriate processing condition considering many processes, process variables, and alloys to obtain good microstructure and properties remains elusive.

Thermal distortion, porosity, and lack of fusion in parts are common occurrences and their mitigation is an important task. Shrinkage of liquid pools contributes to thermal distortion and large liquid pools make a part susceptible to thermal distortion. Printing technique and processing conditions affect the dimensions of the liquid pool (Figure 1.14 (a)) and the susceptibility of parts to thermal distortion represented by thermal strain, can vary for different printing processes (Figure 1.14 (b)). Similarly, the lack of fusion defects depend on the extent of geometric overlap between the neighboring layers and hatches. The choice of process variables is important for mitigating defects. Heat source power, scanning speed, layer thickness, and other variables affect the formation of defects and this matter will be discussed in more detail in Chapter 10.

Later in this book, we will examine the important metallurgical variables such as the cooling rates, temperature gradients, and solidification growth rates that control the microstructure and properties of parts. We will also review how the emerging powerful digital tools of mechanistic models and machine learning are facilitating the use of the most advanced knowledge of metallurgy and data science and empowering engineers to achieve superior microstructure and properties and reduce defects.

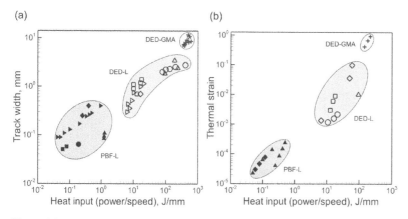

Figure 1.14 (a) The variation of track width with heat input for DED-GMA, DED-L, and PBF-L. (b) Effects of heat input on thermal strain for DED-GMA, DED-L, and PBF-L. All data are for stainless steel 316. Values are either experimental data or computed results. *Source:* T. Mukherjee and T. DebRoy.

1.7 Technological and commercial issues

AM faces several technological challenges (Figure 1.12). Trial and error testing for part qualification is time-consuming and expensive and adversely impacts the cost-competitiveness of parts. Part sizes are limited in some AM processes such as the powder bed fusion because of the size of the powder bed. High deposition rate processes such as the DED-GMA have limitations of feature size and require post-processing to improve surface quality. Complexities in the part geometry often require a support structure that needs to be removed which adds to the cost. Some types of defects such as residual stresses and distortion are not apparent in small coupons but are common in large complex parts [16]. Productivity may become an issue when parts are scaled up. The orientation of the solid model within the build volume is important but not based on any basic principle. It depends on considerations of support structure and recommendations of the machine manufacturer. Different machine manufacturers use various software, scanning pattern, and building sequence that affect product quality and cost. They also add to the machine-to-machine variability of part attributes. Removal of support structure from the base plate, surface cleaning, and other post-processing such as the hot isostatic pressing adversely impact cost-competitiveness. Finally, the handling of fine powders requires special care since metal powders can be both a fire hazard and a health risk.

Several economic issues (Figure 1.12) need to be addressed to expand the market share of AM. In conventional manufacturing such as casting and machining, the cost of parts is high if only a few parts are made because of the initial investment in the tools and part-specific equipment but decreases as more parts are made. Since AM does not incur high tooling and set-up costs, the cost per part does not change significantly with product volume. In many cases, the removal of the support structure, surface finish, and other post-processing such as hot isostatic pressing to reduce porosity and heat treatment add to the product cost. High costs of equipment, maintenance, feedstock, safety practice, and training often limit the adaptation of AM by small and medium-sized companies. In addition, only a handful of commercial alloys can be reliably printed by AM. Lack of available feedstock materials often limits the manufacturing of new products.

The cost of additively manufactured parts varies significantly depending on many factors such as the type of process, equipment, material, and post-processing. First, although the cost of the machine has decreased over the last decade, metal printing equipment remains expensive. For example, powder bed fusion equipment can cost well over a million dollars. The cost of the equipment remains an important part of the overall expense for metal printing. In addition, the cost of equipment maintenance and idle time needs to be factored into the cost. The manufacture of the metal powders/wires used in additive manufacturing is also important. Inert shielding gases and electric power are also needed to make metallic parts. Post-processing operations such as removal of the support structure, improving surface finish, and heat treatment to optimize microstructure, properties, and density need to be considered to estimate the manufacturing cost. An example of estimating the cost of an additively manufactured part is provided in Worked out example 1.8. The following features of additive manufacturing help to make the AM processes cost-competitive.

- Additive manufacturing's capability to produce a complex component in one step often avoids the cost of joining or assembling multiple small parts.
- The ability to produce components on demand reduces inventory costs over conventional manufacturing.
- Production of components close to where they are needed saves transportation costs.

- The use of the same equipment to produce a wide variety of materials saves equipment cost and space.
- By preventing the removal of materials by machining, the process avoids the wastage of materials.
- Improvement of functionality of components enhances quality and durability and reduces failures.
- The production of parts for which the supply chain does not exist allow the repair of expensive equipment avoiding rebuilding a large and expensive plant.

The standards for qualifying parts made by AM are currently being developed by various organizations. Voluntary adaptation of the standards by the stakeholders such as the machine manufacturers, feedstock suppliers, and users will be helpful to make business cases for new products. As the foundational patents expire, the intellectual property landscape will become friendlier to new companies. Improved open-source software will also be beneficial for superior product design. Turnkey machines and cost-sensitive manufacturing have attracted a new source of labor in AM that can benefit from fundamental knowledge of metallurgy and welding. Well-trained personnel can contribute a historical perspective of practice that is important for designing parts for optimum performance.

1.8 Content of this book

This book is divided into 14 chapters.

- The industrial applications of the additive manufacturing processes, common terminologies, its scientific synergy with fusion welding, the role of computers, and the important contemporary scientific and technological issues are discussed in the first chapter.
- In the second chapter, the common AM processes are described with an emphasis on metallic components. The chapter also includes the fabrication processes of powder and wire feedstocks used in AM, their properties, and measurement techniques of those properties.
- The third chapter focuses on the additive manufacturing of polymers and polymer composites.
- AM of ceramic materials is discussed in the fourth chapter.
- The design of parts for additive manufacturing is covered in the fifth chapter.
- The various stages in AM including sensing and control, quality control, and qualification are discussed in the sixth chapter. This chapter describes the applications of various sensors used to control AM processes. In addition, it describes control models used in AM.
- The applications of heat transfer to understand both the AM process and the attributes of AM-built products are presented in the seventh chapter. This chapter provides an in-depth theoretical and quantitative understanding of heat transfer and molten metal flow in AM. Several useful analytical calculations and dimensionless numbers related to heat transfer and fluid flow in AM are introduced.
- The main microstructural features of steels, nickel, titanium, and aluminum alloys are discussed in the eighth chapter. It contains an in-depth discussion of nucleation, grain growth, texture evolution, phase formation, and solid-state phase transformation during AM. In addition, several strategies to control the microstructure of AM parts are described.
- The main mechanical properties, i.e., the strength, hardness, ductility, fracture toughness, fatigue, and creep of important alloys are presented in the ninth chapter. In addition, the corrosion resistance of AM parts is also discussed.
- Common defects in AM such as the lack of fusion, porosity, cracking, balling, surface roughness, and loss of volatile alloying elements are covered in the tenth chapter. This chapter also provides techniques to reduce these defects.

- The evolution of residual stresses and distortion, their measurement techniques, and their effects on the properties and structural integrity of AM parts are discussed in the eleventh chapter. Several strategies to reduce residual stresses and distortion are also provided.
- The various types of mechanistic models of additive manufacturing and their capabilities are discussed in the twelfth chapter. It covers models of heat transfer and fluid flow, grain growth, microstructure evolution, residual stresses and distortion, and defect formation. In addition, the emerging applications of machine learning and digital twins in AM are discussed in the chapter.
- The safety, sustainability, and economic issues in additive manufacturing are discussed in the thirteenth chapter.
- The important trends, possibilities, and future perspectives of AM are evaluated in the last chapter.

Worked out example 1.8

Estimate the cost of manufacturing metal parts, 5 cm long, nearly cylindrical in shape with many unique geometric features and 4 cm^3 volume by powder bed fusion. Two batches, each having 30 specimens were made with the long axis in horizontal and vertical configurations. No support structure was used. Data: The equipment cost, lifetime, and idle time: $1 million, 8 years, and 20%, respectively. Build time: 23 hours when parts were built with orientation parallel to the build plate and 26 hours when the long axis was perpendicular to the build plate. Many more layers were needed to make parts oriented vertically than oriented horizontally. The cost of metal powder: $100/kg, powder density: 7.6 gm/cm^3. Machine setup time and cost: 2 hours at $80/h. Inert gas usage: 500 L of argon for initial purging before building starts and then 10 liters per minute. Cost of argon: $0.01 per liter. Power consumption: 1.5 kW. Electric rate: $0.10/kWh. Post processing: $500.

Solution:

The hourly direct cost of machine usage: $1,000,000/(8 × 365 × 24 × 0.8) = $17.84/hour. This cost does not include any maintenance or indirect costs.
Machine time: $17.84/h × 23 = $410.32 for 23 hours and $463.84 for 26 hours.
Mass of powder to make 30 parts = 30 × 4 × 7.6 = 912 grams
Cost of powder for each batch: 0.912 × 100 = $91.2
Cost of argon gas for 23 hours (h) = (500 + 23 × 60 × 10) liters × $0.01/liter = $143 for 23 hours and $161 for 26 hours.
Cost of electricity: 1.5 kW × 23 h = 34.5 kWh × $0.10/kWh = $3.45 for 23 hours and $3.90 for 26 hours.
Post processing: $500
Cost for 30 specimens printed in horizontal orientation, i.e., parallel to the build plate = $160 (machine set-up) + $410.32 (machine time) + $91.20 (powder) + $143 (argon) + $3.45 (electricity) + $500 (post processing) = $1307.97
Cost for 30 specimens printed in vertical orientation, i.e., perpendicular to the build plate = $160 (machine set-up) + $463.84 (machine time) + $91.20 (powder) + $161 (argon) + $3.90 (electricity) + $500 (post processing) = $1379.94
These costs do not consider any indirect costs and the actual cost would be higher. In each case, the cost of the machine time and post processing are the largest contributors to cost. The electricity is less than 1% of the total cost.

Takeaways

What is additive manufacturing?

- Additive manufacturing or 3D printing is used to fabricate complex parts by progressively adding thin layers of materials to make three-dimensional parts.
- The market revenue, the growing sales of 3D printing machines in recent years, and the number of patents granted globally show the growing acceptance of 3D printing in the industry.

Terminology

- Additive manufacturing or 3D printing of metallic materials is also called metal printing, rapid prototyping, and freeform fabrication.

Uses

- Printed parts are used in aerospace, medical, energy, oil, automotive, marine, and consumer products industries.
- Commonly used alloys include steels, nickel, aluminum, titanium, copper, cobalt-chrome alloys, and several precious metals.
- Printed parts of nonmetallic materials such as ceramics, polymers, and composites are used in medical implants, tools, molds, sculptures, prototypes, and so on.

Scientific synergy with welding

- Fusion welding and the main variants of additive manufacturing of metallic materials share many similarities in their physical processes.
- As an emerging field, additive manufacturing faces many scientific problems and the rich knowledge base of fusion welding is an important resource for their solution.

The role of computers

- Computers are used in part design, making files for printing machines, process monitoring and control and defect reduction using machine learning and artificial intelligence, process simulation, and using digital twins for rapid qualification.

The important scientific issues

- Cooling rates, temperature gradients, and solidification growth rates that affect the structure and properties vary significantly depending on the alloys, process variants, and process variables.
- Avoiding common defects still remains a difficult challenge.

Technological and commercial issues

- Technological issues include part qualification, geometric complexities, scaling up, process design, and post-processing.
- Trial and error optimization of part attributes, expensive machine, and feedstock pose challenges in cost competitiveness of parts.

Appendix – Meanings of a selection of technical terms

<u>Ceramic</u>: Hard, brittle, heat- and corrosion-resistant inorganic, nonmetallic material.

<u>Composite</u>: A composite material is produced from two or more constituent materials. Nonmetallic materials are added with alloys to print metal matrix composites.

<u>Cooling rate</u>: The rate with which the deposited material cools down to room temperature. The cooling rate varies significantly depending on the temperature at which the cooling rate is estimated. Cooling rates affect the microstructure and properties of parts.

<u>Electron beam</u>: A focused beam of electrons used as a heat source in additive manufacturing.

<u>Fusion welding</u>: A manufacturing process that uses a heat source to melt two or more parts to form a sound joint after solidification.

<u>Laser beam</u>: The acronym laser stands for "light amplification by stimulated emission of radiation." A focused beam of laser is often used in additive manufacturing as a heat source.

<u>Polymer</u>: Natural or synthetic substances composed of very large molecules.

<u>Solidification growth rate</u>: The rate with which the solidification front grows during the solidification.

<u>Temperature gradient</u>: Spatial gradient of the temperature field during additive manufacturing. The temperature field is spatially nonuniform and is distributed in three dimensions.

<u>Thermal strain</u>: Strain originated due to the nonuniform temperature distribution during additive manufacturing.

<u>Track width</u>: The width of the deposited track which depends on the molten pool dimensions.

Practice problems

1) It is evident from Table 1.1 that stainless steels are widely used to print parts in various industries. Do a literature review to provide a list of important applications of stainless steels in additive manufacturing.

2) Give examples of some common uses of metallic, ceramic, polymer, and composite materials in additive manufacturing.

3) What are the common defects in metallic components made by additive manufacturing?

4) Based on the information provided in Table 1.1 select an alloy for printing a bone implant at a low material cost.

5) Do a literature search to compile a list of exciting applications of AM parts in the medical industry.

6) Discuss the main technological and commercial challenges faced by additive manufacturing.

7) Discuss the challenges and considerations involved in post-processing additively manufactured parts, such as cleaning, machining, grinding, and polishing.

8) Discuss the role of computer-based digital tools in different stages of lifecycles of additively manufactured metallic components.

9) What are the similarities and differences between wire arc additive manufacturing and multipass gas metal arc welding where filler metals in the form of wires are melted to join metallic components?

10) What are the different costs associated with AM of a part? Provide an example of an AM part and compare the different costs of fabricating the part.

References

1 Grossman, B. https://www.bathsheba.com (accessed on 06 October 2021).

2 DebRoy, T. and Bhadeshia, H.K.D.H., 2021. *Innovations in Everyday Engineering Materials.* Springer.

3 Blakey-Milner, B., Gradl, P., Snedden, G., Brooks, M., Pitot, J., Lopez, E., Leary, M., Berto, F. and du Plessis, A., 2021. Metal additive manufacturing in aerospace: A review. *Materials & Design*, 209, article no 110008.

4 DebRoy, T., Mukherjee, T., Wei, H.L., Elmer, J.W. and Milewski, J.O., 2021. Metallurgy, mechanistic models and machine learning in metal printing. *Nature Reviews Materials*, 6(1), pp.48–68.

5 DebRoy, T., Mukherjee, T., Milewski, J.O., Elmer, J.W., Ribic, B., Blecher, J.J. and Zhang, W., 2019. Scientific, technological and economic issues in metal printing and their solutions. *Nature Materials*, 18(10), pp.1026–1032.

6 Mouzakis, D.E., 2018. Advanced technologies in manufacturing 3D-layered structures for defense and aerospace. *IntechOpen*, 10, pp.89–113.

7 Childerhouse, T. and Jackson, M., 2019. Near net shape manufacture of titanium alloy components from powder and wire: A review of state-of-the-art process routes. *Metals*, 9(6), article no 689.

8 Madara, S.R. and Selvan, C.P., 2017. Review of recent developments in 3-D printing of turbine blades. *European Journal of Advances in Engineering and Technology*, 4(7), pp.497–509.

9 DebRoy, T., Wei, H.L., Zuback, J.S., Mukherjee, T., Elmer, J.W., Milewski, J.O., Beese, A.M., Wilson-Heid, A.D., De, A. and Zhang, W., 2018. Additive manufacturing of metallic components–process, structure and properties. *Progress in Materials Science*, 92, pp.112–224.

10 Hagedorn, Y., 2017. Laser additive manufacturing of ceramic components: Materials, processes, and mechanisms. In *Laser Additive Manufacturing* (pp.163–180). Edited by Milan Brandt. Woodhead Publishing.

11 Schubert, C., Van Langeveld, M.C. and Donoso, L.A., 2014. Innovations in 3D printing: a 3D overview from optics to organs. *British Journal of Ophthalmology*, 98(2), pp.159–161.

12 EWI Report, 2015. *Robotic GMAW-P RP2Z Pipe Welding.* Available at https://ewi.org/18029-2 (accessed on 16 January 2023).

13 Mukherjee, T., DebRoy, T., Lienert, T.J., Maloy, S.A. and Hosemann, P., 2021. Spatial and temporal variation of hardness of a printed steel part. *Acta Materialia*, 209, article no 116775.

14 Wei, H.L., Knapp, G.L., Mukherjee, T., and DebRoy, T., 2019. Three-dimensional grain growth during multi-layer printing of a nickel-based alloy Inconel 718. *Additive Manufacturing*, 25, pp.448–459.

15 Mukherjee, T. and DebRoy, T., 2019. A digital twin for rapid qualification of 3D printed metallic components. *Applied Materials Today*, 14, pp.59–65.

16 Wei, H.L., Mukherjee, T., Zhang, W., Zuback, J.S., Knapp, G.L., De, A. and DebRoy, T., 2021. Mechanistic models for additive manufacturing of metallic components. *Progress in Materials Science*, 116, article no 100703.

2

Feedstocks and Processes for Additive Manufacturing of Metals and Alloys

Learning objectives
After reading this chapter the reader should be able to do the following: 1) Understand the working principles of commonly used additive manufacturing processes for metallic materials. 2) Know several techniques of producing powder feedstocks and their advantages and disadvantages. 3) Select an appropriate powder production technique to achieve desired powder characteristics. 4) Recognize the effects of powder characteristics on part quality. 5) Understand the benefits and limitations of the recycling of powders. 6) Appreciate the roles of wire and sheet feedstocks in additive manufacturing and their production techniques.

CONTENTS

Theory and Practice of Additive Manufacturing, First Edition. Tuhin Mukherjee and Tarasankar DebRoy.
© 2024 John Wiley & Sons, Inc. Published 2024 by John Wiley & Sons, Inc.

2.1 Introduction

Additive manufacturing (AM) is a process of building parts by progressively adding thin layers of materials, sometimes layers thinner than a human hair. A computer program containing a three-dimensional design guides the printing of parts layer-by-layer. A laser beam (L), electron beam (EB), gas metal arc (GMA), or plasma arc (PA) is used to melt a powder or wire feedstock or thin sheets [1,2]. There are two types of main AM processes used in the industry. In directed energy deposition (DED) the feedstock is supplied using a feeding mechanism which is subsequently melted by a heat source. In powder bed fusion (PBF) processes, thin layers of powders are deposited to form a bed and are selectively melted by a heat source. That is why powder bed fusion is often called selective laser or electron beam melting. The selection of an appropriate combination of AM process and feedstock is very important to print high-quality parts. In this section, we discuss commonly used AM processes for metallic materials. In addition, fabrication techniques for powder, wire, and sheet feedstocks are described. Furthermore, the effects of feedstock characteristics on AM processes and the qualities of the printed parts are explained.

2.2 Additive manufacturing processes

In AM of metallic materials, a computer-aided design (CAD) file specifies the part geometry. This file also helps to determine the orientation of the part within the build volume and the support structure that may be needed to build the part. For example, if a part contains overhangs an appropriate supporting structure may be needed to prevent distortion of hot overhangs under their own weight. The CAD file also helps to define the planer layers by slicing the three-dimensional component. The scan path for covering each planer layer is also prepared from the CAD file. The scanning speed of the heat source and the rate of supply of the feedstock material is often selected based on the machine manufacturer's recommendation.

2.2.1 Directed energy deposition processes

In DED-L, schematically shown in Figure 2.1 (a), a stream of metal powder flows through a nozzle and interacts with the laser beam during flight, and forms a molten pool below the nozzle. The molten pool moves with the laser beam along a predetermined track. Multiple adjacent tracks form one layer of deposit and the process is repeated layer-by-layer to form a three-dimensional structure. An inert shielding gas such as argon is used to protect the molten pool from oxidation. The metal deposition rate is an order of magnitude faster than that for the powder bed fusion processes.

An electric arc is used as the heat source in DED-GMA or DED-PA. Filler wires are used as feedstock material. Apart from the power source, a wire feeding system is required to supply the wire in the molten pool. A robot controls the relative motion of the heat source and the build (Figure 2.1 (b)). Wires are more commonly available than powders and their use results in high deposition rates. The surface for DED-GMA and DED-PA often shows wavy features with the distance between the crests and troughs of wave comparable to the layer thickness. After the part is removed from the substrate, the surface often requires machining to achieve a good surface finish.

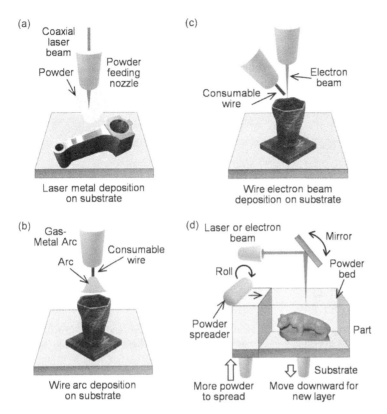

Figure 2.1 A schematic representation of directed energy deposition with (a) powder feedstock and laser heat source, (b) wire feedstock and arc heat source, and (c) wire feedstock and electron beam heat source. (d) Schematic representation of powder bed fusion with a laser or electron beam heat source. *Source:* T. Mukherjee and T. DebRoy.

An electron beam melts a filler wire in DED-EB under vacuum which provides a good environment for preventing oxidation of molten metal (Figure 2.1 (c)). Metal is deposited in layers similar to DED-L. Because the process allows for high deposition rates, it is frequently used to produce larger parts than the powder bed fusion process.

2.2.2 Powder bed fusion processes

In PBF-L, fine powders typically 10–100 micrometers in diameter are packed within a box that defines the maximum dimensions of the part. A laser beam is moved using a mirror attached to a galvanometer to selectively melt the powders (Figure 2.1 (d)). Fusion occurs along the path of a laser beam as it moves at high speed. The powder bed is enclosed in a chamber filled with an inert gas to prevent oxidation. After each layer is deposited, the powder bed is lowered by a short distance equal to the layer thickness, and a thin layer of powder is added by spreading a small amount of powder with a rake. The path of the laser beam does not have to follow the same pattern for each layer. Unidirectional, bidirectional, spiral, zigzag, and other patterns are selected considering the geometry of a given layer and the properties of the alloy deposited. Fusion occurs pass-by-pass and layers-upon-layer to construct the three-dimensional part. The scanning rate for the laser beam is often much faster than what is used in DED. However, because the layer thickness is small, the metal deposition rate is lower than the other AM processes.

An electron beam is also often used as a heat source in powder bed fusion. The deposition chamber is kept at low pressure since the electron beams operate most efficiently under a vacuum.

Table 2.1 Selected attributes of powder bed fusion and directed energy deposition processes [1, 2].

	Powder bed fusion laser (PBF-L) or electron beam (PBF-EB)	Directed energy deposition powder and laser (DED-L)	Directed energy deposition wire electron beam (DED-EB), plasma arc (DED-PA), or gas metal arc (DED-GMA)
Heat source power (W)	50–1000 (up to 4 beams)	400–3000	1000–5000 (DED-GMA, 2000)
Scanning speed (mm s^{-1})	10–1000	6–60	5–50
Deposition rate (cm^{3}hr^{-1})	25–180	20–450	100–>1000
Build size (mm × mm × mm)	Maximum 800 × 400 × 500	Maximum 2000 × 1500 × 1000	Maximum 5000 × 3000 × 1000
Feedstock diameter (μm)	15–60 (PBF-L) 45–105 (PBF-EB)	15–105	900–3000
Dimensional accuracy (mm)	0.04–0.20	0.20–5	1–5
Surface roughness (μm)	7–30 (PBF-L) 20–50 (PBF-EB)	15–60	45–200+, surface needs machining
Post-processing	Heat treatment, hot isostatic pressing, machining	Heat treatment, machining, grinding	Heat treatment, stress relieving, machining
Cooling rate during solidification (Ks^{-1})	10^5–10^7	10^2–10^4	10^1–10^2
Temperature gradient (Km^{-1})	10^6–10^7	10^5–10^6	10^3–10^4
Solidification growth rate (ms^{-1})	10^{-1}–10^0	10^{-2}–10^{-1}	10^{-2}–10^{-1}

The powder spreading mechanism used is the same as that for the PBF-L. The beam rastering is done using electromagnetic coils. Unlike the PBF-L, a two-stage process is commonly used for the PBF-EB. Since electrostatic charging and repulsion create problems in the electron beam processing of powders, sintering at low temperatures is first carried out to reduce particle rearrangement and ejection from the bed. This is followed by rapid scanning of the pre-sintered bed which improves the deposition rate. After the printing of the parts, some solid powders remain unused in both PBF-L and PBF-EB processes and are often reused.

PBF and DED are the most common AM processes. Some of the important attributes of the PBF and DED processes are presented in Table 2.1. Process selections are made based on many factors such as the size of the part, dimensional accuracy, and the time necessary for production. Worked out example 2.1 provides how processes may be selected based on product attributes.

Worked out example 2.1

Which additive manufacturing processes are appropriate for producing (1) a large Ti-6Al-4V aircraft bracket and (2) a small part with intricate internal channels? Please explain why.

Solution:

1) A time-efficient fabrication of a large part like an aircraft bracket requires an additive manufacturing process capable of providing a high deposition rate. Directed energy deposition processes with arc or plasma heat sources (DED-GMA or DED-PA) are suitable for this

> **Worked out example 2.1 (Continued)**
>
> purpose. In addition, aircraft brackets generally do not have complex geometry. Therefore, the deposition of thick layers to achieve a high deposition rate can be used to make brackets in a cost-competitive manner. Parts are machined to achieve the required dimensions.
>
> 2) Parts with small intricate features need to be printed by depositing very thin layers to achieve good surface finish and dimmensional accuracy. It can be made by powder bed fusion process using very fine powders. Since the internal channels cannot be machined after printing, proper selections of layer thickness and powder size are desirable to get the required dimensions and surface finish of the parts. For example, General Electric prints fuel nozzles using PBF-L by depositing thin layers of a Co-Cr alloy from very fine powders.

2.2.3 Emerging processes

Several new additive manufacturing processes are evolving to make metallic components. These processes are still in their initial stages of development and have not yet found many commercial applications.

2.2.3.1 Binder jet process

In a binder jet process, a liquid polymer such as a thermosetting resin is selectively deposited on the powder layers. The liquid polymer or the binder forms an agglomerate of powder particles that defines the shape of the deposit (Figure 2.2 (a)). At the end of each layer, the powder bed is heated using an external heat source to partially cure the binder. This partial curing makes the powder-binder agglomerate strong enough to support the deposition of the next layer of powders. The process continues until the entire part is printed. In the end, the entire part is sintered in an oven to improve its strength. Often, the sintered parts need to be machined to achieve the desired shape and size.

The Binder jet process is used to make prototypes of different shapes and colors, molds, and cores for casting. Since this process does not employ melting and solidification of material, it can avoid solidification shrinkage and cracks. However, the parts need post-process sintering to improve strength which adds cost. Sintering can cause some shrinkage which needs to be considered in part design. In addition, the parts are often not fully dense which may result in poor mechanical properties.

2.2.3.2 Ultrasonic vibration-assisted process

Ultrasonic vibrations have been used to consolidate thin sheets of alloys typically in the thickness range of 0.5 to 1.0 mm under pressure. The sheets are pressed against a metallic substrate using a normal force applied by a sonotrode. The sonotrode also vibrates in the transverse direction at a very high frequency of about 20 kHz. The heat generated by the ultrasonic vibrations softens the metals and they are consolidated under pressure (Figure 2.2 (b)). The sonotrode travels along the scanning direction to form a deposit track. Multiple deposit tracks are created side-by-side for a particular layer. Several sheets are used to create multiple layers. At the end of the process, the final shape is machined to obtain the required geometry and dimensions.

This process has the unique advantage of making parts with varying compositions by using sheets of different alloy compositions at each layer. Additionally, this solid-state process enables the fabrication of graded parts of metallurgically incompatible metals. This process can be applied to soft materials such as aluminum, copper, and magnesium alloys. This process is time-consuming and can only produce parts that can be assembled from thin sheets. It cannot produce parts with complex geometry.

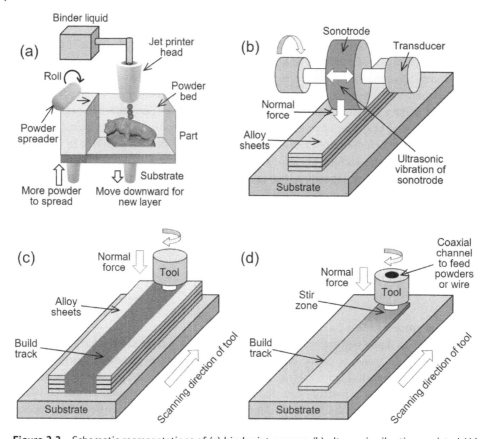

Figure 2.2 Schematic representations of (a) binder jet process, (b) ultrasonic vibration-assisted AM process, (c) friction stir additive manufacturing with sheet feedstock, and (d) additive friction stir deposition of powder or wire feedstock. *Source:* T. Mukherjee and T. DebRoy.

2.2.3.3 Friction stir additive manufacturing

Friction stir additive manufacturing is an emerging process whose working principle is the same as friction stir welding. It is a solid-state process and avoids the complexity of melting and solidification and issues related to them such as solidification shrinkage and hot tearing. In this process, a high-speed rotating tool with a shoulder rubs against the workpiece to generate frictional heat that softens the material. The soft, plasticized material forms the part after cooling down to room temperature.

There are two types of friction stir additive manufacturing processes commonly used. First, friction stir additive manufacturing using sheet feedstock, also known as friction stir cladding, joins thin metallic sheets in a layer-by-layer manner to make the parts (Figure 2.2 (c)). A noncon-sumable tool is used to generate the frictional heat which also applies an axial pressure to join two consecutive sheets. Multiple layers of thin sheets are joined [3] and at the end of the process, the component is machined to get the required dimensions and surface finish. The track width is almost the same as the tool's shoulder diameter [3]. However, multiple passes can be done per layer to make a wider component. Compositionally graded materials can be printed using this process by using sheets of different compositions in different layers. However, this process is slow and is difficult for making parts with complex geometry.

Additive friction stir deposition uses powders or wires as feedstock materials. The feedstock is supplied to the stir zone through a channel coaxial to the rotating tool (Figure 2.2 (d)). The powder or wire feedstock travels through the channel and is deposited at the bottom of the rotating tool. There exists a small gap between the tool and the substrate where the tool rubs the feedstock to generate frictional heat. The frictional heat softens the feedstock which forms the deposit after cooling down to room temperature. The deposit height is controlled by the gap between the tool and the substrate and the feed rate of the feedstock. The width of the track is almost the same as the tool diameter. However, multiple passes can be performed to make a wider component. The movement of the feedstock through the channel inside the rotating tool is significantly affected by the rotational speed of the tool. Very high rotational speed may result in a high centrifugal force on the powder feedstock which may displace the feedstock away from the stir zone. Precise control of the stirring material in the stir zone still remains a challenge and needs further research and development. Tools with internal channels are expensive. Furthermore, extending the life of these tools is an important consideration.

In the friction stir deposition process, often, a consumable tool is used instead of powder or wire feedstocks. The tool rotates at a high speed and rubs against the substrate with an axial force to generate frictional heat. The frictional heat makes the consumable tool soft and linear movement of the tool along the scanning direction creates the deposited track. The deposited track is then machined to obtain the required geometry and surface finish. The track width is almost the same as the diameter of the tool. This process is very slow because the tool is consumed after depositing a very short distance and a new tool needs to be placed. Friction stir additive manufacturing is an emerging process and needs significant research and development to be accepted in the industry for commercial production.

2.2.3.4 Hybrid additive manufacturing

Hybrid manufacturing combines subtractive manufacturing's superior precision and lower cost with the many advantages of additive manufacturing in a single manufacturing step. Some special features of parts such as a threaded hole are difficult to manufacture with precision using additive manufacturing. Dimensional tolerances of printed metal parts are often achieved by machining in a separate operation. Since printed metal parts often need subtractive methods to produce the finished parts, combining an additive manufacturing machine with facilities to perform milling and machining operations make hybrid AM manufacturing more capable and cost-competitive. Since additive manufacturing can add materials where needed, it allows the repair of damaged parts. AM can be combined with surface finishing in a hybrid operation for manufacturing finished parts. Hybrid AM systems often consist of a standard machine tool such as a lathe or a mill together with a DED deposition head that can add material in the form of a wire or powder and a heat source such as a laser beam. In many cases, the additive and subtractive processes can be operated in sequence. After a near-net-shape part is made by the additive process, machining can be used to finish the part. Alternatively, machining can add a feature such as threading after a part is made by an additive process or an internal feature can be machined as a part is printed. Machining capabilities have also been added to powder bed systems to add features in a part or for surface finish. Hybrid systems also allow adding dissimilar metals to a part such as cladding for tailoring surface properties or adding copper for superior heat transfer by AM in specific locations that are carved out by machining (Figure 2.3). Hybrid systems are also available for processing polymers.

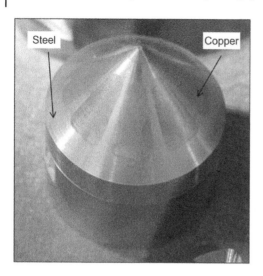

Figure 2.3 A dissimilar metallic part is fabricated by hybrid manufacturing where copper is added to a steel part to increase the thermal conductivity [4]. The figure is taken from [4] with permission from Stephanie Hendrixson, Executive Editor of Additive Manufacturing Media (Photo credit: Barbara Schulz).

2.3 Powder feedstock

Powders are the most commonly used feedstock in additive manufacturing. In DED processes, powders are supplied through one or more nozzles, melted by a laser or electron beam, and solidified to make the part. In the PBF process, thin layers of powders are deposited which are melted by a laser or electron beam to make a part. The commonly used production techniques and the effects of powder attributes on the AM processes and printed parts are discussed below.

2.3.1 Powder production techniques

Ingots of the required metal or alloy are melted, and the molten material is converted to powders. There are four atomization techniques commonly used to make powders for AM, gas atomization, water atomization, centrifugal atomization, and plasma rotating electrode process. They are described below.

2.3.1.1 Gas atomization

All atomization processes begin with the melting of the alloy and transferring the melt to a liquid metal holding equipment called a tundish. A tundish can regulate the flow rate of the molten alloy into an atomizer where the powders are made. The molten alloy enters at the top of the atomization chamber and falls freely through the chamber (Figure 2.4 (a)). The stream of the molten alloy is broken up by impinging the stream of metal with high-pressure jets of inert gas. Since the inert gases have a much lower capacity to absorb heat (low heat capacity) the metal droplets do not freeze instantly. The powders are solidified and collected at a chamber below the atomizer. The powder particle size can be controlled by varying the ratio of the flow rates of the gas to that of the molten alloy. The relation between the gas flow rate and the powder particle size is illustrated in the worked out example 2.2.

2.3.1.2 Water atomization

The water atomization process (Figure 2.4 (b)) is similar to the gas atomization process, with the difference being the use of water instead of inert gas to break up the liquid stream. The use of water makes the process cost-competitive because it does not require expensive inert gases. The

Figure 2.4 Schematic representation of (a) gas atomization method (b) water atomization method (c) centrifugal atomization method, and (d) plasma rotating electrode process. *Source:* T. Mukherjee and T. DebRoy.

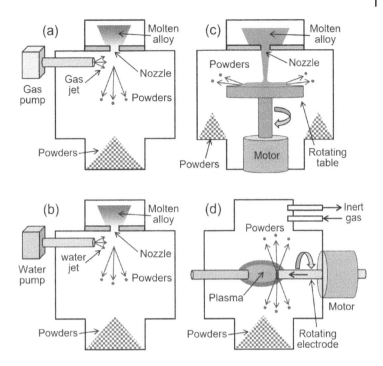

molten alloy enters the atomization chamber from the top and is atomized by water jets. The powders are collected at the bottom of the chamber. The powders need to be dried because of the use of water in the chamber. Powders produced in this way are irregular in shape because of their rapid solidification when the metal stream comes in contact with water. The relation between the nature of the powder and the porosity of the printed part is illustrated in the worked out example 2.3.

2.3.1.3 Centrifugal atomization

An ingot of the alloy is first melted in a furnace and transferred to a tundish. The molten alloy enters from the tundish to the atomization chamber from the top and falls freely through a nozzle on a rotating table (Figure 2.4 (c)). The rotation of the table creates a high centrifugal force that atomizes the molten metal into fine droplets. The droplets are ejected radially outward from the rotating table due to the centrifugal force. The molten droplets solidify before hitting the walls of the chamber forming powders. The fine powders are then collected below. The size of the powder particles can be controlled by adjusting the rotational speed of the table.

2.3.1.4 Plasma rotating electrode process

The plasma rotating electrode process (Figure 2.4 (d)) is used to produce highly spherical particles. A bar feedstock of alloy is used in the process. The bar behaves as a rotating electrode. As the rotating bar enters the atomization chamber plasma torches melt the end of the bar, ejecting fine particulates from its surface. The molten particulates solidify before hitting the walls of the chamber forming powders. The fine powders are then collected in a chamber below the atomizer. The advantages and disadvantages of different powder production techniques are presented in Table 2.2.

Table 2.2 Different powder production techniques and their advantages and disadvantages.

Production techniques	Powder size	Advantages	Disadvantages
Gas atomization	0–500 μm	• Wide ranges of metals and alloys can be processed. • Rapid and less expensive.	• Hard to achieve the spherical morphology of powders. • Entrapment of gas causes internal porosities in powders.
Water atomization	0–500 μm	• High rate of production. • Less expensive experimental setup.	• Post-processing is needed to remove moisture. • Not suitable for reactive metals. • Mostly nonspherical, irregular morphology is produced.
Centrifugal atomization	0–600 μm	• Wide ranges of metals and alloys can be processed. • Uniform powder particle size distribution can be achieved.	• Requires expensive and specialized equipment. • Very fine particles cannot be made unless a very high rotational speed is used.
Plasma rotating electrode process	0–100 μm	• Wide ranges of metals and alloys can be processed. • Very fine particles can be produced. • Powders with spherical morphology can be made.	• Requires expensive and specialized equipment. • Low productivity. • Requires input material in the form of wire which adds cost.

Worked out example 2.2

In the gas atomization process, a desired powder particle size can be obtained by controlling the gas flow rate. Suggest a dimensionless number that can be used to relate the gas flow rate with the powder particle size. If a jet of nitrogen gas is passed through a nozzle of 1 mm diameter to make stainless steel 316 powders of 200 microns size, what should be the necessary gas flow rate in m^3/minute? Data: density of nitrogen: 1.16 kg/m^3, surface tension of stainless steel 316: 1.5 N/m.

Solution:

Weber number can be used to quantify the relation between the gas flow rate and powder particle size. It is defined by the ratio of the inertia force of the gas to the surface tension force of the liquid metal and is expressed as [5],

$$W = \rho U^2 L / \sigma$$

where ρ is the density of gas (kg/m^3), U is a characteristic velocity (m/s) which can be taken as the gas jet velocity for gas atomization process, L is a characteristic length (m) which can be considered as the average particle size, and σ is the surface tension of liquid metal (N/m). For the atomization of fine particles, a high Weber number is required which indicates a higher inertia force of gas compared to the surface tension force of the liquid metal. Under this condition, the inertia force of the gas is capable of disintegrating the liquid into small droplets by overcoming the resistance due to surface tension. It has been reported [5] that a value of Weber number > 100 is needed for atomization of liquid metal using nitrogen gas. Using $W = 100$ and the other values provided, the required velocity of the gas jet = $((100 \times 1.5)/(1.16 \times 200 \times 10^{-6}))^{0.5}$ = 804 m/s.

Worked out example 2.2 (Continued)

Supersonic flows are common in gas atomization. For example, a gas speed of 1700 m/s was used to atomize 15 μm aluminum powder using helium gas [6].

The gas jet is supplied through a nozzle of 1 mm diameter. Therefore, the flow rate of the gas should be = (804 × (π/4) × (0.001)^2) × 60 = 0.038 m^3/minute. The estimations based on dimensionless numbers provide correct trends, but they are approximate and they need to be validated with experimental data.

Worked out example 2.3

Additively manufactured parts made from powder feedstocks often suffer from porosities that affect their properties (see Chapter 10). If the goal is to make a part with low porosity, recommend an atomization technique to produce the powder. Explain your answer.

Solution:

Powders produced by the gas atomization technique can contain gas bubbles entrapped inside the powder particles. These powders when used in additive manufacturing may cause gas porosities. Similarly, oxygen and hydrogen may dissolve in the powder made by the water atomization and form porosities. Powders made by the plasma rotating electrode process or centrifugal atomization process can produce powders free of porosity in printed parts. For example, Figure E2.1 below shows parts made by DED-L using Inconel 718 powders produced by (a) gas atomization and (b) plasma rotating electrode process [7]. The figure shows that the part made from powders produced by the plasma rotating electrode process avoided gas porosities.

Figure E2.1 Parts made by DED-L using Inconel 718 powders produced by (a) gas atomization and (b) plasma rotating electrode process. The figure is taken from [1] with permission from Elsevier.

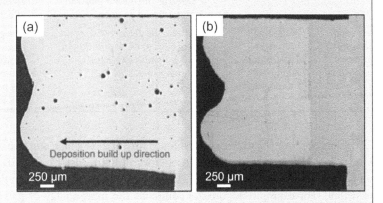

2.3.2 Powder characteristics and their effects on part quality

The attributes of the powder feedstocks affect the microstructure, properties, and susceptibility to defect formation. The important properties of powder feedstock include morphology, size distribution, flowability, chemical composition, and density. Table 2.3 summarizes various powder attributes and their effects on AM process and part qualities.

Table 2.3 Summary of powder attributes and their effects on process and part quality.

Powder attribute	Measurement techniques	Effects on processes and products	Comments
Powder particle morphology	Scanning electron microscopy, computer tomography	• Affects packing density in PBF processes. • Spherical powders are good for heat absorption. • Affects surface roughness.	• Spherical morphology is desired in PBF for a good packing density. • Morphology is quantified by aspect ratio (longest/shortest dimension).
Powder size distribution	Laser diffraction, sieving	• Affects the effective density and thermophysical properties of a powder bed. • Small powders are needed to achieve intricate features. • Finer particles absorb more heat due to the high surface area to volume ratio.	• Typical size range: 10–60 μm (PBF-L), 60–100 μm (PBF-EB), and 10–200 μm (DED-L). • Often follows a Gaussian distribution. • Very fine particles are fire hazards and affect health and safety.
Powder flowability	Hall flow meter	• Poor flowability hinders powder spreader movement in PBF. • Poor flowability often results in nozzle clogging in DED.	• Tiny spherical powders exhibit the best flowability. • Moisture content affects flowability due to capillary forces.
Powder particle composition	X-ray diffraction	• Affects the microstructure and properties of the printed part. • Evaporative loss of volatile alloying elements may make the composition of the part different from that of the feedstock. • Governs the thermophysical properties of powders.	• Uniformity in composition needs to be maintained to reduce inhomogeneity in the microstructure and properties of the part. • The oxygen and nitrogen should be avoided. • The presence of surface-active elements e.g., sulfur in steels affects the molten metal flow.
Powder density and internal porosities	Scanning electron microscopy, computer tomography	• Affects the density of the part. • Gas entrapped/dissolution in powders may cause gas porosity in part.	• The gas-atomization method often causes porosities in powder. • Pre-process sintering can reduce the pores but is time-consuming and expensive.

Powders produced by different techniques vary in their shape and size which govern the packing density in PBF processes and flow properties of the feedstock in DED. The morphology of the powder feedstock can be spherical, elliptical, or irregular depending on the atomization technique used [8,9]. For example, Figure 2.5 compares the morphology of powders produced using the plasma rotating electrode process, gas atomization, and water atomization processes. Spherical powders with smooth surfaces can be produced using the plasma rotating electrode process because the process allows sufficient time for spheroidization. In contrast, powders produced using the water atomization process are rapidly solidified and of irregular shape. Gas-atomized powders are partially spherical with uneven surfaces with small satellite particles attached to their surfaces.

All powder manufacturing processes produce a range of powder sizes. Therefore, a particular sample of powders contains particles of different sizes. Sieves of different mesh sizes are used to

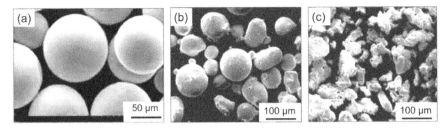

Figure 2.5 Morphology (using scanning electron microscopy) of powders produced using (a) plasma rotating electrode process, (b) gas atomization, and (c) water atomization processes. The figure is taken from [1] with permission from Elsevier.

measure the volume of powder for a particular size. There are ASTM standards of the sieve of different mesh sizes. For example, the "number 200" sieve allows particles of size 75 μm or smaller to pass through it. Therefore, particles of size 75 μm or smaller can be separated from the larger particles. Similarly, the volume of powders for different sizes can be estimated using sieves of different mesh sizes. The volume versus particle size distribution is called powder size distribution that often tends to follow a Gaussian pattern (Figure 2.6 (a)). In the distribution, the mean represents the average particle size and mode corresponds to the particle size of the maximum amount of powders. For symmetric Gaussian distributions, mean and mode are the same. However, the mode is higher than the mean if the sample has more coarse powders (Figure 2.6 (b)). In contrast, for a sample with more fine powders, the mode is less than the mean (Figure 2.6 (c)). However, in practice, powder size is represented using a cumulative distribution which is a plot of the percentage of total volume versus powder size (Figure 2.6 (d)). The powder size is represented by D_X where X% (in volume) of powders are smaller than D_X. Commonly used values of X are 10, 50,

Figure 2.6 Powder size distribution indicating (a) a Gaussian distribution, (b) a negatively skewed Gaussian distribution for a sample with more coarse particles, and (c) a positively skewed Gaussian distribution for a sample with more fine particles. (d) A cumulative distribution of particle size indicating commonly used terminologies of powder size, D_{10}, D_{50}, and D_{90}. *Source:* T. Mukherjee and T. DebRoy.

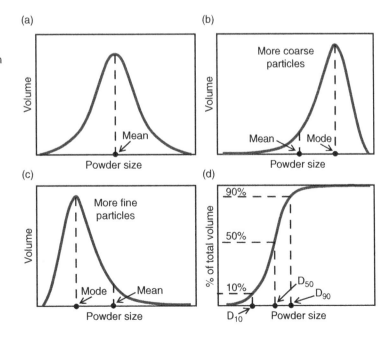

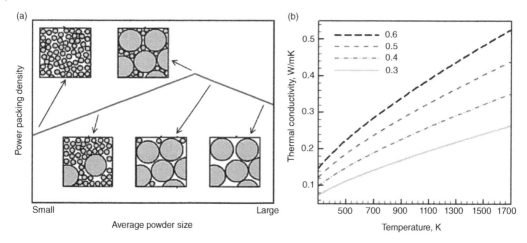

Figure 2.7 (a) Schematic representation of the variation of packing density of a powder bed with average powder particle size. The figure is adapted from [10] with permission from Elsevier. (b) Effect of powder packing density on the effective thermal conductivity of a stainless steel 316 powder bed at different temperatures. *Source:* T. Mukherjee and T. DebRoy.

and 90% (Figure 2.6 (d)). For instance, a powder sample with $D_{50} = 30$ μm represents that 50% (in volume) of the powders are below 30 μm size. An example of how the powder size affects surface roughness is explained in worked out example 2.4.

Powder size is an important variable in both DED and PBF processes. In the DED processes, powder size significantly affects the flow of powders through the nozzle which governs the mass deposition rate. The flow of powders are affected by their size as shown in worked out example 2.5. In PBF, the powder packing depends on the average powder size and the size distribution. Powder packing density is expressed by the ratio of the total volume of powder to the volume of the powder bed. Therefore, packing density varies between 0 and 1. Figure 2.7 (a) shows that for very small powders, the packing density is low because of the gaps among the powder particles. However, for a mixture of large and small powders, small particles can occupy the gaps among the larger particles and increase the packing density [10]. For a powder bed with all very large powders, the packing density is low because of the large gaps among the big particles. Packing density controls the effective thermophysical properties of the powder bed (see Chapter 7). In a powder bed, the properties such as thermal conductivity and specific heat are affected by both the powder size distribution and the shielding gas entrapped in the powder bed. Since gases have lower thermal conductivity than metal powders [11], the effective thermal conductivity of a powder bed significantly decreases with a reduction in packing density as shown in Figure 2.7 (b). A powder bed with a low packing density may be preheated to allow partial sintering to increase the packing density. However, preheating adds an extra cost. Obtaining a powder feedstock with desired properties in a cost-effective manner remains a challenge.

Worked out example 2.4

Surface roughness is an important issue in additive manufacturing (see Chapter 10). Rough surfaces adversely affect the fatigue life of parts. Machining, grinding, and electrochemical polishing are used to achieve smooth surfaces that add cost. Explain which properties of powders are important for reducing surface roughness of parts.

Worked out example 2.4 (Continued)

Solution:

Partially melted powder particles may stick to the deposit and result in rough surfaces [12]. Large powders take more time to melt completely and thus may attach to the deposit and cause surface roughness. Therefore, the use of fine powders is beneficial to produce smooth surfaces.

Worked out example 2.5

In the directed energy deposition processes, poor flowability of powders often results in clogging in the powder feeding nozzle and disruption of the production schedule. Which factors affect nozzle clogging and how?

Solution:

The important factors that affect powder flowability and nozzle clogging are powder flow velocity, nozzle dimensions, powder size, powder density, and carrier gas properties. Their effects are listed below.

- During the flow of powders through a converging nozzle, the velocities of the powder have both axial and radial components. The axial component facilitates the flow of powders. In contrast, the radial component promotes the impingement of the powders on the nozzle wall. Possible adherence of the particles on the wall may cause nozzle clogging.
- Powders are difficult to flow through a nozzle that has a small internal diameter and such nozzles are susceptible to clogging.
- Nozzles may be heated during additive manufacturing because of their proximity to the molten pool. A hot nozzle wall may partially sinter the powders and promote nozzle clogging. Small powders of low-density alloys are easy to sinter and more susceptible to nozzle clogging.
- A high flow rate of the carrier gas maintains the flow of powders through the nozzle and tends to prevent nozzle clogging.

2.3.3 Recycling of powders

In powder-based additive manufacturing processes, not all powders in the deposition chamber are melted and utilized to form the part. The unmelted powders are reused. In DED, a significant portion of the supplied powders is not used to form the part. At the end of the process, the unused powders are collected, blended with fresh powders, and reused to save money. This process is called the recycling of powders and is quite common in AM industry. However, recycled powders often have irregular shapes which affect their flow behavior. For example, Figure 2.8 shows that the use of recycled powders requires more energy to flow through the nozzle during the DED process. Here, the recycled powders are mixed with fresh powders at different proportions. The required energy is higher when the powder mixture contains a high

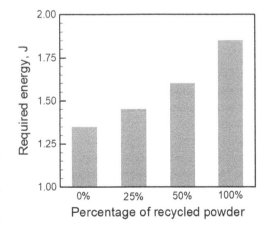

Figure 2.8 Variation in flow energy required to flow the powders through the nozzle in directed energy deposition process for different percentages of recycled powder. The plot is made by T. Mukherjee and T. DebRoy using data reported in [10].

proportion of recycled powders. They often contain dissolved gases such as oxygen which can affect the chemical composition of the part. Gases such as oxygen also affect the way the surface tension of iron alloys changes with temperature. This change in the temperature dependence of surface tension affects the convective heat transfer within the fusion zone and affect the shape of the fusion zone as discussed in worked out example 2.6. In addition, the irregular shape of the recycled powders can contribute to the surface roughness of the parts. Therefore, the extent of powder recycling should be selected considering the cost and the quality of the parts.

Worked out example 2.6

Metallic powders absorb oxygen from the atmosphere when they are used in additive manufacturing at one-atmosphere pressure. Therefore, recycled powders contain more oxygen than fresh powders. How can the use of recycled powders affect the fusion zone geometry, composition, and properties of steel parts?

Solution:

Oxygen is a known surface-active element in many alloys (see Chapter 7) that may change the pattern of convective flow and heat transfer inside the molten pool. For example, during additive manufacturing of steels, the convective flow on the top surface of the molten pool is commonly from the hottest region to the periphery. However, the presence of oxygen may reverse the flow direction (see Chapter 7). Hot liquid under the heat source may flow downward resulting in a deeper and narrower fusion zone. The presence of oxygen may also form oxides that can significantly degrade the mechanical properties and corrosion resistance of parts (see Chapter 9).

2.4   Wire and sheet feedstocks

Wires are commonly used as feedstocks in DED-GMA and DED-PA processes. In addition, wire-based DED with electron beam heat source is also used. Sheet metals are used in ultrasonic additive manufacturing. The commonly used production techniques of wires and sheets and the effects on the AM processes and parts are discussed below.

2.4.1   Wire and sheet production techniques

Wires are commonly manufactured using the wire drawing method. In this method, billets of large diameter are drawn through a converging die under a high tensile force (Figure 2.9 (a)). The diameter of the billet is significantly reduced due to the compression it experiences while passing through the die. Generally, drawing is done in several steps with intermediate heat treatment if needed for stress relieving to reduce the diameter of the billet gradually to the final diameter of the wire. Worked out example 2.7 explains some important issues involved in the wire drawing process. The drawing is often performed by preheating the billet which is called hot drawing. Alloys are softer at high temperatures and require less drawing force. The quality of the wire depends on the preheating temperature, drawing force, initial billet diameter, and design of the die.

Metallic sheets are generally fabricated by the rolling process (Figure 2.9 (b)). Thick slabs of alloy are passed through a small gap between two rotating rollers. The thickness of the slab is reduced due to the compression it experiences while passing through the small gap between the rollers. Rolling is done in several steps to reduce the thickness of the slab gradually to the final thickness of the sheet. The rolling can also be performed by preheating the slab which is called hot rolling. Hot slabs are softer and require less force in the rolling process. The quality of the sheets depends on the preheating temperature, rolling pressure, initial slab thickness, and roller diameter.

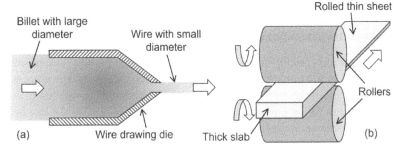

(a) Billet with large diameter — Wire with small diameter — Wire drawing die

(b) Rolled thin sheet — Rollers — Thick slab

Figure 2.9 Schematic representation of (a) wire drawing method to manufacture wire feedstocks and (b) sheet rolling method to fabricate thin sheets for ultrasonic additive manufacturing. *Source:* T. Mukherjee and T. DebRoy.

Worked out example 2.7

A DED-GMA process rewquires stainless steel 304 wires of 2 mm diameter. These wires are to be generated by wire drawing from a 10-mm-diameter cylindrical billet. The wires must have a tensile strength greater than 750 MPa and a ductility of at least 20%. Use the following Figure E2.2 that shows the variations in tensile strength and ductility of stainless steel 304 with percent of cold work and explain how this may be achieved.

Solution:

It is observed from Figure E2.2 that a cold work of at least 10% is needed to achieve a tensile strength of at least 750 Mpa. In addition, the cold work must not exceed 34% to achieve a ductility of at least 20%. However, one-step drawing to reduce the diameter from 10 to 2 mm

requires a cold work of $\left| 1 - \dfrac{\frac{\pi}{4}(2)^2}{\frac{\pi}{4}(10)^2} \right| \times 100\% = 96\%$. Therefore, one step drawing cannot be used

in this case. A possible alternative is to use a multistep wire drawing with 20% cold work in each step along with inter-step heat treatment to make the wires strain free.

In the last step, the wires are made with 20% cold work to achieve a diameter of 2 mm. For this purpose, the wire diameter before the last step should be $= 2 \text{ mm}/(1 - 20/100)^{0.5} = 2 \text{ mm}/0.8944 = 2.24 \text{ mm}$. If the number of steps required is N, then $2 \text{ mm}/(0.8944)^N = 10 \text{ mm}$. By solving, N ~ 15 steps. Therefore, it needs 15 steps of wire drawing with inter-step heat treatment to make 2 mm diameter wires with the desired properties by a cold drawing of a 10 mm diameter cylindrical billet.

Figure E2.2 Variations in tensile strength and ductility of stainless steel 304 with percent of cold work. The plot is made using the data from [13].

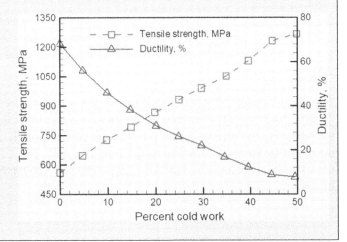

2.4.2   Effects of wire or sheet characteristics on part quality

The microstructure, properties, and susceptibility to defect formation in parts made using wire-based DED or ultrasonic AM are affected by the properties of the wire or sheet feedstocks and the process variables. For wire feedstock, several characteristics are similar to the consumable wire used in fusion welding processes. Like fusion welding processes, the deposition rate of DED-GMA can be controlled by adjusting the wire diameter as explained in worked out example 2.8. Generally, wire and sheet feedstock characteristics are represented by the diameter of wires, the thickness of sheets, surface finish, composition, and density. Wires with large diameters and thick sheets are easy and cost-effective to manufacture. In contrast, very thin wires or sheets are relatively expensive and may break during feeding and therefore are difficult to control. Sheets with rough surfaces are often beneficial because they can generate more frictional heat during ultrasonic additive manufacturing.

The composition of wires plays an important role in affecting the part geometry and properties. For example, wires may contain oxygen that may form oxide scales on the wire feedstocks. These oxides can significantly degrade the microstructure and properties of the parts. Oxide scales on the sheet feedstocks can affect the generation of frictional heat during ultrasonic AM. Therefore, the sheet feedstocks need to be cleaned before usage. The presence of surface-active elements such as sulfur and oxygen in steel wires can change the convective heat transfer pattern inside the molten pool affecting the deposit geometry (see Chapter 7).

2.5   Summary

Powder bed fusion and directed energy deposition are two commonly used additive manufacturing processes used for making metallic parts. Other additive processes such as binder jet and hybrid processes are also used to a limited extent. Lasers, electron beams, electric arcs, and plasma arcs are commonly used as heat sources. Powders and wires are commonly used feedstocks. Metal powders are produced by gas atomization, water atomization, centrifugal atomization, and plasma rotating electrode processes. The shape and size of the powders affect heat absorption, heat conduction through the powder bed, surface roughness of parts, and the ability to make intricate parts. Recycling unused powders help to make the powder bed process cost-competitive and sustainable. Both the wire feed rate and wire diameter affect the deposition rate of the directed energy deposition process.

Worked out example 2.8

Conduct a literature review to investigate how wire diameter and wire feed rate affect deposition rate during directed energy deposition of metal parts, and explain how DED-GMA productivity can be increased. What are the disadvantages of these methods?

Solution:

The productivity of DED-GMA can be enhanced by using wires of larger diameters or increasing the wire feed rate [14]. For example, Figure E2.3 shows that thicker deposits can be achieved by increasing the wire feed rate (for 1.2 mm wire diameter) or by using wires with larger diameters (at 80 mm/s wire feed rate) during DED-GMA of H 13 tool steel. However, both the use of large diameter wire and higher wire feed rate can have detrimental effects as listed below:

Worked out example 2.8 (Continued)

- Deposition of thick layers may result in rough surfaces (see Chapter 10) that need expensive post-process machining or grinding to achieve the desired surface finish.
- At high feed rates, wires may break and interrupt the process.
- Thick layers exhibit considerable shrinkage during solidification and cooling and may result in distortion and cracking (see Chapter 11).

Figure E2.3 Thicker deposits are made by increasing the wire feed rate (for 1.2 mm wire diameter) or by using wires with larger diameters (at 80 mm/s wire feed rate) during DED-GMA of H 13 tool steel. *Source:* T. Mukherjee and T. DebRoy.

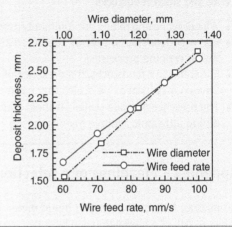

Takeaways

Additive manufacturing processes

- Powder bed fusion and directed energy deposition processes are commonly used to make metallic components.
- Laser, electron beam, electric arc, and plasma arc are the commonly used heat sources.
- Feedstocks are used in the form of powder, wire, and thin sheets.
- Binder jet process, ultrasound-assisted additive manufacturing, friction stir additive manufacturing, and hybrid additive manufacturing are emerging processes.

Powder production techniques

- Commonly used powder production techniques are gas atomization, water atomization, centrifugal atomization, and plasma rotating electrode process.
- Gas and water atomization are widely used because of their low costs. However, they often do not provide high-quality powders.
- The plasma rotating electrode process can produce high-quality powders but is expensive.

Powder characteristics and their effects on part quality

- Powder morphology and size affect heat absorption, heat conduction through a powder bed, surface roughness, and the ability to print intricate geometries.
- Poor flowability of powders through a nozzle during the directed energy deposition can result in nozzle clogging that interrupts the process.
- Powders produced using the gas or water atomization may contain gas or water vapors entrapped inside the particles which may cause porosities in the part.

(Continued)

Takeaways (Continued)

Recycling of powders

- Powders that remain in the chamber after the part is made are often recycled to save money.
- Recycled powders may have irregular geometry and may contain excess oxygen that can adversely affect the properties of the printed parts.

Wire and sheet feedstocks

- Wire drawing and rolling are commonly used to produce wire and sheet feedstocks, respectively.
- For wire feedstocks, several characteristics are similar to the consumable wires used in fusion welding processes.
- For steel wire feedstocks, the presence of sulfur and oxygen affects deposit shape and size. Oxygen may form oxide scales that degrade part properties.
- The presence of oxide scales on sheet feedstocks may affect the generation of frictional heat during ultrasonic additive manufacturing. Therefore, sheets need to be cleaned before usage.

Appendix – Meanings of a selection of technical terms

<u>Atomizing</u>: A process to convert liquid metals and alloys to powders.

<u>Billet</u>: A slab of material that can be rolled to make sheets.

<u>CAD</u>: The acronym of CAD stands for computer-aided design. CAD represents a three-dimensional design of a part.

<u>Die</u>: A specially designed tool for wire drawing. A die is made of very hard material. Wires are drawn through a die to reduce their diameter.

<u>Electric arc</u>: A high-energy arc that is used as a heat source in additive manufacturing.

<u>Electron beam</u>: A focused beam of electrons used as a heat source in additive manufacturing.

<u>Friction stir welding</u>: A manufacturing process to join materials. A rotating tool generates frictional heat that softens the materials to join.

<u>Fusion welding</u>: A manufacturing process that uses a heat source to melt two or more parts to form a sound joint after solidification.

<u>Laser beam</u>: The acronym laser stands for "light amplification by stimulated emission of radiation." A focused beam of laser is often used in additive manufacturing as a heat source.

<u>Plasma</u>: A mixture of ionized, excited, and neutral atoms and electrons that can carry current. A plasma arc is often used as a heat source in additive manufacturing.

<u>Sonotrode</u>: A sonotrode is a device that uses high-frequency sound waves to generate heat. It generates ultrasonic vibrations and applies them to the materials being welded. It is typically made of a hard material such as tool steel or tungsten carbide.

<u>Tundish</u>: A tundish is a container that is used to hold and distribute molten metal where needed in a metallurgical process. It is typically located between a furnace and a casting machine

Practice problems

1) What are the important factors for the selection of an additive manufacturing process for making metallic parts?

2) A dental implant needs to be made using a titanium alloy which requires to have intricate features with high dimensional accuracy to fit in a patient's gum. Which AM process is the best suitable for making the part?

3) Which additive manufacturing process is commonly used to make large metal parts and why?

4) What are the main difficulties in producing metal powders for additive manufacturing by gas and water atomizing?

5) A component is to be made using very fine layers of powders using a material that is reactive to water. What is the most cost-effective atomization technique for making the powders?

6) Among powder and wire-based additive manufacturing processes, which one exhibits higher productivity and why?

7) How do the attributes of the powders affect the quality of metal parts made by additive manufacturing?

8) Two parts, A and B, are made of the same alloy powders and processing conditions. Part A is made of powders produced by centrifugal atomization. Gas-atomized powders are used to print part B. Which part is expected to have a better tensile property and why?

9) How do recycled powders affect the quality of parts made by additive manufacturing?

10) Wire feedstocks are often used in directed energy deposition additive manufacturing. What roles do minor alloying elements such as oxygen and sulfur play in additive manufacturing of steel parts?

References

1 DebRoy, T., Wei, H.L., Zuback, J.S., Mukherjee, T., Elmer, J.W., Milewski, J.O., Beese, A.M., Wilson-Heid, A.D., De, A. and Zhang, W., 2018. Additive manufacturing of metallic components–process, structure and properties. *Progress in Materials Science*, 92, pp.112–224.

2 DebRoy, T., Mukherjee, T., Wei, H.L., Elmer, J.W. and Milewski, J.O., 2021. Metallurgy, mechanistic models and machine learning in metal printing. *Nature Reviews Materials*, 6(1), pp.48–68.

3 Palanivel, S., Sidhar, H. and Mishra, R.S., 2015. Friction stir additive manufacturing: Route to high structural performance. *JOM*, 67(3), pp.616–621.

4 Hendrixson, S., 2019. *AM 101: Hybrid manufacturing*. available at: https://www.additivemanufacturing.media/articles/am-101-hybrid-manufacturing (accessed on 18 January 2023).

5 Antipas, G.S.E., 2013. Review of gas atomisation and spray forming phenomenology. *Powder Metallurgy*, 56(4), pp.317–330.

6 Liu, J., Arnberg, L. and Backstrom, N., 1988. Mass and heat transfer during ultrasonic gas atomization. *Powder Metallurgy International*, 20(2), pp.17–22.

7 Qi, H., Azer, M. and Ritter, A., 2009. Studies of standard heat treatment effects on microstructure and mechanical properties of laser net shape manufactured Inconel 718. *Metallurgical and Materials Transactions A*, 40(10), pp.2410–2422.

8 Sames, W.J., Medina, F., Peter, W.H., Babu, S.S. and Dehoff, R.R., 2014. Effect of process control and powder quality on Inconel 718 produced using electron beam melting. In *8th International Symposium on Superalloys*, 718, pp.409–423.

9 Pinkerton, A.J. and Li, L., 2005. Direct additive laser manufacturing using gas-and water-atomised H13 tool steel powders. *The International Journal of Advanced Manufacturing Technology*, 25(5), pp.471–479.

10 Tan, J.H., Wong, W.L.E. and Dalgarno, K.W., 2017. An overview of powder granulometry on feedstock and part performance in the selective laser melting process. *Additive Manufacturing*, 18, pp.228–255.

11 Mukherjee, T., Wei, H.L., De, A. and DebRoy, T., 2018. Heat and fluid flow in additive manufacturing—Part I: modeling of powder bed fusion. *Computational Materials Science*, 150, pp.304–313.

12 Mumtaz, K. and Hopkinson, N., 2009. Top surface and side roughness of Inconel 625 parts processed using selective laser melting. *Rapid Prototyping Journal*, 15(2), pp.96–103.

13 Milad, M., Zreiba, N., Elhalouani, F. and Baradai, C., 2008. The effect of cold work on structure and properties of AISI 304 stainless steel. *Journal of Materials Processing Technology*, 203(1–3), pp.80–85.

14 Ou, W., Mukherjee, T., Knapp, G.L., Wei, Y. and DebRoy, T., 2018. Fusion zone geometries, cooling rates and solidification parameters during wire arc additive manufacturing. *International Journal of Heat and Mass Transfer*, 127, pp.1084–1094.

3

Feedstocks and Processes for Additive Manufacturing of Polymeric Parts

---

Learning objectives

After reading this chapter the reader should be able to do the following:

1) Understand the working principles of commonly used additive manufacturing processes for polymers.
2) Know several techniques of producing liquid, powder, and wire-based polymer feedstocks and their advantages and disadvantages.
3) Select an appropriate printing technique and polymer feedstocks for desired parts.
4) Recognize the effects of feedstock characteristics on part quality.
5) Understand the benefits and limitations of different additive manufacturing processes for printing polymer composite parts.
6) Know several applications of printed polymeric and composite parts in diverse industries.

CONTENTS

Theory and Practice of Additive Manufacturing, First Edition. Tuhin Mukherjee and Tarasankar DebRoy.
© 2024 John Wiley & Sons, Inc. Published 2024 by John Wiley & Sons, Inc.

3.1 Introduction

Additive manufacturing (AM) of polymeric materials is the process of building parts by progressively adding thin layers of polymers. A computer program containing a three-dimensional design of the part guides the printing of the part layer-by-layer. There are various types of AM processes for polymers including vat photopolymerization, fused deposition modeling, inkjet printing, selective laser sintering, binder jetting, and laminated object manufacturing [1–3]. These processes are used to make various prototypes, tools, molds, and functional parts for the healthcare, automotive, consumer products, and aerospace industries. Polymers in the form of liquids, powders, and wires are commonly used as feedstocks. The selection of an appropriate combination of AM process and feedstock is very important to print high-quality parts. In this chapter, we explain commonly used AM processes and equipment for polymers and polymer composites. In addition, liquid, powder, and wire feedstocks of polymeric materials and their usage in AM are described. Furthermore, different applications of printed polymer and polymer composite parts in various industries are discussed.

3.2 Additive manufacturing processes and equipment

The commonly used AM processes and equipment for polymeric materials significantly vary depending on the shape, size, and complexity of the part, the feedstock used, and productivity. However, for all processes, a computer-aided design (CAD) file helps to determine the orientation of the part within the build volume and the support structure that may be needed to build the part. For example, an appropriate support structure may be needed if a part contains overhangs. The processing conditions and feedstock materials are often selected based on the machine manufacturer's recommendations. This section introduces the commonly used AM processes for polymeric materials.

3.2.1 Vat photopolymerization

Vat photopolymerization is a commonly used additive manufacturing process in which a liquid resin or a photopolymer in a vat is selectively cured by a light source to form solid, long chains of polymers to form a part. This process uses various types of light sources ranging from ultraviolet laser to visible light. The vat photopolymerization process is well-known for producing parts with excellent precision and resolution. Two main types of vat photopolymerization processes, stereolithography, and digital light processing are discussed below. Their main features are compared in worked out example 3.1.

3.2.1.1 Stereolithography

In the stereolithography (SLA) process, a coherent light source, usually an ultraviolet laser, is used to cause the polymerization of photosensitive liquid resins (photopolymers). The polymerization results in the cross-linking of the initially liquid resins and produces solid, hard, long chains of polymers. The light source selectively scans the liquid and makes the solid part (Figure 3.1). After printing a layer, the build platform is lowered by a distance equal to the layer thickness, and a new layer of liquid resin is spread on it. The process continues until the entire part is made.

In SLA, the main time-consuming part is not the laser scanning but the spreading of the new layer of the liquid resin. The layer of the liquid resin should have a uniform thickness with no ripples and bubbles entrapped inside. Therefore, resins with low viscosity (between 0.1 and 10 Pa·s) are often desirable. Generally, nonreactive additives or solvents are used to reduce the viscosity of the liquid photopolymer resin. By varying the spot size of the light source, a high spatial resolution can be achieved. The time necessary to print a layer depends on the scanning speed of the light source. The movement of the light source is usually controlled by a reflector within a Galvano scanner. The maximum allowable layer thickness is dependent on the light penetration depth, which can be controlled by the addition of suitable absorbers to the photopolymer resin.

3.2.1.2 Digital light processing

In digital light processing (DLP), a light source is used to cure a photopolymer in a layer-by-layer manner to build a part. A selectively masked light source from the bottom of the liquid vat cures one entire layer at a time (Figure 3.2). That is why this process is often called dynamic mask photolithography. The information for each layer of the 3D part design is provided in the form of a black-and-white

Figure 3.1 Schematic diagram of the stereolithography process [4]. The figure is taken from an open-access article [4] under the terms and conditions of the Creative Commons Attribution (CC BY) license.

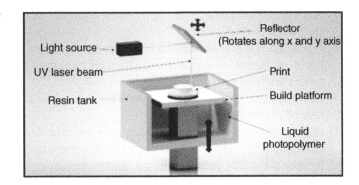

Figure 3.2 Schematic diagram of digital light processing [2]. (a) A vat filled with photopolymer resin, (b) a light source, (c) a reflector within a Galvano scanner, (d) a vertically movable building platform, and (e) a tilting device to spread the resin at the bottom. The figure is taken from an open-access article [2] under the terms and conditions of the Creative Commons Attribution (CC BY) license.

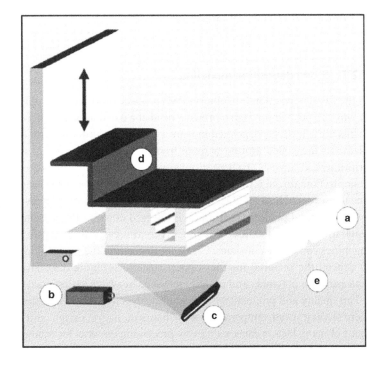

pattern. Such binary patterns are presented via a digital micromirror device that selectively cures the photopolymer. Since an entire layer is made in one step, the productivity of this process is very high. The spatial resolution of the DLP systems depends on the number of pixels/mirrors provided by the digital micromirror device and the light used to project the patterns onto the part building inside the vat. The maximum allowable layer thickness depends on the light penetration depth, which can be controlled by the addition of suitable absorbers to the photopolymer resin.

Worked out example 3.1

Stereolithography (SLA) and digital light processing (DLP) are the two main types of vat polymerization processes. What are the main differences between these two processes?

Solution:

The main differences are listed below:
1) In SLA, the light source selectively scans and cures the liquid photopolymer. In contrast, the entire layer is exposed not point-by-point but all at once with a selectively masked light source in the DLP.
2) SLA uses an ultraviolet laser light source. A digital micromirror device projector projects visible light in DLP process.
3) In SLA, a part is printed from the bottom to the top where the entire part remains inside the liquid vat. In contrast, a part is made from top to bottom in DLP where the part is gradually taken out of the vat as they are made.
4) SLA needs spreading of the new layer of the liquid resin after printing each layer. This is not needed in the DLP process.
5) In SLA, the curing happens on the top surface on the liquid vat and can be affected by the atmosphere. However, the effect of the atmosphere does not affect the DLP part quality because the curing occurs inside the liquid vat.
6) In SLA, productivity depends on the scanning speed of the light source. Since DLP makes one entire layer at a time, the productivity is expressed in mm per hour, that is, the height of the part (number of layers) per unit time.

3.2.2 Fused deposition modeling

In the fused deposition modeling (FDM) process, thermoplastic polymers in the form of a filament or wire are pushed through a printing head at a constant speed in a continuous stream and the filaments are selectively deposited in a layer-by-layer manner to print the part in 3D (Figure 3.3). The filaments are coiled around a spool from which they are fed in a printing head assembly. The printing head consists of a heater and an extrusion nozzle. The heater makes the filament softer and easier to flow through the extrusion nozzle. Once the heated filament is extruded out, it solidifies due to the loss of heat and fuses with the previously deposited layers. After the deposition of a layer, either the build platform (substrate) is lowered down or the printing head is lifted with a distance equal to the layer thickness. The process continues until the entire part is made. Most of the FDM machines contain two nozzles. One is for depositing the material for the part and the other is used to deposit the material for the support structure (Figure 3.3). The support structure is made of water-soluble polymers and is easily removed by water jetting once the printing is completed.

Part quality and productivity are significantly affected by the scanning speed, layer thickness, filament feeding speed, temperature of the filament, nozzle design, and polymer viscosity. The involvement of many factors causes a narrow processing window for many FDM instruments (see worked

Figure 3.3 Schematic diagram of fused deposition modeling [2]. (a) A vertically movable building platform or substrate, (b) a horizontally movable printing head for deposition of (c) build material and (d) support material, and (e) filaments or wires of polymer on a spool. The figure is taken from an open-access article [2] under the terms and conditions of the Creative Commons Attribution (CC BY) license.

out example 3.2). Generally, the polymer filament is heated at a suitable temperature to ensure a complete fusional bonding with the previously deposited layers. An incomplete fusion may result in weak macroscopic surface adhesion, voids, and poor mechanical properties. Printing with thick layers may cause surface roughness (see worked out example 3.3). The FDM process is widely used for printing both prototypes and functional components.

Worked out example 3.2

In a fused deposition modeling process, the printing nozzle has a maximum injection rate of 20 mm^3/s. If the nozzle diameter is 0.5 mm, what is the maximum allowable scanning speed to maintain a continuous deposition process?

Solution:

To maintain a continuous deposition, the injection rate (Q) can be expressed as,

$$Q = \frac{\pi d^2}{4} v \tag{E3.1}$$

where d is the nozzle diameter and v is the scanning speed. Using equation E3.1, the maximum allowable scanning speed to maintain a continuous deposition process is calculated as 101.9 mm/s. This scanning speed is similar to what is generally used in fused deposition modeling [5].

Worked out example 3.3

In the fused deposition modeling process, surface roughness due to the staircase effect is a major issue. Calculate the average surface roughness of an inclined plane with an oblique angle of 45 degrees. The thickness of the layers is 100 micrometers.

Solution:

The average surface roughness (Ra) due to the staircase effect (see Figure E3.1 below) can be expressed as,

$$Ra = t\cos\theta \tag{E3.2}$$

where t is the layer thickness and θ is the oblique angle of the surface. Using equation E3.2, the average surface roughness is calculated as 70.7 micrometers.

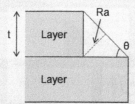

Figure E3.1 Surface roughness due to the staircase effect during fused deposition modeling. *Source:* T. Mukherjee and T. DebRoy.

3.2.3 Inkjet printing

In the inkjet printing process, a liquid photopolymer ink is selectively jetted onto the build platform (substrate) and then is cured by an ultraviolet light source to form the solid deposit (Figure 3.4). A horizontally movable inkjet printing head has several hundred nozzles to eject small droplets of liquid photopolymer. After the selective deposition of a layer of ink, an ultraviolet light source flashes to cure the liquid ink. The process is repeated until the entire part is made. An inkjet printing head can deposit both the building material and the material for the support structure. The support structures are made of water-soluble polymers and are removed by water jetting at the end of the process.

The main advantage of this process is its capability of printing 3D multi-material or multicolor structures by ejecting various types of liquid polymers through different nozzles. Parts consisting of multiple materials with different optical or mechanical properties can be printed without an additional assembly step. However, this process has not been used to make functional components. In addition, the processing window for the available polymer inks is very narrow with strict requirements regarding the density, viscosity, and surface tension of inks. A dimensionless Ohnesorge number is often used to ensure that the ink flows continuously during inkjet printing (see worked out example 3.4).

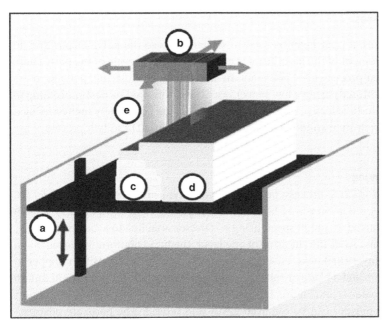

Figure 3.4 Schematic diagram of inkjet printing [2]. (a) A vertically movable building platform or substrate, (b) a horizontally movable multi-nozzle inkjet printing head, (c) layers of support material, (d) layers of building material, and (e) an ultraviolet light source attached to the inkjet printing head. *Source:* The figure is taken from an open-access article [2] under the terms and conditions of the Creative Commons Attribution (CC BY) license.

Worked out example 3.4

During inkjet printing, a continuous flow of the polymer ink through the inkjet nozzle is desired. Inks that have either very low or very high viscosity cannot maintain a continuous flow. Propose a dimensionless number to indicate the continuity of the flow of the ink during inkjet printing. For a typical inkjet printing, the radius of the nozzle is 0.3 mm. If the surface tension, density, and viscosity of the ink are 0.046 N/m, 1110 kg/m^3, and 0.016 Pa.s, respectively, can this ink flow continuously?

Solution:

The continuity of the flow can be indicated by Ohnesorge number [1] defined as:

$$Oh = \frac{\eta}{\sqrt{\gamma \rho r}} \tag{E3.3}$$

where r is the radius of the inkjet nozzle, γ, ρ, and η are the surface tension, density, and viscosity of the ink, respectively. For a continuous flow of ink during inkjet printing, the value of $1/Oh$ should be between 1 and 10.

For the parameters provided, the calculated value of $1/Oh$ is 7.73. Therefore, the ink can flow continuously during this inkjet printing process.

3.2.4 Powder bed processes

In the powder bed processes, a laser beam is used to selectively fuse the polymer powders to print a part in 3D. The movement of the laser beam is guided by the 3D design of the part. There are two types of powder bed processes for polymers. In selective laser sintering, polymer powders are partially melted (sintered) using a low-power laser beam followed by post-processing to achieve the desired strength. In selective laser melting, powders are completely melted using a high-power laser beam which then solidifies to produce the 3D part. These two processes are discussed below.

3.2.4.1 Selective laser sintering

In the selective laser sintering (SLS) process, polymer powders typically 10–100 micrometers in diameter are selectively sintered in a layer-by-layer manner to print 3D parts (Figure 3.5). Within a layer, the laser beam is moved at a high speed using a reflector attached to a galvanometer to selectively sinter the powders. After the sintering of one layer, the build platform is moved down by a distance equal to the layer thickness. Powders are supplied from the deposition hopper to make a new layer of powders and a blade or roller distributes the powder uniformly to maintain a constant height of the powder layer (Figure 3.5). These steps are repeated until the entire part is made. In the end, loose powder can be removed easily and reused. The parts are post-processed to remove any interparticle voids and to achieve the desired strength. In this process, loose powder particles remain on the build platform to serve as a support structure. Therefore, parts with intricate and complex structures can be printed without the need for support structures.

3.2.4.2 Selective laser melting

Unlike selective laser sintering, in selective laser melting (SLM), polymer powders are completely melted using a high-power laser beam. Fine powders typically 10–100 micrometers in diameter are packed within a box that defines the maximum dimensions of the part. A laser beam is moved using a mirror attached to a galvanometer to selectively melt the powders. Fusion occurs along the path of a laser beam as it moves at high speed. The powder bed is enclosed in a chamber filled with

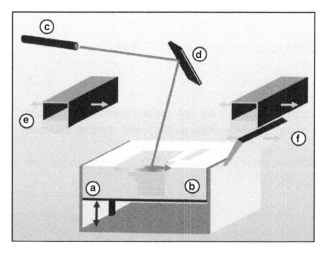

Figure 3.5 Schematic diagram of selective laser sintering [2]. (a) A vertically movable building platform or substrate, (b) powder bed, (c) laser source, (d) reflector with Galvano scanner, (e) powder feedstock and deposition hopper, and (f) blade for powder distribution and leveling. The figure is taken from an open-access article [2] under the terms and conditions of the Creative Commons Attribution (CC BY) license.

an inert gas to prevent oxidation. After each layer is deposited, the powder bed is lowered by a short distance equal to the layer thickness, and a thin layer of powder is added by pushing a small amount of powder. The path of the laser beam does not have to follow the same pattern for each layer. Unidirectional, bidirectional, spiral, zigzag, and other patterns are selected considering the geometry of a given layer and the properties of the deposit. The process continues until the entire part is made. Selective laser melting can print 3D complex parts with intricate geometry and high-dimensional accuracy without the need for post-processing.

3.2.5 Binder jetting

In a binder jetting process, a liquid thermosetting polymer is selectively deposited on the layers of fine polymer powders. The liquid polymer or the binder forms an agglomerate of the polymer powders that defines the shape of the part (Figure 3.6). At the end of each layer, the powder bed is heated using an external heat source to partially cure the binder. This partial curing makes the powder-binder agglomerate strong enough to support the deposition of the next layer of powders. The process continues until the entire part is printed. In the end, the entire part is sintered in an oven to improve its strength. Often, the sintered parts need to be machined to achieve the desired shape and size.

The Binder jet process is used to make prototypes of different shapes and colors, molds, and cores for casting. Since this process does not employ melting and solidification of material, it can avoid solidification shrinkage and cracks. Unused powder particles serve as support materials and upon completion, they may be recovered and reused. However, the parts printed by this process often lack the precision that can be achieved by SLS and SLM. Furthermore, the parts need post-process sintering to improve strength which adds cost. Post-process sintering can cause some shrinkage which needs to be considered in part design. In addition, the parts are often not fully dense which may result in poor mechanical properties.

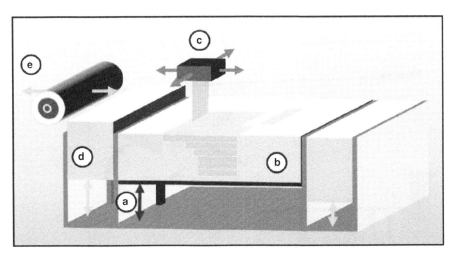

Figure 3.6 Schematic diagram of binder jetting [2]. (a) A vertically movable building platform or substrate, (b) powder bed, (c) horizontally movable printing head for deposition of binder material, (d) storage for the powder feedstock, and (e) roller for powder distribution and leveling. The figure is taken from an open-access article [2] under the terms and conditions of the Creative Commons Attribution (CC BY) license.

3.2.6 Laminated object manufacturing

In the laminated object manufacturing (LOM) process, 3D objects are made by stacking and laminating thin sheets of polymeric material (Figure 3.7). The adhesive-coated polymer sheets are fed through rollers to the building platform. A laser beam is used to selectively cut the contours in two-dimensional slices guided by the 3D design of the part. The movement of the laser beam is controlled by a reflector attached to a Galvanometer. The contoured area is attached to the building part by the adhesive. A fresh sheet is passed through the rollers and the process is repeated until the entire part is made. The surrounding sheet acts as the support structure. However, this process involves a large amount of material wastage. In addition, this process is limited to a few polymers for which thin sheets can be made. The vertical resolution of the part depends on the sheet thickness and the use of thick sheets may result in rough surfaces. Since the layers are bound by the adhesive, the part may exhibit delamination with time. This process has been used to make nonfunctional prototypes.

3.2.7 Comparison among different processes

The commonly used AM processes for polymeric materials vary widely depending on the shape, size, and complexity of the part, the feedstock used, and productivity. These processes are compared in Table 3.1. The part attributes significantly vary depending on the process used. For example, parts made using fused deposition modeling exhibit rougher surfaces than those printed using stereolithography and selective laser sintering (Figure 3.8). The information provided in the table for the selection of an appropriate process for making a part.

3.2.8 Available equipment

The additive manufacturing of polymeric materials has been developed over a long period of time. Currently, there are many manufacturers worldwide making various types of AM machines for polymeric materials. Table 3.2 summarizes a selection of widely used AM equipment and their manufacturers, maximum build size, and minimum layer thickness that affects the dimensaional accuracy of parts.

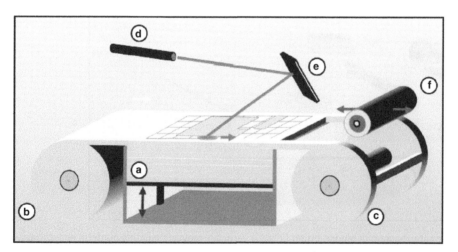

Figure 3.7 Schematic diagram of laminated object manufacturing [2]. (a) A vertically movable building platform or substrate, (b) adhesive-coated polymer sheet rolls, (c) residual sheet collection, (d) a laser source, and (e) a reflector for the laser movement for cutting layer contours and crosshatch pattern. The figure is taken from an open-access article [2] under the terms and conditions of the Creative Commons Attribution (CC BY) license.

Table 3.1 Comparison among different AM processes for polymers. Adapted from [1, 2].

AM process	Type of polymer used	Spatial resolution, μm	Benefits	Limitations
Stereolithography	Liquid	50–100	Excellent surface quality and precision	Poor mechanical properties
Digital light processing	Liquid	25–100	High productivity Low initial vat volume Good surface quality	Low-viscosity polymer needed Poor mechanical properties
Fused deposition modeling	Wire or filament	100–150	Inexpensive machines Cheap materials	Rough surfaces High-temperature process
Inkjet printing	Liquid	50–100	Allows multi-materials Room temperature process	Low-viscosity ink required Needs post-processing
Selective laser sintering	Powder	50–100	Less anisotropic properties Good for intricate parts	Slow productivity Needs post-processing
Binder jetting	Powder	50–100	High productivity Low-temperature process Allows multi-materials	Rough surfaces Needs post-processing
Laminated object manufacturing	Sheet	200–300	High productivity Support structure not needed	Limited materials Low resolution High anisotropy

Figure 3.8 Comparison of parts made by fused deposition modeling (FDM), stereolithography (SLA), and selective laser sintering (SLS) [6]. The figure is reprinted with permission from Manufactur3D Magazine.

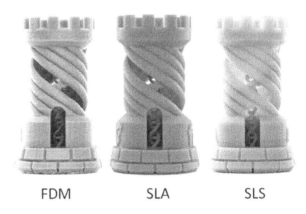

FDM   SLA   SLS

Table 3.2 A selection of commercial machines for additive manufacturing of polymers. Adapted from [1].

AM process	Manufacturer	Maximum build size (mm³)	Minimum layer thickness (μm)
Stereolithography	Stratasys	145 × 145 × 175	25
	3D Systems	1500 × 750 × 550	25
Digital light processing	EnvisionTEC	140 × 79 × 100	25
	Prodway	800 × 330 × 400	25

(Continued)

Table 3.2 (Continued)

AM process	Manufacturer	Maximum build size (mm^3)	Minimum layer thickness (μm)
Fused deposition modeling	Stratasys	$223 \times 223 \times 305$	100
	Makerbot	$1005 \times 1005 \times 1005$	100
Inkjet printing	Stratasys	$294 \times 192 \times 148$	16
Selective laser sintering	EOS	$340 \times 340 \times 600$	60
	3D Systems	$1400 \times 1400 \times 500$	60
Binder jetting	Voxeljet	$1800 \times 1000 \times 700$	260
	ExOne	$4000 \times 2000 \times 1000$	100
Laminated object manufacturing	EnvisionTEC	$160 \times 210 \times 135$	100
	Katana	$180 \times 280 \times 150$	100

3.3 Feedstocks

In additive manufacturing, polymer feedstocks are used in the form of liquid, powders, wires or filaments, and sheets. Table 3.3 summarizes the commercially available feedstocks.

3.3.1 Liquid polymers

Liquid photopolymers (also known as liquid photo-curable resins) are used both in the vat photopolymerization processes as well as in inkjet printing. However, different types of photopolymers are used in these two processes (Table 3.3). In inkjet printing, long chains of liquid polymers are used that have very good thermal stability and yet cure rapidly when exposed to light. In contrast, photopolymers used in the vat photopolymerization processes generally contain a mixture of liquid monomers and photoinitiators. Because of the light, the photoinitiators undergo a chemical transformation and produce free radicals or ions. These free radicals or ions initiate the polymerization reaction in which the monomer molecules are combined to form a polymer chain. The long and cross-linked polymer chains are solid and hard materials that form the printed part.

The polymerization reaction occurs by using either free radicals or ion-based photoinitiators. Table 3.3 provides a list of commonly used free radical and ion-supported photopolymers. For example, a mixture of benzophenone and benzyl dimethylamine is used as a free radical-based photoinitiator to initiate the polymerization reaction for the vinyl monomers such as urethane acrylate, bisphenol A acrylate, and poly (ethylene glycol) acrylate. Figure 3.9 explains the reactions to produce free radicals. The nitrogen in the benzyl dimethylamine has a lone pair of electrons in the 2s orbital. The carbon attached to the benzene ring can take advantage of that lone pair, release an H$^+$ ion in the solution, and become a free radical. The presence of light can put the benzophenone in an excited state. Under that condition, one of the two covalent bonds between carbon and oxygen can break and oxygen is negatively charged (O$^-$). Then it can react with the H$^+$ ion in the solution and produce a free radical. Thus, the mixture of benzophenone and benzyl dimethylamine when activated by a light source can produce two free radicals. These free radicals can initiate polymeric chain reactions. Engineers have found several such photoinitiators for initiating the polymeric chain reactions during the vat photopolymerization processes.

Table 3.3 Different polymers used in additive manufacturing. Adapted from [1, 3].

Polymer type	Name of polymers	AM processes
Liquid	Photopolymers such as poly (diethyl amino) ethyl acryl amide, poly (diethyl amino) ethyl methacrylate, poly (morpholine) ethyl methacrylate, poly (dimethyl amino) ethyl methacrylate, Poly (oligo (ethylene glycol) methacrylate), and poly (diethyl acryl amide)	Inkjet printing
	Free radical-supported photopolymers such as urethane acrylate, bisphenol A acrylate, and poly (ethylene glycol) acrylate	Stereolithography, digital light processing
	Ion-supported photopolymers such as bisphenol A diglycidyl ether and trimethylolpropane triglycidyl ether	
Powder	Polycarbonate, nylons, acrylic styrene, polyamides, polystyrenes, thermoplastic, elastomers, polyaryletherketones	Selective laser sintering, selective laser melting, binder jetting
Wire or filament	Thermoplastics such as acrylonitrile styrene acrylate, acrylonitrile butadiene styrene, polycarbonate, polyetherimide, polylactic acid, high-impact polystyrene, thermoplastic polyurethane, aliphatic polyamides, and high-performance plastics such as polyether ether ketone, and polyetherimide	Fused deposition modeling
Sheet	Acrylonitrile butadiene styrene copolymers, polylactide, polycarbonate, and polyamides	Laminated object manufacturing

Figure 3.9 Generation of the free radicals from the benzophenone and benzyl dimethylamine photoinitiators. The "dot" indicates the free radical state. The reaction is taken from [2]. *Source:* T. Mukherjee and T. DebRoy.

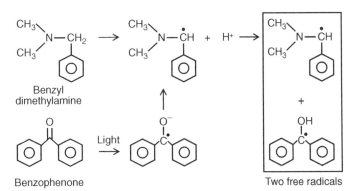

3.3.2 Powder feedstocks

Polymer powders are used in the selective laser sintering, selective laser melting, and binder jetting processes. Although various polymers are used to make powders (Table 3.3), polyamides and poly-styrenes are the most commonly used polymers for powder production. Table 3.4 summarizes several polyamide and polystyrene powders and their suppliers. Polymer powders are primarily made using three techniques as described below. In addition, a comprehensive comparison of these processes based on their relative advantages and disadvantages is provided in Table 3.5.

Cryogenic grinding: Several polymeric pellets are cooled well below 0°C and then crushed bet-ween two counter-rotating discs to produce powders. Crushed particles of different sizes are sepa-rated by sieving to achieve a narrow size distribution. Pellets are cooled because the heat generated

Table 3.4 A selection of polyamide and polystyrene powders for additive manufacturing. Here, PA: Polyamides and PS: Polystyrenes. Adapted from [1].

Polymer name	Trade name	Supplier
PA-12	Duraform PA	3D Systems
PA-12	NyTek 1200 PA	Stratasys
PA-12	Orgasol Invent Smooth	Arkema
PA-12	PA 2201	EOS
PA-12	PA650	ALM
PA blend	DuraForm EX	3D Systems
PA-12	Duraform GF	3D Systems
PA-12	NyTek 1200 GF	Stratasys
PA-12	PA 615-GS	ALM
PA-12	PA 616-GS	ALM
PA-12	PA3200 GF	EOS
PA	Windform GT	CRP Technologies
PA	Windform LX 2.0	CRP Technologies
PA-12	CarbonMide	EOS
PA-12	NyTek 1200 CF	Stratasys
PA	Windform SP	CRP Technologies
PA	Windform XT 2.0	CRP Technologies
PA-12	Alumide	EOS
PA-11	NyTek 1100	Stratasys
PA-11	PA 1101	EOS
PA-11	PA-850_NAT	ALM
PA-6	Sinterline	Solvay
PS	CastForm PS	3D Systems
PS	PrimeCast101	EOS

Table 3.5 Different polymer powder production techniques and their advantages and disadvantages. Adapted from [1].

Production techniques	Advantages	Disadvantages
Cryogenic grinding	• Simple process, widely used • Suitable for most polymers • Can produce a blend of different polymers	• Irregular powder shape (Figure 3.10 (a)) • Poor powder flowability • Particle rounding or post-processing treatment may be required
Thermally-induced phase separation	• Potato-shaped powder (Figure 3.10 (b)) • Good powder flowability	• Complex and time-consuming procedure • Needs special types of solvents
Melt emulsification	• Spherical particle shape (Figure 3.10 (c)) • Good powder flowability • Continuous production possible	• Requires expensive and specialized equipment • Complex and time-consuming procedure

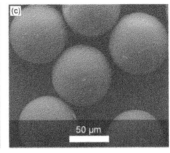

Figure 3.10 (a) Polyamide (PA-11) powders made by cryogenic grinding [2]. (b) Polyamide (PA-12) powders precipitated from an ethanol solution during the thermally-induced phase separation process [2]. (c) Polystyrene powders produced by melt emulsification [2]. *Source:* The figure is taken from an open-access article [2] under the terms and conditions of the Creative Commons Attribution (CC BY) license.

during the grinding or crushing process may result in an agglomeration of the particles. Powders produced using this process are nonspherical with rough edges (Figure 3.10 (a)) and thus exhibit poor flowability. In addition, parts produced with these powders yield low density and poor mechanical properties. However, this is an inexpensive and rapid process. In addition, this process can also be used to blend different polymers.

Thermally-induced phase separation: In this process, powders are produced by controlled cooling of a polymer solution in ethanol. Below a certain temperature, particles start precipitating from the solution. This process yields powders of potato-like shapes (Figure 3.10 (b)) that exhibit good flowability and thermal absorptivity. However, the need for a special type of solution hinders the application of this process for many polymers.

Melt emulsification: Since oil is insoluble in water, oil droplets in water take a spherical shape. The melt emulsification process uses a similar method to produce powders. Spherical particles (Figure 3.10 (c)) can be obtained by emulsion polymerization of water-insoluble monomers. The process is based on the preparation of a monomer solution at elevated temperature and pressure (e.g., a mixture of ethanol and water in the case of lauryl lactam). Precipitation of polymeric material starts as soon as the temperature or pressure drops below a certain limit. Powder size and size distributions can be controlled by adjusting the pressure and temperature during the process.

3.3.3 Wire and sheet-based polymer feedstocks

Wires are commonly used as feedstocks in the fused deposition modeling process. Different polymers used to make wires are provided in Table 3.3. Generally, thermoplastics are used because of their rigidity, high strength, high-temperature tolerance, and stability. Worked out example 3.5 indicates an important issue of wire feeding during the fused deposition modeling. The wires are made by extruding polymer pellets. This process is done using extruders that push the heated and soft polymer through holes in a die to produce the wires. Figure 3.11 schematically explains the working principle of the extrusion process. In the extrusion process, die temperature, roller pulling speed, spindle speed, and inlet polymer temperature affect the wire diameter and quality. Sheet feedstocks are used in the laminated object manufacturing process. This process was originally developed by bonding together adhesive-coated paper sheets and producing wood-like parts. Currently, laminated object manufacturing uses sheets of several thermoplastics and polymer-based composites (Table 3.3).

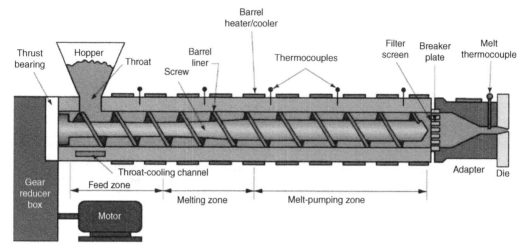

Figure 3.11 Schematic diagram of polymeric wire manufacturing using the extrusion process [7] The figure is taken from an open-access article [7] under the terms and conditions of the Creative Commons Attribution (CC BY) license.

Worked out example 3.5

In the fused deposition modeling process, the wire (filament) is supplied from a coiled spool to a heater-nozzle assembly. The heater makes the wire soft so that it can flow easily through the nozzle. The wires should be ductile enough so that they can be coiled in a spool. In addition, the wires should have sufficient strength to prevent buckling between the spool and the heater-nozzle assembly during feeding. For a filament of 1.5 mm diameter and elastic modulus of 1.8 GPa, calculate the critical strength required to prevent buckling. Assume the distance between the spool and the heater-nozzle assembly is 50 mm.

Solution:

The critical strength to prevent buckling can be estimated from the buckling stress assuming that the filament is a long column.

$$\sigma = \frac{\pi^2 E d^2}{16 L^2} \tag{E3.4}$$

where d and E are the diameter and elastic modulus of the filament wire, and L is the length of the filament between the spool and the heater-nozzle assembly. The equation above is for calculating the buckling stress of a long slender column and its derivation can be found in any mechanics of materials textbook. Using the equation above, the critical strength to prevent buckling is 0.998 MPa.

3.4 Additive manufacturing of polymer composites

Additively manufactured polymer parts often do not have the required strength for their usage as functional components. Therefore, hard and brittle reinforcement materials are often added in the form of fibers, whiskers, long wires, and nano and micro-particles to increase strength. The addition of reinforcement materials to low-strength polymer employs the plastic flow of the polymeric material under stress to transfer the load to the reinforcement materials, resulting in a high-strength composite. For example, Figure 3.12 shows that the strength of an epoxy ink part made by fused deposition modeling can be significantly increased by adding SiC whiskers and carbon fibers. Commonly used reinforcement materials include carbon fiber, silicon nitride, aluminum oxide, silicon carbide, tungsten, and glass fiber.

Figures 3.13 (a-f) show that the shape, size, and alignment of the reinforcement materials in polymer composites vary widely. The different types of composites exhibit a wide variety of mechanical properties. For example, Figure 3.14 compares the tensile strength of fiber-reinforced polymer composites (Figure 3.13 (c) and (d) made by laminated object manufacturing and fused deposition modeling) with that made by conventional processes. The tensile strength of composites increases with fiber volume fraction. However, even at high fiber volume fractions, the tensile strength of composites is often less than conventionally made composites. Composites with long and unidirectional fibers (Figure 3.13 (a)) exhibit better mechanical properties. However, their tensile strength along the direction perpendicular to the fibers is significantly less that that along the direction parallel to the fibers. The anisotropy in mechanical properties can be reduced by using bidirectional (Figure 3.13 (b)) or network structure (Figure 3.13 (e)) fiber reinforcement. Calculations of effective mechanical properties of fiber reinforced polymer composite are described

Figure 3.12 Comparison of engineering stress versus strain plots of additively manufactured epoxy ink polymer part and SiC whiskers and carbon fiber reinforced epoxy ink polymer composite part. The figure is made by T. Mukherjee and T. DebRoy using data reported in [8].

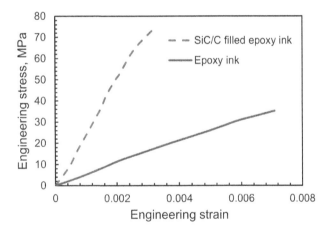

Figure 3.13 Fiber-reinforced polymer composites with (a) unidirectional long fibers, (b) bidirectional or multidirectional long fibers, (c) aligned short fibers, (d) randomly oriented short fibers, and (e) fibers woven in a honeycomb-like network structure. (f) Polymer composites with nano or microparticles randomly dispersed in the polymer matrix. *Source:* T. Mukherjee and T. DebRoy.

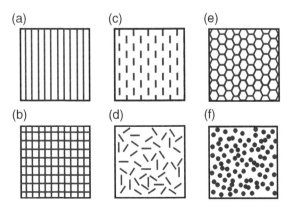

in worked out examples 3.6 and 3.7. Polymer composites are printed using vat photopolymerization, fused deposition modeling, powder bed fusion, and laminated object manufacturing processes which are discussed in Table 3.6. Worked out example 3.8 discusses important considerations in the additive manufacturing of polymer composites.

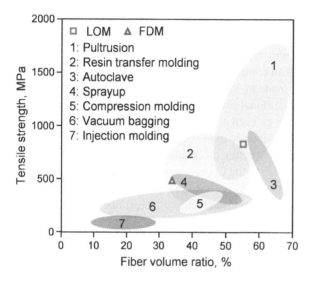

Figure 3.14 Tensile strength of polymer composite (carbon fibers in an epoxy resin polymer matrix) parts produced using laminated object manufacturing (LOM) and fused deposition modeling (FDM) are compared with that of parts made using conventional methods. The figure is adapted from [9]. The figure is taken from an open access article [9] under the terms and conditions of the Creative Common Attribution (CC BY) license. Discussions on the conventional methods are available in other textbooks [10].

Table 3.6 Comparison among different AM processes for printing polymer composites.

AM process	Fiber orientation	Advantages	Disadvantages
Vat photopolymerization	Random short fiber or particle orientation	• Random alignment of fibers to achieve isotropic mechanical property • Fine resolution can be obtained	• Sedimentation of fiber in liquid resin • Increased resin viscosity with the addition of fibers • Fibers may obstruct the light from penetration
Fused deposition modeling	Unidirectional long fiber or short fiber aligned along the scanning direction	• Inexpensive machines • Can print diverse polymer and reinforcement materials	• Degradation of the nozzle due to the abrasive effect of reinforcement particles • At a high concentration of reinforcement particles, the nozzle may clog
Powder bed fusion	Random short fiber or particle orientation	• Complex and intricate parts can be made • Easy to control fiber volume ratio	• Not possible to fabricate composites with long fibers • Slow and expensive process • Difficult to maintain uniformity of fiber or particle distribution
Laminated object manufacturing	Generally, unidirectional long fiber orientation	• Low cost • Support structures not needed	• High material wastage • Fibers need to align during the feedstock sheet preparation

Worked out example 3.6

A long and unidirectional glass fiber-reinforced (Figure 3.13 (a)) polyester resin composite consists of 40 vol.% of glass fibers having a modulus of elasticity of 69 GPa. The modulus of elasticity of polyester resin is 3.4 GPa. (a) Compute the modulus of elasticity of this composite in the direction parallel to the fibers. (b) If a stress of 50 MPa is applied to the composite in the direction parallel to the fibers, calculate the magnitude of stresses carried by each of the fiber and polymer matrix phases. Assume that the fiber–matrix interfacial bond is very good so that the deformation of both the polymer matrix and fibers is the same.

Solution:

(a) The total force (F_c) experienced by the composite is equal to the sum of the forces on the polymer matrix phase (F_m) and the fiber phase (F_f).

$$F_c = F_m + F_f \tag{E3.5}$$

If σ_c, σ_m, and σ_f are the stresses and A_c, A_m, and A_f are the cross-sectional area of the composite, matrix, and fiber, respectively,

$$\sigma_c A_c = \sigma_m A_m + \sigma_f A_f \tag{E3.6}$$

Since the composite, polymer matrix, and fiber have the same lengths, the volume fractions of the matrix (V_m) and fiber (V_f) can be written as, (A_m/A_c) and (A_f/A_c), respectively. Therefore,

$$\sigma_c = \sigma_m V_m + \sigma_f V_f \tag{E3.7}$$

Since we assume that the deformation of both the polymer matrix and fibers is the same, the strains along the fiber in the composite, matrix, and fiber are equal. Therefore, by dividing the both sides of equation E3.7 by strain, we get:

$$E_c = E_m V_m + E_f V_f \tag{E3.8}$$

where E_c, E_m, and E_f are the modulus of elasticity of the composite, matrix, and fiber, respectively. The values of E_m and E_f are provided. $V_f = 0.4$ and $V_m = 1 - 0.4 = 0.6$. Using these values, $E_c = 30$ GPa.

(b) The strain in the composite = σ_c/E_c = 50 MPa/30 GPa = 1.67×10^{-3} = strain in matrix and fiber
   Therefore, stress in the matrix = 3.4 GPa $\times 1.67 \times 10^{-3}$ = 5.67 MPa
   Stress in the fiber = 69 GPa $\times 1.67 \times 10^{-3}$ = 115.23 MPa

Worked out example 3.7

A long and unidirectional fiber-reinforced (Figure 3.13 (a)) polymer composite is under a stress (σ) applied in the direction perpendicular to the fibers. (a) Derive an expression to represent the modulus of elasticity of the composite (E_c) in the direction of stress as a function of the modulus of elasticity of the polymer matrix (E_m) and fiber (E_f) and the volume fraction of fibers (V_f). Assume an elastic deformation. (b) If the magnitudes of the modulus of elasticity of the fiber and the polymer matrix are 70 GPa and 5 GPa, respectively, plot and explain the variation in the modulus of elasticity of the composite as a function of the volume fraction of fibers.

(Continued)

Worked out example 3.7 (Continued)

Solution:

(a) Both the fiber and the polymer matrix experience the same stress (σ) as the composite in the direction perpendicular to the fibers.

However, the volumetric deformation of the fiber (δ_f) and matrix (δ_m) are different and their addition provides the total deformation of the composite (δ_c). Therefore,

$$\delta_c = \delta_m + \delta_f \tag{E3.9}$$

Dividing both sides by the volume of composite (vol_c),

$$\delta_c / vol_c = \delta_m / vol_c + \delta_f / vol_c \tag{E3.10}$$

or,

$$\delta_c / vol_c = \frac{\delta_m \times vol_m}{vol_c \times vol_m} + \frac{\delta_f \times vol_f}{vol_c \times vol_f} \tag{E3.11}$$

Here vol_m and vol_f are the volume of matrix and fiber, respectively. From Eq. E3.11 we get the following:

$$\varepsilon_c = \varepsilon_m V_m + \varepsilon_f V_f \tag{E3.12}$$

Here ε_c, ε_m, and ε_f are the strains in the composite, matrix, and fiber, respectively. V_m and V_f are the volume fractions of matrix and fiber, respectively, where $V_m = 1 - V_f$. From Eq. E3.12 we get the following:

$$\frac{\sigma}{E_c} = \frac{\sigma}{E_m}(1 - V_f) + \frac{\sigma}{E_f}V_f \tag{E3.13}$$

or,

$$\frac{1}{E_c} = \frac{1}{E_m}(1 - V_f) + \frac{1}{E_f}V_f \tag{E3.14}$$

By rearranging the terms, one gets the following:

$$E_c = \frac{E_m E_f}{(1 - V_f)E_f + V_f E_m} \tag{E3.15}$$

(b) Figure E3.2 shows the variation in the modulus of elasticity of the composite as a function of the volume fraction of fibers. At zero volume fraction which indicates a pure polymer part, the modulus of elasticity is 5 GPa. The modulus of elasticity of the composite gradually increases with the fiber

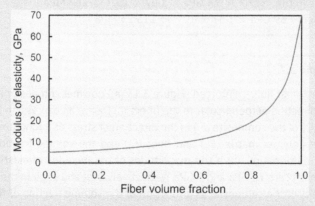

Figure E3.2 Modulus of elasticity of the composite as a function of the fiber volume fraction. *Source:* T. Mukherjee and T. DebRoy.

Worked out example 3.7 (Continued)

volume fraction. This nonlinear variation is significantly different from the linear variation expressed by Eq. E3.8 in worked out example 3.6 because of the differences in fiber orientations.

Worked out example 3.8

Based on a literature search, discuss the most important considerations in additive manufacturing of fiber-reinforced polymer composites.

Solution:

Fiber-polymer matrix bonding: Improper bonding between the fibers and polymer matrix may result in a separation between them when a load is applied resulting in a degradation in the mechanical properties of the composite.

Fiber homogeneity: The tensile strength of a composite may reduce due to poor stress transfer from matrix to fibers caused by inhomogeneity of fibers inside the polymer matrix. Proper mixing methods should be used to produce homogenous composite materials during additive manufacturing.

Fiber alignment: Various fiber alignments (Figure 3.13) can result in significantly different mechanical properties. Additive manufacturing offers precise control of the fiber alignment within the polymer matrix to achieve the desired mechanical properties.

Printability of fiber-polymer combination: Not all combinations of polymers and reinforcement particles can be used to print composites. For example, for liquid polymers, the density and viscosity of the polymer should be high enough to achieve a uniform distribution of the fibers.

3.5 Applications of additively manufactured polymer and composite parts

Additively manufactured polymer parts are used as prototypes, tools, dies, molds, and cores for injection molding and investment casting, and parts for electronic appliances, textiles, implants, and interior components of cars and airplanes [11]. For example, 3D Systems prints a prototype of a complex duct for an automobile made by selective laser sintering of polyamide (Figure 3.15 (a)). Sacrificial patterns for investment casting are made by stereolithography (Figure 3.15 (b)). A prototype of a shock absorber was printed at once with all moving parts together (Figure 3.15 (c)). These examples indicate that AM of polymeric materials can be used to print complex parts with intricate geometries. The aerospace industries require parts with lower weight and high strength which can be manufactured on demand. Such parts are now being made by AM. NASA has designed rovers that contain about 70 additively manufactured polymer parts [3] including flame-retardant vents and housings, large pod doors, front bumpers, camera mounts, and complex electronics connectors.

Printed polymeric parts are used in biomedical applications for making tools, medicines, organs, dental implants, prosthetics, and cranical implants (Figure 3.16). The polymers are both biodegradable and nonbiodegradable. Biodegradable polymers are used where the tissues grow and the printed part inside the body is not needed after some time. These parts degrade inside the body. Nonbiodegradable polymers are used to print structural implants. For the potential to improve complexity and functionality, printed polymeric parts are used in textile industries. In addition, these parts allow novel designs and better performance. Customized and flexible clothing is printed using polymers (Figure 3.17). The primary aim of these printed parts is to achieve important textile properties such as softness, strength, flexibility, and porosity.

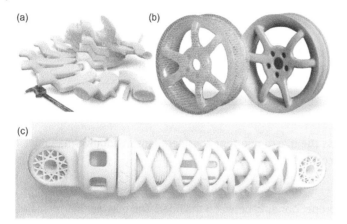

Figure 3.15 (a) Complex air ducts made by selective laser sintering of polyamide. (b) Sacrificial patterns of castable plastics made for investment castings by stereolithography. Figures (a) and (b) are provided by 3D Systems. (c) A prototype of a shock absorber is printed using different polymers and different processes [12]. Figure (c) is reprinted with permission from Elsevier.

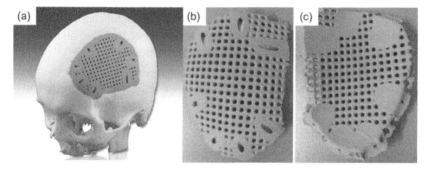

Figure 3.16 A cranial implant made by selective laser sintering of a polymer powder. a) The location of the implant, and (b and c) the printed implants [13]. The figure is reprinted with permission from Elsevier.

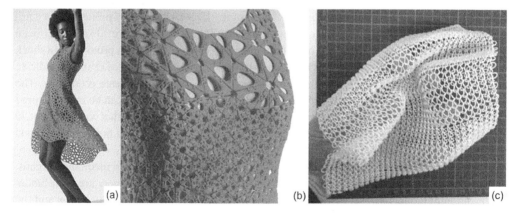

Figure 3.17 Flexible textiles made by additive manufacturing. (a and b) The "Kinematics Dress" printed by Nervous System using the SLS process. (c) A polyamide textile printed by Weft using selective laser sintering [2]. The figure is taken from an open-access article [2] under the terms and conditions of the Creative Commons Attribution (CC BY) license.

Figure 3.18 Photograph of a honeycomb structure printed using epoxy ink polymer with 1 vol.% carbon fibers [14]. The black fibers can be seen inside the polymer matrix. Image credit: Brett Compton, Harvard SEAS. Permission obtained from Paul Karoff (Associate Dean for Communications & Strategic Priorities, Harvard SEAS).

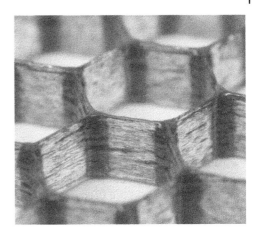

Additively manufactured polymer composites are used lightweight, high-strength structural parts used in the aerospace, energy, and automotive industries. For example, Figure 3.18 shows a photograph of a honeycomb structure printed using epoxy ink with 1 vol.% carbon fibers. The unidirectional black fibers are visible within the transparent polymer matrix walls throughout the structure. The structure is used in a wind turbine blade panel to achieve a combination of light weight and strength. Spacecraft components such as wing and fuselage panels and rocket motor casings, automotive components such as bumpers and exhaust systems, sporting goods such as golf clubs, tennis rackets, and fishing rods, and biomedical applications such as prosthetic limbs, dental implants, and surgical instruments are just a few of the major products.

3.6 Summary

In the last few decades, significant research and development have made the AM processes for parts made by polymers and polymer composites more reliable. High-quality prototypes and functional parts can now be produced for specialized applications. However, AM of polymers is still restricted mostly to printing components for specialized applications. This is primarily because of the lack of availability of feedstocks for a wide variety of polymers or the difficulties in producing cost-competitive parts. In addition, additively manufactured polymeric parts exhibit a wide variety of defects and poor mechanical properties depending on the AM process, process variables, and the feedstock used. A better understanding of the science and technology of the AM processes for polymers is needed for its continued expansion.

Takeaways
Additive manufacturing processes
• Commonly used additive manufacturing processes for polymers include vat photopolymerization, powder bed processes, inkjet printing, binder jetting, fused deposition modeling, and laminated object manufacturing.
• Two commonly used vat photopolymerization processes are stereolithography and digital light processing.

(Continued)

Takeaways (Continued)
• Two commonly used powder bed processes are selective laser sintering and selective laser melting.
• Fused deposition modeling is the most commonly used additive manufacturing process for polymers.
Polymer feedstocks
• Polymer feedstocks are available in the form of liquid, solid, wire, and sheet.
• Vat photopolymerization and inkjet printing use liquid polymers.
• Powder feedstocks are used in selective laser sintering, selective laser melting, and binder jetting.
• Wires are used in fused deposition modeling. The laminated object manufacturing uses polymer sheets as feedstocks.
• Liquid feedstocks contain photoinitiators that help to start the polymeric chain reaction when activated by light during the vat photopolymerization processes.
• Polymer powders are made by cryogenic grinding, thermally induced phase separation, and melt emulsification.
• Wires or filaments are made by the extrusion process.
Additive manufacturing of polymer composites
• Hard and brittle reinforcement materials are added to polymers to make composites.
• Reinforcement materials are added in the form of fibers, whiskers, long wires, and nano and microparticles.
• Polymer composites have higher strength than polymer parts.
• Polymer composites are printed using vat photopolymerization, fused deposition modeling, powder bed fusion, and laminated object manufacturing processes.
Applications of additively manufactured polymer and polymer composite parts
• Additively manufactured polymer parts are used as prototypes, functional prototypes, tools, dies, molds, and cores for injection molding and investment casting, and parts for electronic appliances, textiles, implants, and interior components of cars and airplanes.
• Additively manufactured polymer composites are used lightweight, high-strength structural parts used in the aerospace, energy, and automotive industries.

Appendix – Meanings of a selection of technical terms

Atomizing: A process to convert liquid to powders.

CAD: The acronym CAD stands for computer-aided design. It represents a three-dimensional design of a part.

Delamination: Separation of layers typically observed in printed parts. It is an undesirable defect.

Die: A specially designed tool for wire drawing and extrusion. A die is made of very hard material. Wires are drawn through a die to reduce their diameter.

Extrusion: Extrusion is a material-forming process in which hot and soft material is forced to flow through a die to reduce its cross-section.

Free radical: A free radical is an atom, molecule, or ion that has at least one unpaired valence electron that makes the free radicals highly chemically reactive.

Laser beam: The acronym laser stands for "light amplification by stimulated emission of radiation." A focused beam of laser is often used in additive manufacturing as a heat source.

Plasma: An electrically neutral gas consisting of electrons, excited atoms, and positively charged ions. A plasma arc is often used as a heat source in additive manufacturing.

Resin: Resin is a liquid, solid, or highly viscous polymeric substance.

Ultraviolet: Ultraviolet is a form of electromagnetic radiation with a wavelength shorter than that of visible light, but longer than X-rays.

Practice problems

1) Explain why stereolithography is a good process to fabricate patterns for investment casting?
2) Derive an expression to correlate the maximum allowable scanning speed with the mass flow rate through the nozzle during fused deposition modeling.
3) Why is fused deposition modeling more suitable for medical scaffold architectures compared with selective laser sintering-fabricated scaffolds made from a similar material?
4) Using an Internet search, find the recommended range of processing parameters for selective laser sintering of polyamide powders. Which parameters could be adjusted to increase the build rate?
5) Is it possible for a part made by selective laser sintering to have the same mechanical properties as one made by selective laser melting? Please explain why.
6) Based on a literature search, make a list of commercially available inks for inkjet printing together with their room temperature density, viscosity, and surface tension.
7) In the fused deposition modeling process, why is the selection of an appropriate nozzle diameter essential to reduce defects and improve the properties of parts? Explain.
8) Sheet feedstocks are used in both laminated object manufacturing and ultrasonic additive manufacturing (Chapter 2). What are the similarities and differences between these two processes?
9) Some photopolymers can react with the atmospheric oxygen during the photopolymerization process which degrades the quality of the printed parts. Which vat photopolymerization process is suitable for such photopolymers and why?
10) What is the smallest diameter nozzle that could be used in inkjet printing for the ink that has the following properties: surface tension: 0.04 N/m, density: 1200 kg/m^3, and viscosity: 0.015 Pa.s.

References

1 Tan, L.J., Zhu, W. and Zhou, K., 2020. Recent progress on polymer materials for additive manufacturing. *Advanced Functional Materials*, 30(43), article no 2003062.

2 Ligon, S.C., Liska, R., Stampfl, J., Gurr, M. and Mülhaupt, R., 2017. Polymers for 3D printing and customized additive manufacturing. *Chemical Reviews*, 117(15), pp.10212–10290.

3 Saleh Alghamdi, S., John, S., Roy Choudhury, N. and Dutta, N.K., 2021. Additive manufacturing of polymer materials: progress, promise and challenges. *Polymers*, 13(5), article no.753.

4 Nath, S.D. and Nilufar, S., 2020. An overview of additive manufacturing of polymers and associated composites. *Polymers*, 12(11), article no.2719.

5 Xia, H., Lu, J., Dabiri, S., and Tryggvason, G., 2018. Fully resolved numerical simulations of fused deposition modeling. Part I: fluid flow. *Rapid Prototyping Journal*, 24(2), pp.463–476.

6 Manufactur3D, February 13, 2018. *3D Printing technology choice: FDM v/s SLA v/s SLS.* Available at: https://manufactur3dmag.com/3d-printing-technology-choice-fdm-v-s-sla-v-s-sls (accessed on 26 January 2023).

7 Kristiawan, R.B., Imaduddin, F., Ariawan, D. and Arifin, Z., 2021. A review on the fused deposition modeling (FDM) 3D printing: filament processing, materials, and printing parameters. *Open Engineering*, 11(1), pp.639–649.

8 Wang, X., Jiang, M., Zhou, Z., Gou, J. and Hui, D., 2017. 3D printing of polymer matrix composites: a review and prospective. *Composites Part B: Engineering*, 110, pp.442–458.

9 Wickramasinghe, S., Do, T. and Tran, P., 2020. FDM-based 3D printing of polymer and associated composite: A review on mechanical properties, defects and treatments. *Polymers*, 12(7), article no.1529.

10 Astrom, B.T., 1997. *Manufacturing of Polymer Composites*, 1st Edition. CRC Press, London.

11 Prakash, K.S., Nancharaih, T. and Rao, V.S., 2018. Additive manufacturing techniques in manufacturing-an overview. *Materials Today: Proceedings*, 5(2), pp.3873–3882.

12 Holmes, M., 2019. Additive manufacturing continues composites market growth. *Reinforced Plastics*, 63(6), pp.296–301.

13 Berretta, S., Evans, K. and Ghita, O., 2018. Additive manufacture of PEEK cranial implants: Manufacturing considerations versus accuracy and mechanical performance. *Materials & Design*, 139, pp.141–152.

14 Harvard School of Engineering and Applied Sciences News, June 25, 2014. *Carbon-fiber epoxy honeycombs mimic the material performance of balsa wood.* Available at: https://seas.harvard.edu/news/2014/06/carbon-fiber-epoxy-honeycombs-mimic-material-performance-balsa-wood (accessed on 26 January 2023).

4

Feedstocks and Processes for Additive Manufacturing of Ceramic Parts

Learning objectives
After reading this chapter the reader should be able to do the following: 1) Appreciate the importance of the additive manufacturing of ceramics. 2) Know about the commonly used ceramic feedstocks. 3) Understand the working principles of commonly used additive manufacturing processes for ceramics. 4) Appreciate the properties of the additively manufactured ceramic parts. 5) Understand the common defects in the ceramic parts. 6) Recognize the common applications of additively manufactured ceramic parts.

CONTENTS

4.1 Introduction

Metals and polymers are the most widely used materials in additive manufacturing. However, their strong reactivity at high temperatures, high density, and other properties make them unsuitable for many applications. Ceramic materials provide low density, good corrosion resistance at elevated temperatures, low coefficient of thermal expansion, and other desirable properties. They also have good thermal and electrical resistance, high hardness, and wear resistance. In addition,

Theory and Practice of Additive Manufacturing, First Edition. Tuhin Mukherjee and Tarasankar DebRoy.
© 2024 John Wiley & Sons, Inc. Published 2024 by John Wiley & Sons, Inc.

many of the mechanical properties of ceramic materials are stable over a wide range of temperatures. Some ceramic materials are biocompatible, strong, and have a low density, making them suitable for various types of prostheses. Their unique properties make components of ceramic materials attractive for many applications in aerospace, automotive, machine tools, biomedical, and electronic applications. Because of the market demands, availability of ceramic materials, and additive manufacturing processes to make components, they are increasingly used for a variety of applications, particularly in shapes that cannot be easily made by conventional processes.

A variety of ceramic materials are used as feedstock depending on the application. These include alumina, zirconia, silicon carbide, silicon nitride, silicon carbonitride, and calcium phosphate. The form of the feedstock varies depending on the additive manufacturing process. For example, fine powders without any additives are used for the selected laser melting, selected laser sintering, and directed energy deposition processes. Binder liquids are added in binder jetting, slurries containing a ceramic powder and a liquid polymer are used for robocasting, and monomers are added to ceramic powders in stereolithography.

Most industrial ceramics are characterized by their high melting points and this poses a challenge in their additive manufacturing via the liquid state. A variety of processing techniques have been developed to print ceramic parts. Similar to the powder bed fusion process for the additive manufacturing of metallic parts from powders, one-step powder bed fusion of ceramic powders without any addition of low melting binders has been tried for several ceramic powders. However, the applications of ceramic parts made by complete melting have so far been very limited because of the generation of defects such as porosity, rough surface finish, and cracking [1]. As a result, the bulk of the additive manufacturing of ceramic parts is made by a multistep process in which a lower melting polymer or other binder is added to form a green part which is heated to remove the binder and subsequently the structure is sintered at high temperatures. Most of the additively made commercial ceramic components of different shapes and sizes are made using stereolithography, binder jetting, robocasting, direct inkjet printing, extrusion, laminated object manufacturing, selected laser melting, selected laser sintering, and directed energy deposition. Of these, the selected laser melting/sintering and the directed energy deposition are single-step processes. The remaining multistep processes take more time than the one-step processes. However, more work is needed to make the single-step powder melting process industrially viable.

In this chapter, the commonly used ceramic feedstocks are discussed, the additive manufacturing processes are examined, the properties of the additively manufactured parts are reviewed, and the common applications of additively manufactured ceramic parts are studied.

4.2 Feedstocks

Industrial ceramics are inorganic compounds often referred to as glasses, refractories, cement, clays, and whiteware. They may be amorphous or polycrystalline. They are characterized by high strength, hardness, stiffness, and good resistance to corrosion and oxidation. The constituent atoms are bound by ionic and covalent bonds. The feedstock is often an oxide, nitride, carbide, or other compounds of metals. However, in many processes, the powder is mixed with a polymer to make a slurry which is used as a feedstock. Because ceramic materials are hard and brittle, melting and solidifying them to make an intricate part often results in defects. A monomer in the ceramic slurry may be heated by a laser beam or another heat source to polymerize and make a part of the same shape as the final part. The part may be heated at a low temperature to remove the polymer and subsequently sintered to improve its density. A variety of inorganic materials are commonly used to make ceramic parts by additive manufacturing as discussed below.

Alumina: Alumina or aluminum oxide (Al_2O_3) is a commonly used structural material for additive manufacturing because of its good mechanical strength, stiffness, low density, excellent thermal and corrosion resistance, high melting point, and relatively low cost. Lightweight high-purity alumina components are used in aerospace and biomedical industries because of their good mechanical properties, thermal stability and corrosion resistance, and comparable coefficient of thermal expansion compared with titanium alloys. Alumina is also used in the electronic industry because of its insulating properties. Alumina components have been produced by several additive manufacturing processes including stereolithography.

Zirconia: Zirconia or zirconium oxide (ZrO_2) is lightweight, tough, has good resistance to wear and corrosion, and can withstand high temperatures. Zirconia is often mixed with 3–8 weight % yttria to improve properties. Additions of yttria stabilize the crystal structure and also enable the yttria-stabilized zirconium to conduct oxygen ions which makes yttria-stabilized zirconia well suited for application as solid electrolyte in solid oxide fuel cells. In the biomedical industry, zirconia is used to create orthopedic and dental implants. Zirconia is biocompatible and bone tissue grows around the implant [2, 3]. It is also used to make prosthetic limbs. Components of zirconia are made using stereolithography where a photopolymer resin is selectively cured by a laser, and the final product is sintered at high temperatures to improve density.

Silica: Silica (SiO_2) is a commonly used inexpensive, hard glassy material and has been used to make various 3D-printed parts. For example, a silica-based ceramic core is used in the manufacture of high-temperature gas turbine blades. 3D-printed ceramic cores can improve the quality of cast superalloy blades. Although photopolymerization and extrusion processes have been used for making silica-based parts, they are still considered to be an emerging technology [4, 5].

Silicon carbide: Silicon carbide, SiC, is a hard, strong, and creep-resistant material that has good oxidation and corrosion resistance and high thermal conductivity. Silicon carbide is used in a variety of applications such as abrasives, cutting tools, heat exchangers, and high-power electronic devices. They are also used in car brakes, car clutches, and ceramic plates in bulletproof vests. The additive manufacturing methods used to print parts of silicon carbide commercially include powder bed, binder jetting, and photopolymerization. In order to build intricate 3D geometries, these techniques combine SiC powder with a polymer as a binder [6].

Silicon nitride: Silicon nitride (Si_3N_4) has an excellent combination of strength, toughness, wear resistance, corrosion resistance, and biocompatibility [7]. Because of these desirable properties additively manufactured silicon nitride parts are widely used in aerospace, automotive, and medical industries. Two common processes used to make silicon nitride parts are binder jetting and stereolithography. In binder jetting, a liquid binder is selectively deposited onto a bed of silicon nitride powder, layer by layer, to form the part. In stereolithography, a laser cures a liquid resin containing silicon nitride particles, also layer by layer. Both processes require post-processing to make dense parts.

Calcium phosphate-based biomaterials: Amorphous bioglasses and crystalline inorganic materials, often called bioceramics, include a wide range of calcium phosphates based on their chemical compositions, particularly the Ca/P molar ratios [8]. For example, amorphous calcium phosphates (Ca/P: 1.2–2.2), α-tricalcium-phosphate (Ca/P: 1.5), β-tricalcium-phosphate (Ca/P: 1.5), Hydroxyapatite (Ca/P: 1.67), tricalcium silicates, and bioactive glasses have been used as biomaterials [8]. Hydroxyapatite (HA) or hydroxyl calcium phosphate has a chemical formula $Ca_{10}(PO_4)_6(OH)_2$ and occurs naturally in teeth and bones. It is inexpensive and biocompatible and used for custom-made bone implants and scaffolds for bone tissue engineering. The 3D printing process for HA involves stereolithography. The process is done in a layer-by-layer fashion and the layers are then hardened with a laser.

4.3   Additive manufacturing processes

There are several AM processes used to print ceramic parts. Fusion-based AM processes such as selective laser sintering/melting and directed energy deposition often encounter defects such as cracking and porosity. In addition, a very high-energy heat source is needed to melt high-melting point ceramics. Therefore, one-step fabrication of ceramic parts using AM like that used for metals and polymers is not always feasible. A comparison of additive manufacturing of metallic, polymeric and ceramic materials is presented in worked out example 4.1. In general, ceramic particles are mixed with binders to print parts (Figure 4.1). For example, ceramic powders are mixed with photocurable monomers which are cured during stereolithography process (Chapter 3). Ceramic powders are bound by selectively depositing binder liquid in a layer-by-layer manner to make 3D parts using the binder jet method (Chapter 2). The importance of adding an appropriate amount of binder is explained in the worked out example 4.2. Sheets made from a mixture of ceramic particles, polymeric fibers, binders, and adhesives, known as preceramic sheets [9], are selectively cut using a laser beam and bound in a layer-by-layer manner to make parts using laminated object manufacturing (Chapter 3). Robocasting [10] and fused deposition-based processes involve the deposition of a ceramic slurry in which ceramic powders and liquid binders are mixed. The slurry is dispensed through a computer-controlled nozzle onto a build platform in a layer-by-layer fashion. These processes have similarities with the fused deposition modeling process for polymers as discussed in Chapter 3. Table 4.1 compares the common AM processes used to print ceramic parts.

As-printed parts made by binder jetting, stereolithography, laminated object manufacturing, fused deposition process, and robocasting are green parts with very low strength. These parts need to be heated in a furnace to burn out the binder (Figure 4.1). In this process, the organic binder burns out by reacting with atmospheric oxygen. The brown parts after binder removal are sintered in a high-temperature furnace to achieve high density (Figure 4.1). The sintering temperature and time affect the density and mechanical properties of parts as discussed in the next section.

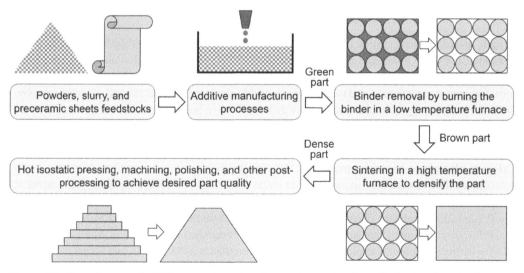

Figure 4.1   Flow diagram of additive manufacturing of ceramics starting from the feedstock to the final part. *Source:* T. Mukherjee and T. DebRoy

Table 4.1 Comparison between different additive manufacturing processes for ceramics.

AM process	Feedstock	Processing steps	Advantages and limitations
Stereolithography	Ceramic powders of alumina, zirconia, silicon carbides, and nitrides mixed with photocurable monomers	• Apply light to selectively cure monomers guided by 3D design • Removal of parts from the substrate • Densify the part using post-process sintering	• Can produce high-resolution complex parts • Good dimensional accuracy and low surface roughness • Wide range of design flexibility • A support structure is often needed • There is a limit to how much ceramic powders may be contained in a liquid resin
Binder jetting	Powders of alumina, zirconia, silicon carbides and nitrides, and calcium phosphate	• Spray binder liquid on a powder bed to selectively bind powders guided by 3D design • Take out the part from the bed and remove the loose powders • Densify the part using post-process sintering	• Wide range of ceramic materials • Support structure not required • Removal of loose powders can damage the green part • Printing of large parts is challenging
Robocasting	A slurry in which ceramic powders and liquid binder are mixed	• Ceramic slurries are deposited through a nozzle on a heated substrate • Post-process drying and sintering to densify the parts	• Designed for printing ceramics • Rapid and cost-effective • Difficult to make slurries with desired flow properties such as low viscosity • Nozzle may clog
Fused deposition processes	A slurry in which ceramic powders and liquid binder are mixed	• Slurries are deposited through a nozzle on a heated substrate • Post-process sintering to densify the parts	• High productivity can be achieved • Complex parts can be made • Nozzle clogging may affect the process • Difficult to control the flow
Laminated object manufacturing	Preceramic sheets [9] made from a mixture of ceramic particles, polymeric fibers, binders, and adhesive	• Sheets are selectively cut using a laser beam and bound in a layer-by-layer manner • Densify the part using post-process sintering • Post-process machining to achieve the desired shape and size	• Processing dense structures with good mechanical properties • Cost-effective • Preceramic sheets may not be available for different ceramics • Needs post-process machining to get the final shape
Selective laser sintering	Powders of alumina, zirconia, silicon carbides, and nitrides	• Preheat the ceramic powder bed • A laser beam selectively scans the powder bed to sinter the powder • The part is removed from the substrate	• Complex geometries can be printed • Support structure not needed • Printing large parts is often challenging • Slow and expensive process • Post-processing is still needed to make dense finished parts

(Continued)

Table 4.1 (Continued)

AM process	Feedstock	Processing steps	Advantages and limitations
Powder bed fusion	Powders of alumina, zirconia, silicon carbides, and nitrides	• A laser beam scans the powder bed to selectively melt the powder • Parts are formed after the solidification of materials • The process continues to make parts in a layer-by-layer manner	• Complex geometries can be printed • Support structure not needed • Printing large parts is often challenging • Slow and expensive process • Careful control of process parameters is needed to avoid cracking due to large temperature gradient
Directed energy deposition	Powders of alumina, zirconia, silicon carbides, and nitrides	• Deposition of ceramic powders using a nozzle • Melting of powders using a coaxial laser or electron beam • The nozzle moves guided by the computer to print parts in a layer-by-layer manner	• Graded compositions can be achieved • As-printed parts have dense structures • Support structures are often needed • Nozzle clogging

---

Worked out example 4.1

Additive manufacturing of metallic materials, polymers, and ceramics are discussed in Chapters 2, 3, and 4, respectively. Compare the processes used, feedstocks, processing temperature, applications, cost, and market share of additive manufacturing of these three classes of materials.

Solution:

Table E4.1 provides a comparison of additive manufacturing of metallic materials, polymers, and composites.

Table E4.1 Comparison of AM of metallic materials, polymers, and composites.

	Metals and alloys	Polymers	Ceramics
AM processes used	Powder bed fusion with laser or electron beam, directed energy deposition with laser or electron beam or electric arc, and binder jetting	Stereolithography, selective laser sintering, fused deposition modeling, laminated object manufacturing, binder jetting, and powder bed fusion	Stereolithography, selective laser sintering, robocasting, laminated object manufacturing, binder jetting, and powder bed fusion
Feedstocks	Metal powders, wires, and sheets	Liquids, powders, wires, and sheets	Powders, slurries, and preceramic sheets
Processing temperature	High	Low	Higher than that for metals in fusion-based processes. Other processes are done at low temperatures

(Continued)

Table E4.1 (Continued)

	Metals and alloys	Polymers	Ceramics
Applications	Diverse components in aerospace, automotive, healthcare, energy, consumer products, and marine industries	Prototypes, tools, dies, molds, parts for electronic appliances, textiles, implants, and interior components of cars and airplanes	Different parts for dentistry, healthcare, aerospace, automotive, construction, consumer products, and electronics
Strength and durability	High strength and durability, suitable for harsh environments	Less strong and durable compared to metals	High strength and durability, suitable for high-temperature applications
Need for post-processing	Hot isostatic pressing, machining, grinding, heat treatment	Sanding, polishing, and painting	Sintering, binder burnout, hot isostatic pressing, machining, and grinding
Cost	Higher cost compared to polymers and ceramics	Lower cost compared to metals and ceramics	Higher cost compared to polymers
Market size	Largest and rapidly growing	Medium	Much smaller than metals and polymers

Worked out example 4.2

Binder jetting is a common process for additive manufacturing of ceramic materials. In this process, a binder liquid is supplied at a constant volumetric flow rate (V) on the powder bed to selectively bind the powders. Derive an expression to represent the volumetric flow rate required for a complete homogeneous binding of powders to produce a layer of thickness "t" and width "w" at a scanning speed of "S." Assume that the packing density of the powder bed is "η."

Solution:

The volume of the track printed per unit time = track width × layer thickness × scanning speed = w t S

The volume of binder needed per unit time = volume of the track printed per unit time – volume of powders = (1 – packing density) × total track volume = (1 – η) w t S

For a complete homogeneous binding of powders, the volume of binder needed should be equal to the volumetric flow rate of the binder (V). Therefore, V = (1 – η) w t S

In practice, the process parameters are carefully adjusted to achieve a complete homogeneous binding of powders during binder jetting [11].

4.4 Defects and properties

The evolution of properties and defects of ceramic parts made by stereolithography, binder jetting, robocasting, and laminated object manufacturing are affected by the sintering process. The sintering process involves placing the green parts in a furnace which is heated to a temperature between 1200°C and 1800°C. The parts are subjected to a gradual increase in temperature over several hours. During sintering, the ceramic particles (Figure 4.2 (a)) gradually fuse with each other

(Figure 4.2 (b) and (d)). At the end of the process, all ceramic particles are fused to make a dense structure (Figure 4.2 (c)). The effect of particle size on the kinetics of sintering is explained in worked out example 4.3. In this process, the as-printed green parts shrink in size and become denser and stronger. After sintering, the parts are cooled down gradually to prevent a rapid temperature change, and the final product is inspected for defects. Sintering for a longer duration provides the particles sufficient time to completely fuse and improve the part density (Figure 4.3 (a)). Higher temperatures accelerate the sintering process, reduce porosity, and increase density (Figure 4.3 (b)).

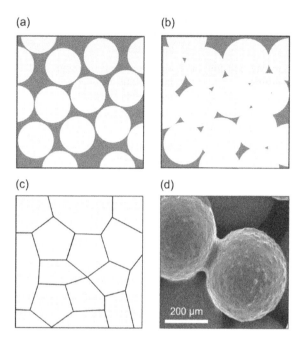

Figure 4.2 (a-c) Schematic representation of particle sintering process. The ceramic particles in figure (a) start to fuse with each other gradually as shown in figure (b). At the end of the sintering process, all ceramic particles are fused to make a dense structure, as shown in figure (c). (d) SEM image of partially sintered two powder particles [12]. Source for figures (a-c): T. Mukherjee and T. DebRoy. Figure (d) is taken from [12] with permission from Elsevier.

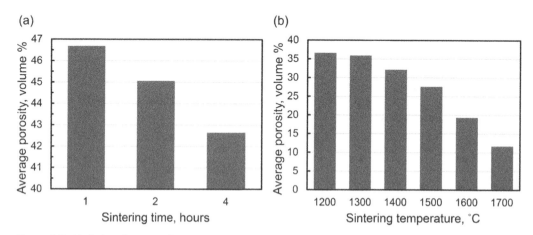

Figure 4.3 Variations in pore volume with (a) sintering time and (b) sintering temperature. For figure (a), the sintering temperature was 1300 °C. For figure (b), the parts were heated at 3 °C/min to reach the specified temperature and then cooled down to room temperature at 3 °C/min. Figures (a) and (b) are for hydroxyapatite and alumina, respectively. Figures (a) and (b) are made by T. Mukherjee and T. DebRoy using the data reported in [13] and [14], respectively.

Worked out example 4.3

Figure 4.2 (d) provides an SEM image of two partially sintered particles which is schematically depicted in the figure below (Figure E4.1). It is assumed that two adjacent spherical particles of radius "R" are joined during sintering. As shown below, the particles overlap a distance of "2h" and the contact radius is "a." If the rate of sintering can be approximately quantified as the relative velocity between particles, dh/dt, as has been suggested in the literature [15], show that the sintering rate is inversely proportional to the particle radius at the beginning of sintering.

Solution:

From the figure, we can write approximately: $(R - h)^2 + a^2 \approx R^2$
  or, $R^2 - 2Rh + h^2 + a^2 \approx R^2$ or, $a^2 \approx 2Rh - h^2$ or, $2a\, da/dt \approx 2\, dh/dt\, (R - h)$
  or, $dh/dt \approx a/R\, da/dt$, simplifying $R - h = R$ since $h \ll R$
  From the expression, it is evident that the smaller particles show faster rate of sintering (dh/dt) which is also consistent with the results reported in the literature [15].

Figure E4.1 Schematic representation of two partially sintered particles. *Source:* T. Mukherjee and T. DebRoy.

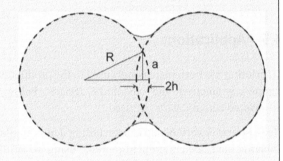

Apart from porosity, cracking is another major concern. Cracking during sintering is usually caused by the constraint that prevents the sintering material from shrinking in one or more directions and the resulting tensile stresses. Pores can also act as crack initiators. Ceramics are brittle and are susceptible to cracking when subjected to stress. Also, care must be taken when removing support structures or handling green parts. Any sudden or excessive force applied to the part can cause it to crack. The presence of both cracking and porosities significantly degrades the mechanical properties of 3D-printed ceramic parts. For example, Figure 4.4 shows that the flexural strength of

Figure 4.4 Variation in flexural strength of a 3D-printed alumina part with porosity. Flexural strength resists failure due to bending. The plot is made by T. Mukherjee and T. DebRoy using the data reported in [16].

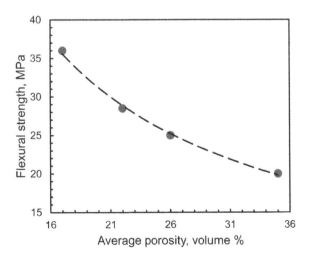

a 3D-printed alumina part decreases significantly at a higher volume of pores. The relation between porosity and strength is explained in worked out example 4.4. A proper selection of sintering temperature and time is needed to ensure a high density of parts.

Worked out example 4.4

It has been reported [17] that the strength (σ) of additively manufactured ceramic parts is inversely proportional to the exponential of pore fraction (p). That is why printed and sintered parts often need to post-processed by hot isostatic pressing. How much strength can be improved if the part density is increased from 90% to 98% by hot isostatic pressing?

Solution:

As stated in the problem, the strength is related to porosity as $\sigma \propto 1/\exp(p)$
 Therefore, improvement in σ is $= \exp(1 - 0.9)/\exp(1 - 0.98) = 1.08$ times
 Therefore, the strength can be improved by 8% if the part density is increased from 90% to 98%.

4.5 Applications

3D printing has been increasingly used in the production of ceramic parts due to its ability to create complex geometries and customize designs. Below are some examples of industries where 3D-printed ceramic parts are used.

Healthcare: A variety of parts, including dental implants (Figure 4.5), hearing aids, and surgical implants have been attempted to make using 3D printing [18, 19]. 3D-printed ceramic parts are often biocompatible, and they can be designed to mimic natural bone structures, promoting better integration with the human body. AM processes such as stereolithography and inkjet printing have been explored [19] to make parts of biocompatible ceramics including zirconia and calcium phosphate. This technology has the potential to produce customized implants and prosthetics, which are tailored to each patient's specific needs, thereby enhancing patient outcomes. Additionally, the use of 3D printing can reduce the cost of producing these ceramic parts, making them more accessible to patients in need of them.

Construction: 3D printing of houses [20–22] is an important application (see Chapter 14). Cement is a common ceramic material for 3D-printed houses. Two-third of the mass of cement is calcium silicates ($3CaO.SiO_2$ and $2CaO.SiO_2$) and the remaining amount is aluminum and iron-containing phases and other compounds. The MgO in cement may not exceed 5 wt.%. Fused deposition-based processes are used in the construction industry [20]. 3D printing can reduce the time required to

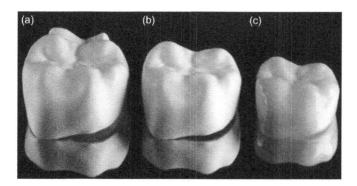

Figure 4.5 Zirconia tooth printed using stereolithography (a) as-printed green state, (b) brown state after polymeric binder burnout, and (c) final sintered part [18]. The figure is taken from an open-access article [18] under the terms and conditions of the Creative Commons Attribution (CC BY) license.

build a house from months to days [21], making it an ideal solution for emergency housing to cope with natural disasters or for rapid urbanization [21, 22]. Ceramic materials are also resistant to heat and corrosion, which makes them perfect for use in harsh environmental conditions. 3D printing technology reduces the waste generated during the manufacturing process, thus making it an eco-friendly solution. The use of 3D-printed ceramic parts can significantly reduce construction time, labor costs, and material waste making it a cost-effective solution for the construction industry [20]. 3D printing of ceramic tiles is also gaining attention [23].

Aerospace: Alumina, zirconia, and silicon carbides and nitrides have been attempted to print using stereolithography to produce components for gas turbines and jet engines, parts for spacecraft and satellites including heat shields and insulation [24]. 3D-printed ceramic parts can withstand high temperatures and extreme pressure, which shows their potential use in the harsh environment of outer space. Alumina has been tried to print electrical and thermal insulators for spacecraft [24]. Silica-based ceramics with added alumina and zirconia have a very low thermal expansion, high porosity, and exceptional surface quality and have been explored to print cores [24]. These cores can be used to cast single crystal blades for aeroengine [24].

Consumer products: 3D-printed ceramic objects are gaining attention in the consumer products industry, where they can offer new opportunities for creating intricate and unique designs. 3D-printed ceramic parts have been explored for a variety of applications, including sculptures [25], home décor [26], and even functional objects like vases and bowls [27]. Generally, clay or glass-based ceramics are printed using fused deposition processes and the green part is sintered to make the final parts [27]. The result is a piece of art or design that is not only visually stunning but also practical for everyday use.

Electronics: Silicon carbides and nitrides and other ceramics have shown their potential [28] to be printed using selective laser sintering and stereolithography to make capacitors, insulators, and other parts with complex shapes and geometries (Figure 4.6). Ceramic materials have high thermal conductivity and can withstand high temperatures, making them ideal for use in electronic circuits that generate heat. Additionally, 3D-printed ceramic parts are highly durable and resistant to wear and tear and can ensure the reliability of electronic devices.

Automotive: Stereolithography and inkjet printing of alumina have been attempted to 3D print camshafts, intake valves, brake discs, and exhaust manifolds [30]. Ceramic parts can be very useful in automotive applications because of their high strength, heat resistance, and durability.

Figure 4.6 (a) Photograph and (b) SEM image of a printed 3D structure of lead zirconate titanate [29]. This type of fine intricate parts can be used to make electronic devices. The figure is taken from [29] with permission from Elsevier.

4.6 Outlook

Low density, excellent corrosion and oxidation resistance, low coefficient of thermal expansion, good mechanical properties at elevated temperatures, and many other unique properties make ceramic materials attractive for many practical applications in various industries. The design freedom of additive manufacturing, low volume production of components, and the ability to make complex parts on demand are important considerations for the additive manufacturing of ceramic parts over conventional manufacturing. In recent years there has been significant progress in the development of new ceramic products made by additive manufacturing. However, the development of additive manufacturing of ceramic materials has also faced significant challenges. The one-step printing of metallic parts by melting and solidification route is difficult to replicate for ceramics because the melting and solidification of ceramics often result in unacceptable defects in parts. The industrial solution has been to use multiple-step processes discussed earlier in the chapter that avoided the melting and solidification of ceramic materials. These processes include binder jetting, stereolithography, and other processes that require several steps involving the manufacture of green parts and their subsequent sintering. Such processes take more time than the one-step processes and are expensive. While there has been significant progress in research and development, the development of printed ceramic parts has not kept pace with those of metal and polymers.

As progress is made, a wider range of printed ceramic components is continuing to be developed. Some of the ceramic printing processes such as stereolithography can print high-quality parts with good surface finish. The parts have to be sintered after the green parts are made and optimizing the time and temperature of post-processing can reduce porosity. As a result, new parts that are difficult to make using conventional processing are continuing to be developed for various industries using stereolithography, binder jetting, and other multistep processes using readily available and relatively inexpensive raw materials. However, it would be desirable to be able to print high-quality parts in one step as is done for printing metallic and polymeric parts and work is continuing to achieve this goal.

4.7 Summary

The need for the additive manufacturing of ceramic materials is supported by their unique properties, availability of a variety of affordable feedstocks, and several well-tested additive manufacturing processes. Additively manufactured components of alumina, zirconia, silicon carbide, silicon nitride, and calcium phosphate are in significant demand and the industry is expanding. However, compared to metallic and polymeric parts, the market for printed ceramic components is currently much smaller, mostly due to flaws like cracking and porosity in the one-step production of final parts. Printed ceramic parts made by binder jetting and stereolithography are increasingly used in healthcare, construction, aerospace, and other industries. These processes produce parts that resemble the final part in shape but necessitate extensive post-processing, such as sintering. It would be ideal to be able to print high-quality ceramic parts in one step, as is done for metallic and polymeric components, but significant challenges remain to be overcome to attain this goal.

Takeaways

Need for additive manufacturing

- The high market demand, availability of affordable feedstock materials, and a variety of additive manufacturing processes for ceramics make their printing an attractive undertaking.
- While there is considerable demand for additively manufactured ceramic components in a variety of industries, the market value of printed ceramic components remain much smaller than those for metal and polymers because of the serious challenges in the printing of ceramics.

Feedstocks and manufacturing processes

- Feedstocks include alumina, zirconia, silicon nitride, silicon carbide, and a variety of calcium phosphate-based compounds.
- Ceramic powders or slurries containing powders and a liquid such as an organic compound are used.
- Two commonly used additive manufacturing processes are binder jetting and stereolithography that produce parts that are similar in shape as the final part but require considerable post processing such as sintering.
- One step additive manufacturing of finished ceramic components face significant challenges because of the presence of defects such as porosity and cracking.
- Unlike additive manufacturing of metallic components, most ceramic additive manufacturing involves the preparation of an unfinished intermediate component that requires subsequent sintering to improve mechanical properties and reduce porosity.

Status and outlook

- Additively made ceramic components are increasingly used in healthcare, construction, aerospace, and other industries.
- It would be ideal to be able to print high-quality ceramic parts in one step, as is done for metallic and polymeric components, but significant challenges remain to be overcome.

Appendix – Meanings of a selection of technical terms

Biocompatible: Not toxic or injurious to living systems.

Flexural strength: It is the stress in failure during the bending of a part.

Fuel cells: A device that converts fuels to electricity.

Glassy: Glass-like. Glass is an amorphous, transparent, or translucent material.

Monomer: It is commonly an organic molecule that can react with other molecules to form large molecules or polymers.

Photocurable: An organic liquid that forms a solid when exposed to light or other radiation.

Polymers: A compound consisting of repeating units of molecules.

Sintering: A process of consolidating small particles of solids by heating to reduce porosity.

Resin: Resin is a liquid, solid, or highly viscous polymeric substance.

Ultraviolet: Ultraviolet is a form of electromagnetic radiation with a wavelength shorter than that of visible light, but longer than X-rays.

Practice problems

1) Why is it important to make ceramic components by additive manufacturing? What ceramic materials are candidates for 3D printing?
2) Based on a literature search, provide a few examples of biocompatible ceramics used in 3D printing.
3) Do a literature search on the market size of 3D-printed ceramic components and compare the market size with that for metallic materials.
4) What are the commonly used processes for the additive manufacturing of ceramic parts? What are their advantages and disadvantages?
5) What are the one-step 3D printing processes for making finished ceramic parts? What are the challenges in the one-step printing of finished ceramic parts?
6) Give three examples of 3D-printed commercial ceramic parts and explain how they are made.
7) Which additive manufacturing process is best suited for making a silica casting core?
8) What type of post-processing techniques are widely used for the 3D printing of ceramic materials and why?
9) Why is the fusion-based additive manufacturing of ceramics challenging?
10) Explain how the sintering time and temperature may affect the strength of sintered parts.
11) What are the common defects in the 3D-printed ceramic parts and how are the defects mitigated?
12) Additively manufactured ceramic parts often contain porosity. How does porosity affect part strength?

References

1 Lakhdar, Y., Tuck, C., Binner, J., Terry, A. and Goodridge, R., 2021. Additive manufacturing of advanced ceramic materials. *Progress in Materials Science*, 116, article no.100736.

2 Mansfield, B., Torres, S., Yu, T. and Wu, D., 2019, June. A review on additive manufacturing of ceramics. In *International Manufacturing Science and Engineering Conference* (Vol. 58745, p. V001T01A001). American Society of Mechanical Engineers.

3 Boissonneault, T., 22 December 2021. *The state of technical ceramic materials for laser stereolithography*. Available at https://www.3dprintingmedia.network/the-state-of-technical-ceramic-materials-for-laser-stereolithography (accessed on 17 February 2023).

4 Zhang, H., Huang, L., Tan, M., Zhao, S., Liu, H., Lu, Z., Li, J. and Liang, Z., 2022. Overview of 3d-printed silica glass. *Micromachines*, 13(1), article no.81.

5 Wang, X., Zhou, Y., Zhou, L., Xu, X., Niu, S., Li, X. and Chen, X., 2021. Microstructure and properties evolution of silicon-based ceramic cores fabricated by 3D printing with stair-stepping effect control. *Journal of the European Ceramic Society*, 41(8), pp.4650–4657.

6 Sher, D., 23 August 2019. Silicon carbide 3D printing is hot. *Let's find out why*. Available at https://www.3dprintingmedia.network/silicon-carbide-3d-printing-is-hot-lets-find-out-why (accessed on 18 February 2023).

7 International Syalons, 25 April 2018. *Introducing additive manufacturing for Silicon nitride ceramics*. Available at https://www.syalons.com/2018/04/25/introducing-additive-manufacturing-for-silicon-nitride-ceramics (accessed on 17 February 2023).

8 Kumar, A., Kargozar, S., Baino, F., and Han, S.S. 2019. Additive manufacturing methods for producing hydroxyapatite and hydroxyapatite-based composite scaffolds: a review. *Frontiers in Materials*, 6, article no.313.

9   Travitzky, N., Windsheimer, H., Fey, T., and Greil, P. 2008. Preceramic paper-derived ceramics. *Journal of the American Ceramic Society*, 91(11), pp.3477–3492.

10 Peng, E., Zhang, D. and Ding, J., 2018. Ceramic robocasting: recent achievements, potential, and future developments. *Advanced Materials*, 30(47), article no.1802404.

11 Du, W., Ren, X., Ma, C. and Pei, Z., November 2017. Binder jetting additive manufacturing of ceramics: A literature review. In *ASME International Mechanical Engineering Congress and Exposition* (Vol. 58493, p.V014T07A006). American Society of Mechanical Engineers.

12 He, H., Lou, J., Li, Y., Zhang, H., Yuan, S., Zhang, Y., and Wei, X. 2018. Effects of oxygen contents on sintering mechanism and sintering-neck growth behaviour of FeCr powder. *Powder Technology*, 329, pp.12–18.

13 Monmaturapoj, N. and Yatongchai, C., 2010. Effect of sintering on microstructure and properties of hydroxyapatite produced by different synthesizing methods. *Journal of Metals, Materials and Minerals*, 20(2).

14 Safonov, A., Chugunov, S., Tikhonov, A., Gusev, M. and Akhatov, I., 2019. Numerical simulation of sintering for 3D-printed ceramics via SOVS model. *Ceramics International*, 45(15), pp.19027–19035.

15 Carazzone, J.R., Martin, C.L. and Cordero, Z.C., 2020. Crack initiation, propagation, and arrest in sintering powder aggregates. *Journal of the American Ceramic Society*, 103(9), pp.4754–4773.

16 Li, Q., An, X., Liang, J., Liu, Y., Hu, K., Lu, Z., Yue, X., Li, J., Zhou, Y. and Sun, X., 2022. Balancing flexural strength and porosity in DLP-3D printing Al2O3 cores for hollow turbine blades. *Journal of Materials Science & Technology*, 104, pp.19–32.

17 Pan, Y., Li, H., Liu, Y., Liu, Y., Hu, K., Wang, N., Lu, Z., Liang, J. and He, S., 2020. Effect of holding time during sintering on microstructure and properties of 3D printed alumina ceramics. *Frontiers in Materials*, 7, p.54.

18 Schweiger, J., Edelhoff, D. and Güth, J.F., 2021. 3D printing in digital prosthetic dentistry: an overview of recent developments in additive manufacturing. *Journal of Clinical Medicine*, 10(9), p.2010.

19 Sertoglu, K., 02 July 2020. *XJet and Straumann partner to scale up ceramic additive manufacturing for dentistry*. Available at https://3dprintingindustry.com/news/xjet-and-straumann-partner-to-scale-up-ceramic-additive-manufacturing-for-dentistry-173094 (accessed on 21 February 2023).

20 Xu, W., Huang, S., Han, D., Zhang, Z., Gao, Y., Feng, P. and Zhang, D., 2022. Toward automated construction: the design-to-printing workflow for a robotic in-situ 3D printed house. *Case Studies in Construction Materials*, 17, article no.e01442.

21 Massie, C., 28 January 2015. *China's WinSun Unveils Two New 3D printed buildings*. Available at https://www.architectmagazine.com/technology/chinas-winsun-unveils-two-new-3d-printed-buildings_o (accessed on 10 February 2023).

22 Tomlinson, C., 21 March 2022. *First complete 3D-printed home can withstand hurricanes*. Available at https://www.houstonchronicle.com/business/columnists/tomlinson/article/3-D-printed-house-offers-resiliency-and-beauty-17007516.php (accessed on 10 February 2023).

23 Aysha, M., 12 October 2020. *Studio RAP will 3D print 4,000 ceramic tiles for a new architecture project*. Available at https://www.3dnatives.com/en/studio-rap-ceramic-tiles-121020205/# (accessed on 21 February 2023).

24 Allan, S., August 2021. *Additive manufacturing aerospace ceramics*. Available at https://www.aerospacemanufacturinganddesign.com/article/additive-manufacturing-aerospace-ceramics (accessed on 21 February 2023).

25 Halliday, A., 23 April 2021. *3D Print 18,000 famous sculptures, statues & artworks: Rodin's Thinker, Michelangelo's David & More*. Available at https://www.openculture.com/2021/04/3d-print-18000-famous-sculptures-statues-artworks-rodins-thinker-michelangelos-david-more.html (accessed on 21 February 2023).

26 Ganea, S., 05 September 2019. *The future is now – 3D printed home accessories.* Available at https://www.homedit.com/3d-printed-home-accessories (accessed on 21 February 2023).

27 Vialva, T., 03 April 2019. *Andrea Salvatori and WASP create 3D printed ikebana vases.* Available at https://3dprintingindustry.com/news/andrea-salvatori-and-wasp-create-3d-printed-ikebana-vases-152618 (accessed on 21 February 2023).

28 Hanaphy, P., 23 June 2022. *Synteris awarded $2.7M grant to develop new method of 3D printing ceramic packaging for electronics.* Available at https://3dprintingindustry.com/news/synteris-awarded-2-7m-grant-to-develop-new-method-of-3d-printing-ceramic-packaging-for-electronics-211213 (accessed on 21 February 2023).

29 Lewis, J.A., 2002. Direct-write assembly of ceramics from colloidal inks. *Current Opinion in Solid State and Materials Science*, 6(3), pp.245–250.

30 Davies, S., 06 May 2022. *XJet ceramic 3D printing technology used to produce ultracar engine piston.* Available at https://www.tctmagazine.com/additive-manufacturing-3d-printing-news/ceramic-and-exotic-additive-manufacturing-news/xjet-ceramic-3d-printing-technology-used-to-produce-ultracar/#:~:text=XJet's%20ceramic%203D%20printing%20technology,which%20is%20currently%20under%20development (accessed on 21 February 2023).

5

Design for Additive Manufacturing

Learning objectives

After reading this chapter the reader should be able to do the following:

1) Understand what is design for additive manufacturing (DFAM) and how it differs from the design principles for conventional manufacturing.
2) Appreciate the unique capabilities of additive manufacturing that are important in the design for additive manufacturing.
3) Comprehend the important constraints in the design for the additive manufacturing.
4) Understand the principles of material selection in the design for additive manufacturing.
5) Appreciate the important steps in the design for additive manufacturing.
6) Understand the roles of computers and the emerging digital tools in the design for additive manufacturing.
7) Appreciate how sustainability is considered in the design for additive manufacturing.

CONTENTS

Theory and Practice of Additive Manufacturing, First Edition. Tuhin Mukherjee and Tarasankar DebRoy.
© 2024 John Wiley & Sons, Inc. Published 2024 by John Wiley & Sons, Inc.

5.1   Introduction

Engineers work to improve the lives of people through the products they design and manufacture. The engineering design of a part considers its purpose, function, economics, sustainability, and aesthetics. The manufacture of parts of a selected material by additive manufacturing starts with and is guided by a digital file known as the computer-aided design (CAD) file. With a good design file, additive manufacturing enables the creation of complex geometries and shapes that may be difficult or impossible to make using traditional manufacturing methods. The design could be used for a variety of purposes, including prototyping, custom production of one or more legacy components, and manufacturing.

The design for additive manufacturing is important for several reasons. First, when compared to traditional subtractive manufacturing, a good design for additive manufacturing improves the efficiency of the utilization of valuable feedstock by reducing materials wastage. Another powerful aspect of the design is the ability to create unique customized products. Also, for the production of a small batch of products or prototypes, well-designed additively manufactured components may be produced faster than traditional manufacturing methods. Parts made by additive manufacturing may be cost-competitive, especially for complex and customized parts that cannot be easily produced by traditional manufacturing. Also, well-designed additively manufactured products can be more sustainable than that produced by conventional manufacturing. As a result, the design is important to take advantage of these benefits and optimize it for a specific additive manufacturing process.

The design starts with a computer-aided design (CAD) file that includes geometric data such as the size, shape, and orientation, the additive manufacturing process, process parameters to be used, and other data. Depending on the specific additive manufacturing machine and the software, different types of CAD files are used. Table 5.1 shows a selection of several important file formats, their features, capabilities, and limitations. However, this table does not include all CAD file formats used in additive manufacturing. The selection of a specific file format depends on the printing software and hardware, as not all formats are supported by all equipment manufacturers.

Of the various design files, the STL format was developed specifically for 3D printing and it has been used since the late 1980s. The OBJ files are often used to represent more complex models, such as those with curved surfaces or intricate details. The AMF file was developed to improve the capabilities of the STL file by including color and material properties, as well as information about the object's internal structure. The 3MF (3D Manufacturing Format) file format has been developed as a more advanced alternative to those developed earlier. This format can store a wide range of information about an object's geometry, materials, and other properties, and they are designed to be more flexible than the other formats. The development of the design file formats was aimed to improve the description of the product features for the manufacture of improved products. Selection of the right CAD file format for a 3D printing project ensures that the manufacturing process runs smoothly and produces high-quality parts.

Apart from the file format, there are several other important issues to consider when designing for additive manufacturing. The choice of materials is important because material properties

Table 5.1 File format, features, capabilities, and limitations of several additive manufacturing design file formats.

File Format	Features	Capabilities	Limitations
STL (STereoLithography)	Simple format designed for additive manufacturing, widely used	The model is represented as a collection of flat, triangular facets	Limited support for color and texture
OBJ (object)	Similar to STL, but includes information about color and texture	Represent 3D model as a collection of facets	May require more computing power and memory
3MF (3D Manufacturing Format)	Developed by Microsoft as a flexible alternative to STL and OBJ formats	Supports color and texture, and metadata. Can be shared and exchanged between different software and hardware systems	Supported by a growing, but limited, number of 3D printing software and machines
SLDPRT (SolidWorks Part File) or DWG (Drawing, by Autodesk)	Developed by software companies for their products	Includes features and capabilities not found in more general-purpose formats	Supported by a limited number of 3D printing software and machines
Process-specific formats such as G-code (geometric code), AMF (Additive manufacturing file)	G-code was originally used with computer numerical control machines. AMF is designed to be an efficient and reliable format	May include specialized data and commands specific to the process	Supported by a limited number of 3D printing software and machines

have to be appropriate for applications. Another important issue is the design of the object itself, as the geometries and structures that are possible with additive manufacturing differ from those that can be achieved with traditional manufacturing methods. Additionally, there are considerations around the overall production process, such as the speed and cost of production, as well as post-processing steps like surface finishing and assembly. So the design includes, apart from the geometry, the build orientation, the support structure (Figure 5.1), the material

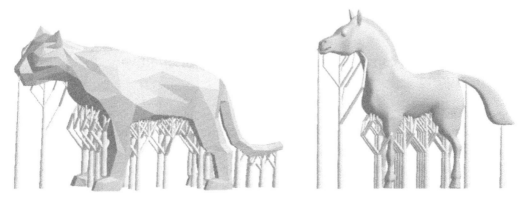

Figure 5.1 Support structures are shown by the yellow treelike lines. The figure is taken from [1] with permission from Elsevier.

properties, the uniqueness of the product, and the design constraints of the chosen additive manufacturing technology. The unique aspects of DFAM that separate it from conventional DFM are discussed in worked out example 5.1.

Worked out example 5.1

Discuss the unique aspects of design for additive manufacturing (DFAM) that separate it from design for manufacturing (DFM).

Solution:

Design for manufacturing (DFM) and design for additive manufacturing (DFAM) are two approaches to product design that consider the manufacturing process during the design phase.

DFM is typically used for traditional manufacturing methods such as injection molding, casting, and machining, which involve removing material from a raw material to create a finished product. On the other hand, DFAM is a design approach that specifically considers the use of additive manufacturing techniques in the production process. Additive manufacturing allows for more design freedom and the ability to create complex geometries and internal features that may not be possible with traditional manufacturing methods. However, it also requires a different set of design considerations, such as ensuring that the product can be printed with minimum support structures and that the design is optimized for the specific properties of the chosen additive manufacturing process.

One key difference between DFM and DFAM is the level of customization that is possible. Traditional manufacturing methods are typically geared toward mass production, with a focus on producing a large number of identical parts at a low cost. As such, DFM aims to design parts that can be easily and cheaply reproduced. In contrast, additive manufacturing allows for a higher level of customization, as each product can be built to order rather than being mass produced. This opens up new possibilities for design, such as personalized products or small batch production, but also requires a different set of design considerations to ensure that the product can be efficiently produced in small quantities.

Another difference between the two approaches is the level of material usage. In traditional manufacturing methods, material is often wasted as it is removed to create the finished product. For products that are made from expensive materials, the wastage can be expensive. In contrast, additive manufacturing typically has a lower material wastage as the product is built up layer by layer. The cost of additive manufacturing can be offset by the ability to create complex internal features and shapes that would be difficult or impossible to produce using traditional manufacturing methods.

5.2   Uniqueness considerations in the design for additive manufacturing

The unique capabilities of additive manufacturing, shown in Table 5.2, are important for design. The importance of these capabilities and their impact are also shown in the table.

Design Freedom: The design freedom provided by additive manufacturing enables the creation of new and innovative designs that were difficult or impossible to achieve before using conventional manufacturing methods. The complex geometries including overhangs, internal cavities, and intricate details of a part can now be included in the design.

Table 5.2 The unique capabilities of additively manufactured parts, their importance in design, and their impact.

Uniqueness	Importance	Impact
Design freedom	Complex geometries include overhangs, internal cavities, and intricate details in a single part.	Innovative designs and improved part performance; only possible using AM.
Parts consolidation	Merging multiple parts into a single, complex part; avoids assembly and improves overall performance.	Reduces assembly time and improves the overall performance of the product.
Weight reduction	Reduced part density; high specific strength and stiffness; particularly useful in the aerospace and automotive industries.	Improves the overall performance and efficiency of the part where weight reduction is important.
Compositionally graded materials	Avoids abrupt change in composition and mechanical properties.	Improves mechanical properties, reliability, and serviceability of dissimilar joints.
Reduced waste	Lower waste compared to conventional manufacturing; this feature makes it environmentally friendly and helps to make it cost-competitive.	Cost savings and environmentally friendly.
Customization	Produces customized products on demand; can produce only a few pieces based on customer needs.	Sensitive to customer needs.
Faster speed	Faster than traditional manufacturing products with complex geometries; does not require tooling.	Faster production improves process efficiency and reduces lead times.
Functional integration	May integrate sensors and electronics into a part reducing the need for additional components.	Improves performance and functionality.

Part consolidation: Additive manufacturing also enables the integration of multiple parts into a single part, thus eliminating assembly time and improving product performance. For example, a fuel nozzle of a jet engine that used to be made by assembling twenty parts is now made by powder bed fusion in one piece [2]. GE found the new nozzle to be more reliable making jet engines safer.

Weight reduction: The unique ability of additive manufacturing for the creation of complex, three-dimensional structures with precise control over the shape, size, orientation, and properties of each layer allows for the creation of lightweight structures that are optimized for specific applications. Designing parts with local lightweight parts while maintaining mechanical strength and stiffness is possible with additive manufacturing. One example of part weight reduction using additive manufacturing is the creation of lightweight structural components in automotive and aerospace industries to improve fuel efficiency, reduce emissions, and lower operating costs. Similarly, parts made from multiple materials can be used to create lightweight, high-quality equipment that improves athlete performance.

Compositionally graded components: These can avoid abrupt changes in the chemical composition of parts such as dissimilar metal joints that are susceptible to degradation of mechanical properties. In joints between steel and nickel alloys, carbon often diffuses away from the joint at elevated temperatures thus causing degradation of the creep resistance. This degradation can be mitigated by creating compositionally graded joints using directed energy deposition.

Reduced waste: Well-designed additively manufactured parts generate minimal waste compared to conventional manufacturing methods, as they only use the exact amount of material needed thus making these parts environmentally friendly and saving cost. Additive manufacturing does not mandate the production of a minimum number of parts of a given design, making it viable to produce even a single customized product when needed by a customer.

Customization: One of the unique attributes of additive manufacturing is its ability to customize the design considering the manufacturing process to meet the requirements of the product. Additive manufacturing allows for almost any design to be created within the physical limits of the printer. Products can be designed with intricate details and complex geometries that would not be possible with traditional manufacturing methods. The creation of customized products that are tailored to the specific needs and requirements of the user through the use of different materials, intricate designs, or manufacturing process is unique to additive manufacturing.

Fast speed: One way that additive manufacturing achieves fast speed is through its use of digital files as a guide for the manufacturing of prototypes and parts. With traditional manufacturing, the production of a part often requires physical tools and molds, which can take a significant amount of time. In contrast, additive manufacturing uses digital files, which can be easily updated, thus allowing for rapid changes to the design and production process. This uniqueness enables the testing of new designs when needed and also the development and marketing of new parts quickly.

Functional integration: Additive manufacturing allows functional integration within a part, which refers to the ability to combine multiple functions or components into a single, integrated structure rather than having to create separate components and then assemble them to achieve the desired functionality. An example of this attribute is the creation of parts with embedded sensors or electronics that can be integrated directly into the structure during the printing process. It allows the creation of complex, multifunctional structures that are not possible with other manufacturing techniques.

These unique capabilities are important in the design of parts because of the unprecedentedly ambitious designs enabling the creation of complex, customized, and innovative products that were not possible before. They also allow faster production of more capable environmentally friendly products and reduced waste, making additive manufacturing an attractive option for a wide range of industries. However, a good design needs to consider these advantages together with the limitations of additive manufacturing and the selection of appropriate materials.

5.3 Constraints in the design for additive manufacturing

Constraints play a crucial role in the design of parts for additive manufacturing because they affect the quality, functionality, and cost of the parts. Additive manufacturing allows the manufacture of parts of complex geometries that are impossible to make by conventional processes. However, this flexibility also means that the designer must be mindful of the various constraints that can impact the success of the manufacturing process. Some of the key constraints that need to be considered include material selection, the need for the support structure, specifications of layer thickness, part orientation, and wall thickness, the cost and time, post-processing, and the available build volume. These constraints, their important attributes, and their impact on design are listed in Table 5.3 and discussed below. These constraints can significantly impact the properties and serviceability of the part and must be considered to produce high-quality, functional, and cost-effective parts.

Material selection: Parts must have appropriate properties to be serviceable. A major constraint in the design of additively manufactured components is the material selection among the available

Table 5.3 The constraints, their important attributes, and their impact on the design for additively manufactured parts.

Constraint	Attributes	Impact on design
Material selection	The availability of feedstock materials and their attributes and printability affect the design	Products may require redesign to use a suitable material
Support structures	The necessity of support structures to maintain overhanging or cantilevered parts	May require redesign to minimize support structures and lower time and cost
Layer thickness	Specification of minimum and maximum layer thicknesses affects the surface finish and mechanical properties	May require redesign to achieve desired surface finish or strength, and dimensional accuracy
Part orientation	Part orientation affects strength, surface finish, and support requirements	Optimization of part orientation affects machine time and cost of feedstock and post-processing
Cost and time	The price of the selected material, machine time, and post-processing affect production rate and cost-competitiveness	Product cost and productivity are important factors in the design
Post processing	Post-processing is needed to remove support structure, improve surface finish and microstructure, and achieve high density	The design needs to minimize post-processing to save time and money
Build volume	The build volume restricts the maximum size of parts in some processes.	Some components may have to be designed for the assembly of multiple parts

feedstocks. It is important to consider an appropriate material for a specific application and consider factors such as mechanical and chemical properties, cost, and production time. Additive manufacturing allows for a wide range of materials to be used in the printing process, including metals, plastics, ceramics, and even living cells. Each of these materials has its unique properties that must be considered when selecting the appropriate material for a part. For example, metals are strong and durable, but may be expensive and have a longer production time. Polymers are lighter than metals but are not as strong as metals. Some materials may have high strength and stiffness but may be brittle and prone to cracking. Other materials may have good fatigue resistance, but may not be suitable for elevated temperature applications. As a result, material selection for additively manufactured components is a trade-off between required performance and available materials. Because of its importance, materials selection is discussed separately after the discussion on constraints.

Support structures: Manufacturing of parts with intricate geometry containing overhangs, cantilevers, and other features may require the use of support structures to ensure that the object does not collapse or distort during the printing process. These support structures can be difficult to remove. When they are removed, they may leave marks on the surface of the part. Some AM processes require the use of support structures to hold some portions of the part in place. The designer must explore the possibility to minimize their use or be able to remove them as easily as possible after printing. Some AM processes have difficulty printing overhangs, and the designer may have to either minimize these features or use support structures when these features need to be printed.

Layer thickness: The thickness of each layer deposited during the printing process is typically limited by the resolution of the printer. Thinner layers can provide higher resolution, but may also be more prone to defects and require a longer time to make a part. The layer thickness also affects the surface finish of the printed part. Thicker layers tend to result in a rougher surface finish, while thinner layers produce a smoother finish. Thus, thin layers are needed when a smooth surface finish is required. Thus, the layer thickness is an important factor that must be carefully considered in the design since it has a direct impact on the surface finish, and printing speed of the final printed product. The choice of layer thickness in the design needs to consider the requirements and the constraints of the application.

Part orientation: Part orientation must be considered in the design because the orientation of the part determines how different layers of material are deposited during the printing process. Different orientations can result in different scanning paths, support structures, different amounts of material requirement, and different properties of parts. Also, orienting a part in a particular way can affect the surface finish and speed of production. Appropriate choice of part orientation can reduce the number of required support structures, which may result in reduced material usage and higher productivity. Appropriate part orientation can lead to an efficient printing process and a superior product.

Cost and time: Additive manufacturing can be more expensive than traditional manufacturing methods in some cases, particularly for low-volume production runs. The cost of additive manufacturing can be influenced by several factors, including the material used, the size and complexity of the part, the required post-processing steps, and the chosen 3D printing technology. Careful consideration of these factors can help to minimize the overall cost of the manufacturing process. Time is another important factor that must be considered when designing for additive manufacturing. The printing process can take significantly longer than traditional manufacturing methods, particularly for larger or more complex parts. Cost and time can significantly impact the feasibility of a manufacturing project, and must be carefully balanced with the specific requirements and constraints of an application.

Post-processing: Post-processing is a critical constraint because it can significantly impact the microstructure, density, dimensional accuracy, and performance of the printed part. The printed parts may have rough or uneven surfaces, and support structures that need to be removed through post-processing. Post-processing may include sanding, polishing, painting, or other steps. These steps can improve the surface finish, appearance, and overall performance of the part. For example, smooth surfaces may improve the fatigue resistance of parts. However, post-processing adds additional time and cost to the overall manufacturing process.

Build volume: Build volume is a limiting factor because it determines the maximum size of the parts that can be printed on a specific 3D printer, and it affects the overall efficiency and the cost of the manufacturing process. The build volume of a 3D printer is determined by the size of the printer's build chamber or print bed and may vary significantly depending on the specific machine. For example, a desktop 3D printer may have a relatively small build volume, while an industrial 3D printer may have a much larger build volume. It ensures that the part can be printed in a single build, rather than having to be printed in multiple pieces and then assembled.

5.4 Materials selection in the design for additive manufacturing

In additive manufacturing, a wide variety of materials including metals and alloys, polymers, ceramics, and composites are used for fabricating parts. The same part can be manufactured using different materials. However, all materials cannot provide the desired properties or cannot produce parts economically. Therefore, the selection of an appropriate material for a given AM process to print a specific part is a significant step in the design for AM. In addition, there is scope for developing more reliable and cost-effective materials for existing products as well as realizing new products through novel materials that show distinct properties. In this section, we summarize the principles for material selection in AM, review the commonly used methods for selecting materials, and describe the available materials database for AM that can help in material selection.

5.4.1 Materials selection criteria

The selection of materials in AM is done by evaluating their suitability for the specified application, required aesthetics, mechanical and thermal properties, their ability to reduce defects, their availability as feedstocks, and costs. Table 5.4 summarizes these criteria and explains their effects on materials selection in AM.

Table 5.4 Common materials selection criteria for additive manufacturing.

Criterion	Description	Effect on materials selection
Application	Materials need to have the attributes required for the application. For example, a prototype may need to have good aesthetics, but a product should have durability, functionality, reliability, and serviceability.	Parts often need biocompatibility, sterilization capabilities, toxicity certifications, chemical resistance, or other functional requirements. The selected material needs to satisfy the engineering requirements of the product.
Aesthetics	AM parts and prototypes often need good aesthetics including color, surface finish, and cosmetic appearance.	Although better aesthetics can be achieved by post-process finishing services, some materials provide better aesthetics than others before and after post-processing.
Mechanical properties	Important mechanical properties include tensile strength, yield strength, elastic modulus, fracture toughness, and fatigue life.	Materials need to withstand the loads during manufacturing and in service. Mechanical properties of parts can often be improved by post-processing.
Thermal properties	Parts undergo repeated heating and cooling during AM. Thermal properties such as thermal conductivity, diffusivity, and specific heat affect how the materials behave during thermal cycling.	Materials with poor thermal properties may cause undesired microstructure, poor mechanical properties, and defects. For example, a material with a low thermal conductivity accumulates more heat and results in distortion and warping.
Printability	Printability or the ability of a material to be successfully converted to a product considers both the susceptibilities to common defects as well as metallurgical and mechanical properties.	Materials selection should be done by evaluating printability that helps to select an appropriate AM process–material combination and can be helpful to reduce common defects and improve microstructure and properties.
Feedstock availability	Many potentially printable materials often cannot be used in AM because of their unavailability as powder or wire feedstock.	Some products may require the design of new materials. However, their adaptation needs a strong business case to pay for testing and qualifying the part.
Cost	The cost of the feedstock material can significantly affect the overall cost of the part.	The cost of the material is an important factor in materials selection.

5.4.2 Materials selection methodologies

In AM, metals and alloys, polymers, ceramics, and composites are used for fabricating parts. Although there has been significant research progress in understanding the use of different materials in AM, currently, there is no well-accepted method for materials selection for AM. Materials selection for AM is generally done using materials selection charts and materials selection indexes described below.

5.4.2.1 Use of materials selection charts

Ashby's materials selection charts [3] are often used to make a selection of available materials for additive manufacturing of a product. These charts are plots of several mechanical and thermo-physical properties of materials that affect the performance of a product. The properties include density, Young's modulus, strength, toughness, damping coefficient, thermal conductivity, diffusivity, and thermal expansion coefficient. The charts plot one property against another and map out the fields in the property space occupied by each material class. For example, Figure 5.2 shows Ashby's chart indicating Young's modulus versus density for different materials. These charts condense a large volume of information into a compact but accessible form. In addition, they reveal correlations between important material properties that help in checking and estimating data for materials selection. Ashby's materials selection charts are well-accepted in materials selection in both traditional manufacturing as well as additive manufacturing.

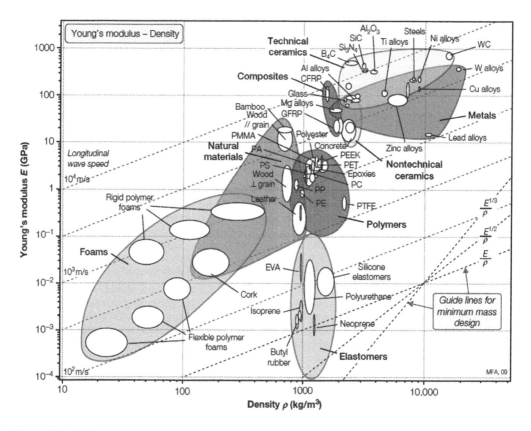

Figure 5.2 Ashby's materials selection chart shows Young's modulus versus density. Reproduced with permission from Professor Mike Ashby of Cambridge University.

5.4.2.2 Use of materials selection indexes

Material selection is a type of engineering decision-making. Many analytical methods based on several materials selection indexes have been introduced into engineering design to help with the evaluation process and in making a rational decision. These indexes are widely used in design for AM as discussed below.

a) Weighted property index

Design for AM often requires the optimization of multiple material properties. For example, the selection of materials for a printed bracket for aircraft needs optimization of weight and fatigue strength. Typically, each material property is assigned a weighting factor based on its importance. Individual weighted property values are then added together to produce a comparative weighted property index. However, the weighted property index frequently has to deal with properties that have different units. The best method is to use a scaling factor to normalize the property values. Each property is scaled so that its highest value does not exceed 100. The scaled value = (Value of property × 100)/Highest value of property in the list. Properties for which a low value is the best, such as weight, mass loss, and wear rate, are scaled as: The scaled value = (Lowest value of property in the list × 100)/Value of property. A subjective rating is required for properties that cannot be expressed numerically, such as printability. Once each material property has been assigned a scaled value, the weighted property index for the material is simply the sum of the scaled value multiplied by the weighting factor. The use of the weighted property index for materials selection is explained in worked out example 5.2.

b) Cost per unit property index

Often in materials selection for AM, one property stands out as the most dominant service requirement. In this case, the cost per unit property index is a suitable materials selection index that determines how much material would cost to provide that property requirement. For example, strength is the dominant material property in the design of additively manufactured frames for an automobile. To achieve higher strength the frames may need to have more materials which increases cost. Therefore, cost per unit strength is the most suitable index for designing such automobile frames. The use of the cost per unit property index for materials selection is explained in worked out example 5.3.

c) Printability index

The ability of a material to be successfully converted to a product by a given AM process is referred to as printability. It considers both susceptibilities to common defects and metallurgical and mechanical properties. Several indexes are available in the literature to compare the susceptibility of different materials to common AM defects (see Chapter 10). Materials should be chosen by assessing printability using those indexes that aid in the selection of an appropriate AM process-material combination and can help to reduce common defects and improve microstructure and properties.

Worked out example 5.2

An aircraft landing gear part needs to be designed for manufacturing using AM. There are seven alloys (A to G) that can be used to make the component. Their important mechanical and thermophysical properties at room temperature are provided in Table E5.1 below. Use the weighted property index method to select the most suitable alloy.

(Continued)

Worked out example 5.2 (Continued)

Table E5.1 Mechanical and thermophysical properties of seven alloys at room temperature.

Alloys	Toughness, J	Yield strength, MPa	Young's modulus, GPa	Density, g/cc	Thermal expansion, 10^{-6}/K	Thermal conductivity, cal/s-cm-K	Specific heat, cal/g-K
A	75.5	420	74.2	2.8	21.4	0.37	0.16
B	95	91	70	2.68	22.1	0.33	0.16
C	770	1365	189	7.9	16.9	0.04	0.08
D	187	1120	210	7.9	14.4	0.03	0.08
E	179	875	112	4.43	9.4	0.016	0.09
F	239	1190	217	8.51	11.5	0.31	0.07
G	273	200	112	8.53	19.9	0.29	0.06

Solution:

The weighted property index method starts with pairwise comparison where each property is compared with each of the other properties. If property 1 is more important than property 2, property 1 is assigned "1" and property 2 is assigned "0." If there are N properties, the total number of comparison is equal to N(N − 1)/2. Here, we have seven properties and thus the total number of comparisons is equal to 21. The comparisons are shown in Table E5.2. For example, in comparison 1, toughness and yield strength are compared; toughness is assigned 1 and yield strength is assigned 0.

Table E5.2 Pairwise comparison table.

Property	Comparison number																				
	1	2	3	4	5	6	7	8	9	10	11	12	13	14	15	16	17	18	19	20	21
Toughness	1	1	1	1	1	1															
Yield strength	0						1	0	0	1	1										
Young's modulus		0					0					0	0	0	1						
Density			0					1				1				1	1	1			
Thermal expansion				0					1				1			0			1	1	
Thermal conductivity					0					0				1			0		0		0
Specific heat						0					0				0			0		0	1

A landing gear component often suffers from a high impact load during takeoff and landing. Therefore, toughness is the most important property for their design. The comparison table shows that toughness is always assigned 1 when compared to other properties. Density is the second important property since aircraft components are always designed to minimize weight. Table E5.3 provides the weighting factors for each property. Weighting factors are calculated as number of 1 divided by the total number of comparisons. For example, for toughness, weighting factor is equal to 6/21 = 0.28.

Worked out example 5.2 (Continued)

Table E5.3 Weighting factors for each property.

Property	Total number of 1	Weighting factor
Toughness	6	0.28
Yield strength	3	0.14
Young's modulus	1	0.05
Density	5	0.24
Thermal expansion	4	0.19
Thermal conductivity	1	0.05
Specific heat	1	0.05
Total	**21**	**1.00**

Table E5.4 shows the calculations of weighted property index and ranking of alloys based on that index. The scaled values of each property and weighted property index for each alloy are calculated as described in Section 5.4.2.2. Please note that toughness, yield strength, and Young's modulus need to maximized and other properties should be minimized and scaling is done accordingly. For example, the scaled value of toughness of "C" is 100 because "C" has the highest toughness. In contrast, the scaled value of density of "B" is 100 because "B" has the least density. Weighted property index of "A" is equal to ($10\times0.28 + 30\times0.14 + 34\times0.05 + 96\times0.24 + 44\times0.19 + 4.3\times0.05 + 38\times0.05$) = 42.2. The alloy with a higher weighted property index is more suitable. Here, alloy "C" is the most suitable for designing the component.

Table E5.4 Calculated weighted property index and ranking of alloys.

Alloys	Toughness	Yield strength	Young's modulus	Density	Thermal expansion	Thermal conductivity	Specific heat	Weighted property index	Ranking
A	10	30	34	96	44	4.3	38	42.2	5
B	12	6	32	100	43	4.8	38	40.1	6
C	100	100	87	34	56	40	75	70.9	1
D	24	82	97	34	65	53	75	50.0	4
E	23	64	52	60	100	100	67	59.8	2
F	31	87	100	30	82	5.2	86	53.3	3
G	35	15	52	30	47	5.5	100	35.9	7

Worked out example 5.3

Connecting rods between the cylinder piston head and the crankshaft in an automobile engine can be fabricated using AM. Since the connecting rods face high stress during the operation of the automobile, yield strength is the dominant property in their design. Explain how the cost per unit property index can be used for selecting an appropriate material for designing the connecting rod. Assume that the cross-sectional area of the connecting rod is constant and does not vary with length.

(Continued)

Worked out example 5.3 (Continued)

Solution:

The mass of the connecting rod can be written as: average area × length × density
To prevent failure, the yield strength of material should be higher than or equal to
  maximum allowable force/average area
By combining the above equations,
  mass = (maximum allowable force/yield strength) × length × density
    = maximum allowable force × length × (density/yield strength)
Material cost is directly proportional to the mass of the connecting rod. Therefore,
  cost ∝ maximum allowable force × length × (density/yield strength)
The design requires that the length and force are specified and cannot be changed. They are
constraints and remain constant. Therefore, cost ∝ density/yield strength
  Here, cost/(density/yield strength) is a cost per unit property index for designing the
connecting rod. The value of (density/yield strength) of an appropriate material can be found
from the Ashby chart showing strength versus density data.

5.4.3   Materials databases for additive manufacturing

AM researchers all over the world have recognized the need for a materials database for AM. Such a database is immensely useful in selecting an appropriate material for an AM process to fabricate a desired product. Attempts have been made to create such databases and Table 5.5 provides a few examples. However, significant research and development are needed to create a database that includes all materials and AM process combinations.

Table 5.5   A selection of currently available materials database for additive manufacturing.

Database	Developer and URL	Description and features
Additive Manufacturing Materials Database (AMMD)	National Institute of Standards and Technology (NIST) https://ammd.nist.gov	Open access data: "AMMaterial" provides vendor material information, "AMMachine" gives machine information, "AMBuild" supplies the information on a specific AM part and part design, pre-process, in-process, and post-process information, and testing.
Database of Additive Manufacturing Machines & Materials	Senvol http://senvol.com/database	Two open access modules: Machine Search and Materials Search. Contains data for different materials feedstock and various AM machines.
Database on mechanical properties of AM materials	Massachusetts Institute of Technology (MIT) http://apt.mit.edu/ am-material-data	Open access mechanical properties database of additively manufactured parts of polymers and metals and alloys.
Wohlers Report materials database	Wohlers Associates https://wohlersassociates.com/ product/wohlers-report-2022	The materials database is available in part 2 of the report. The database is on materials for different AM processes and products.

Table 5.5 (Continued)

Database	Developer and URL	Description and features
Additive Manufacturing Standards Database	Society of Manufacturing Engineers (SME) https://www.sme.org/ technologies/medical-additive-manufacturing/am3dp	Data can be found by area (design, testing, and materials), process (powder bed fusion, binder jetting, and directed energy deposition), material (metal, polymers, and ceramic), material form (powder and liquid), and development organization.
beAM Database	Beamler https://www.beamler.com/ 3d-printing-material-database	Database for AM of metals, alloys, polymers, and ceramics. Available for purchase.
Materials database for AM	The National Aeronautics and Space Administration (NASA) https://ammo.ncms.org/ nasa-material-database-for-dod	Material database of test data and results on AM processes and properties on Inconel 718. Users need to register on NASA's Materials and Processes Technical Information System.
Materials database book for AM	InssTek http://www.insstek.com/core/ board.php?bo_table= technical_news&wr_id=28	Materials database for AM of tool steels, stainless steels, Ti-6Al-4V, Inconel alloys, Co-Cr alloys, and Al-Bronze. This is an open-access database.

5.5 Topology optimization

Topology is a branch of mathematics that studies the properties of objects of various shapes. Two objects are topologically equivalent if they can be deformed into each other by stretching, bending, and twisting, but without tearing. Topology optimization in additive manufacturing is concerned with various shapes and orientations of the part or structure, the distribution of material within it, and the size and placement of openings and voids while meeting certain performance requirements and constraints. These requirements may include a specified load-bearing ability, stiffness, weight restriction, or other constraints. The goal of topology optimization may be to design a part that is structurally safe and efficient, lightweight, satisfy the functional specifications, and can be manufactured by the AM process in a cost-competitive manner. For example, Figure 5.3 shows that topology optimization can reduce the need for support structures and minimize the material waste. Depending on the application, the topology optimization may create either a simple structure or a complex part that cannot be made by conventional manufacturing techniques. Topology optimization involves the use of mathematical optimization techniques to identify the optimal distribution of material within the part.

5.5.1 Benefits

There are many benefits of the topology-optimized design of additively manufactured parts. It allows designers to create complex and intricate structures that would be impossible to fabricate using traditional manufacturing because AM can build intricate structures with internal voids and complex features. Also, topology optimization in AM can help to reduce waste and promote sustainability. Unlike traditional manufacturing processes that often waste a significant amount of excess material, AM produces parts with minimal waste since the material can be added in the exact amount needed to make the desired shape.

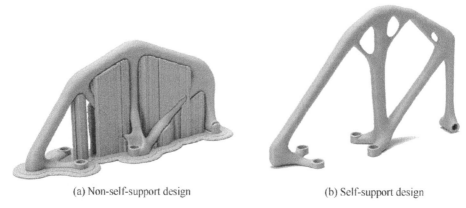

(a) Non-self-support design   (b) Self-support design

Figure 5.3 Topology optimization in the design to reduce the need for support structures [4]. The figure is taken from an open-access article [4] under the terms and conditions of the Creative Commons Attribution (CC BY) license.

5.5.2 Specifications

In order to realize the aforementioned benefits, topology optimization needs to comply with the various design specifications. Examples of several of these constraints are indicated below.

Part performance: Load bearing ability, ability to bear external forces that the part or structure are be subjected to during use, such as gravity, wind, or other applied loads. Fatigue strength is also a consideration in many situations since many material failures occur by metal fatigue. Also, the ability to withstand elevated temperatures is important in many applications.

Part geometry: These are constraints related to the overall shape and size of the part or structure, as well as any specific features or details that must be included. They also include minimum and maximum dimensions, wall thickness, and overhang specifications. The specific AM process, build chamber dimensions, and other factors may affect the constraints of part geometry.

Material properties: The available materials affect what can be built and the specification of the part. The properties of the available materials, such as strength, stiffness, and density, will also impact the design process.

Manufacturing specifications: These are constraints related to the specific AM process. The size and shape of the build envelope, the layer thickness, and the minimum feature size can be produced depending on the specific AM process and the process parameters.

Cost: The cost of the part or structure, will be affected by the cost of materials, the cost of manufacturing including the machine time, and the post-processing cost such as the cost of removal of the support structure, machining the surface and heat treatment or hot isostatic pressing of the part. Depending on the application, the overall cost may be a constraint.

5.5.3 Steps in topology optimization

An efficient, cost-effective, and sustainable part may be made by considering these and other constraints, including the necessary performance requirements. The steps that are involved in the topology optimization process are the following:

Define performance requirements and constraints: The first step in topology optimization for AM is identifying the performance criteria that the structure should meet, such as strength, stiffness, weight, or other specifications. It also involves specifying the constraints, such as the size and shape of the structure, the materials to be used, and the AM manufacturing process.

Modeling the mechanical behavior of the structure: A finite element model (FEM) of the structure would enable the analysis of its behavior under different external forces. The FEM should include all relevant boundary conditions, such as the applied loads as well as the material properties. Structural analysis can be undertaken to determine the performance of the structure under the specified load conditions. The results of the structural analysis may be used to evaluate the performance of the structure and identify any areas for possible improvement.

Optimize the material distribution: Optimizing the material distribution within the structure subject to the desired performance criteria requires the selection of an optimization algorithm. such as gradient-based or evolutionary algorithms. The optimization algorithm will determine the optimal material distribution that meets the performance criteria and satisfy the design constraints.

Building plan: After the material distribution is optimized, the AM build plan is made. This involves dividing the structure into layers and defining the placement of the material within each layer. The AM build plan should be designed to minimize both the amount of material used and the number of supports required during the AM process.

Building, post-processing, and testing: After the part is made, the support structure is removed and the part is tested to ensure that it satisfies the desired performance specifications such as strength, stiffness, and weight.

5.5.4   Topology optimization software

Optimization software can help designers create efficient and superior parts that are stronger and lighter that can be made using AM. Table 5.6 explains a selection of several optimization software and their main features.

Table 5.6   A selection of available topology optimization software and their salient features.

Software	Salient features	Comments
FreeFEM http://www.freefem.org	Open-source software for topology optimization, solution of differential equations, geometry and mesh generation, and data visualization	Active user community and wealth of resources; requires an understanding of the underlying mathematical principles
ToPy https://github.com/williamhunter/topy	Open-source topology optimization software written in Python	Maintained by a group of volunteers
Altair Inspire https://www.altair.com/products/inspire	User-friendly optimization capabilities, support for multiple materials and manufacturing processes	Previously developed OptiStruct is now integrated into Altair Inspire
nTopology https://ntopology.com/topology-optimization-software	Enables generative design to design high-performing parts iteratively; enables weight reduction while satisfying performance specification	Demo version available; automated smoothening of parts

(Continued)

Table 5.6 (Continued)

Software	Salient features	Comments
Tosca https://www.3ds.com/ products-services/ simulia/products/tosca	Topology, bead, shape, and sizing optimization based on FEA; it optimizes the structure to speed up the development process	Works with widely used software Abaqus; no trial version is available
Cero https://www.ptc.com/en/ products/creo	A CAD solution to design, optimize, and validate models. Generative topology optimization considers the restrictions, specifications, materials, and manufacturing processes	Free trial available; relatively inexpensive
ANSYS topology optimization https://www.ansys.com/ applications/ topology-optimization	Optimization of shape and weight; performance simulation with CAD Interoperability	Wide-ranging capabilities; free student downloads

5.6   Process selection in the design for additive manufacturing

Additive manufacturing (AM) enables the fabrication of complex structures ranging from the overall geometry of parts to the topology of its internal architecture to the spatial distribution of its material composition. The freedom of fabrication using AM cannot be achieved without the knowledge of the design for AM. Design for AM encompasses not only designing the part but also designing how the part is going to be built to overcome the inherent limitations of the selected AM process. The build orientation, process layout, process parameters, and post-processing need to vary for each AM process, and they can dramatically impact the quality, performance, and cost of the part. Sending a computer-aided design (CAD) file to a machine specifying only the material and dimensional tolerances is not sufficient. In this section, the seven tasks commonly practiced in the design of AM are discussed. They include the generation of a 3D computer-aided design (CAD) model, converting computer-aided design (CAD) file to stereolithography (STL) file, slicing 3D design into layers for layer-by-layer manufacturing, materials selection, AM process and machine selection, process planning and computer-aided manufacturing (CAM), and specifying sensing and control if required for process monitoring. Designers should be involved with all of these steps because they may necessitate a redesign of the part to manufacture high-quality parts in a time-efficient and cost-effective manner.

5.6.1   Generating a 3D computer-aided design (CAD) model

Although there are many 3D printing processes, the majority of them rely on computer-aided design (CAD). Before a manufacturing company can use a 3D printer to build a prototype or product, the object must first be digitally designed on a computer. The process of designing an object with computer software is known as CAD. There are various types of CAD software to design prototypes and products (Table 5.7). CAD software is used to create three-dimensional object models with precise shapes and sizes. Most 3D printing processes necessitate the use of CAD software, which provides the instructions required to construct a prototype or product. Rules or instructions included in a CAD file govern 3D printers. They use the information in a CAD file to determine how much material needs to be deposited and where it should be deposited.

During AM, support structures are required to support overhangs and loose powder layers. For example, support structures are required in fused deposition modeling processes for polymers to prevent overhanging features from sagging due to gravity. Support structures are required in metal laser-based powder bed fusion to support overhanging structures to the substrate to prevent distortion, warping, and jamming of the recoater blade. Support structures must be designed and included in the part's 3D model. In many AM processes, part orientation has a significant impact on the need for support structures. For example, Figure 5.4 explains the effect of build orientation on the need for a support structure during AM of a cylinder. A support structure is not needed when the build direction is along the axis of the cylinder vertically upward from the substrate (Figure 5.4 (a)). However, support structures are required when the axis of the cylinder is either parallel to the substrate (Figure 5.4 (b)) or at an angle with the substrate (Figure 5.4 (c)).

Table 5.7 Commonly used computer-aided design (CAD) software.

Software type	CAD software	Company	Salient features
Commercial software	AutoCAD	Autodesk	Available from: www.autodesk.com/products/autocad/overview For Windows, macOS, iOS, and Android operating systems; suitable for both 2D and 3D design
	CATIA	Dassault Systèmes	Available from: www.3ds.com/products-services/catia For Windows and Linux operating systems; suitable for computer-aided design, computer-aided manufacturing, 3D modeling, and product lifecycle management
	SolidWorks	Dassault Systèmes	Available from: www.solidworks.com For Windows operating system; suitable for solid modeling, computer-aided design, and computer-aided engineering applications
	ProE/PTC Creo	Parametric Technology Corporation (PTC)	Available from: www.ptc.com/en/products/creo/pro-engineer For Windows operating system; suitable for solid modeling, computer-aided design, and computer-aided manufacturing
	TurboCAD	IMSI Design LLC	Available from: www.turbocad.com For Windows and macOS operating systems; suitable for 2D and 3D design and drafting

(Continued)

Table 5.7 (Continued)

Software type	CAD software	Company	Salient features
Open-source software	Shapr3D	Shapr3D	Available from: www.shapr3d.com For Windows, iOS, and macOS operating systems; suitable for 2D and 3D design, visualization, validation of concepts of 3D printed prototypes
	Fusion360	Autodesk	Available from: www.autodesk.com/products/fusion-360/overview Available for Windows and macOS with simplified applications available for Android and iOS; paid version is available for commercial use
	Blender	Blender Foundation	Available from: www.blender.org Available for Windows, Linux, and macOS operating systems; useful for designing, modeling, and rendering
	BricsCAD Shape	Bricsys	Available from: www.bricsys.com/en-us/bricscad-shape Available for Windows, Linux, and macOS operating systems; allows creating 3D concepts and designs for prototypes for architecture, mechanical assemblies, and 3D printing
	DesignSpark	Ansys	Available from: www.rs-online.com/designspark/home Available for Windows operating system; enables users to create solid models in a 3D environment and create files for use with 3D printers
	FreeCAD	FreeCAD	Available from: www.freecadweb.org Available for Windows, Linux, and macOS operating systems; used for 3D computer-aided design modeling with finite element method support
	Onshape	Parametric Technology Corporation (PTC)	Available from: www.onshape.com/en Makes use of cloud computing, with compute-intensive processing and rendering performed on internet-based servers; users are able to interact with the system via a web browser or the iOS and Android operating systems
	SelfCAD	SelfCAD	Available from: www.selfcad.com Used for the 2D sketch to a fully-fledged 3D design, sculpting, rendering, and creating design files for 3D printing; both web and PC versions are available
	Ultimaker Cura	Ultimaker	Available from: https://ultimaker.com/software/ultimaker-cura Available for Windows, Linux, and macOS operating systems; used for 3D design and slicing applications for 3D printing
	TinkerCAD	Autodesk	Available from: www.tinkercad.com Web-based 3D computer-aided design platform
	LibreCAD	LibreCAD	Available from: https://librecad.org Available for Windows, Linux, and macOS operating systems; used for 2D designing and sketching

(a)   (b)   (c)

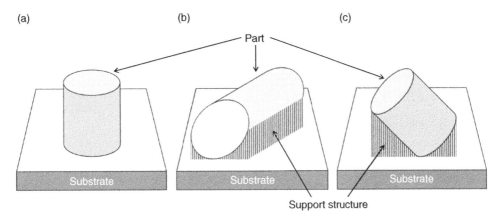

Figure 5.4 Effect of build orientation on the need for support structure during AM of a cylinder. The axis of the cylinder is (a) perpendicular to the substrate, (b) parallel to the substrate, and (c) at an angle with the substrate. *Source:* T. Mukherjee and T. DebRoy.

5.6.2 Converting computer-aided design (CAD) file to stereolithography (STL) file

Before being sent to the AM machine, the CAD file is converted to a stereolithography (STL) file. The STL file format is made up of a triangulated mesh that approximates the 3D solid model's boundaries and surfaces (Figure 5.5). Tessellation is the process by which the surfaces of a solid model are approximated into triangles. The resulting STL file contains vertices for each triangle and the normal vectors, which indicate which face points outward and which faces point inward. The calculations of normal vectors are explained in worked out example 5.4. The surface deviation, as shown in Figure 5.5, defines the maximum allowable difference between the tessellation and the actual part surface. The deviation determines the accuracy and smoothness of the geometry in the STL file. The highest resolution is obtained by using the smallest deviation, but this is often not practical due to the large number of triangles created, and the large file size. When converting CAD models to STL files, designers must maintain a balance between accuracy/resolution and file size. STL files can be saved in both ASCII and binary formats. ASCII files are readable by humans, but binary files are about one-sixth the size of the same file saved in ASCII format.

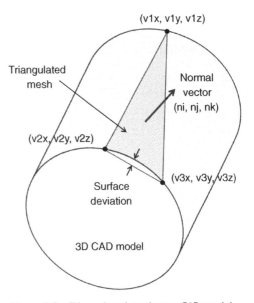

Figure 5.5 Triangulated mesh on a CAD model of a cylinder needed for the STL file. *Source:* T. Mukherjee and T. DebRoy.

Worked out example 5.4

CAD files of 3D design of parts are converted to STL files that are made up of triangulated mesh (Figure 5.5). If the three vertices of a triangulated mesh and their coordinates are A (1,2,5), B (5,7,9), and C (3,2,−1), find out the normal vector.

(Continued)

Worked out example 5.4 (Continued)

Solution:

The normal vector can be found from the cross product of two vectors $\overrightarrow{AB}$ and $\overrightarrow{AC}$

$$\overrightarrow{AB} = (5-1)\,\hat{i} + (7-2)\,\hat{j} + (9-5)\,\hat{k} = 4\,\hat{i} + 5\hat{j} + 4\,\hat{k}$$

$$\overrightarrow{AC} = (3-1)\,\hat{i} + (2-2)\,\hat{j} + (-1-5)\,\hat{k} = 2\,\hat{i} - 6\,\hat{k}$$

Therefore, $\vec{n} = \overrightarrow{AB} \times \overrightarrow{AC} = \begin{vmatrix} \hat{i} & \hat{j} & \hat{k} \\ 4 & 5 & 4 \\ 2 & 0 & -6 \end{vmatrix} = -30\,\hat{i} + 32\,\hat{j} - 10\,\hat{k}$

Note that the cross product of two vectors, A and B is given by the following expression:

$$\vec{A} \times \vec{B} = \begin{vmatrix} \hat{i} & \hat{j} & \hat{k} \\ a & b & c \\ x & y & z \end{vmatrix} = (bz - cy)\,\hat{i} - (az - cx)\,\hat{j} + (ay - bx)\,\hat{k}$$

A commonly used ASCII STL format is shown below:

```
solid name
facet normal ni nj nk
outer loop
vertex v1x v1y v1z
vertex v2x v2y v2z
vertex v3x v3y v3z
endloop
endfacet
endsolid name
```

The file always begins with the "solid name" syntax followed by "facet" syntax with its normal vector "n" (refer to Figure 5.5). The "outer loop" contains the vertices data "v." The vertices are ordered by the right-hand rule of the triangles using a three-dimensional Cartesian coordinate system (refer to Figure 5.5). The "n" and "v" are floating numbers with exponential number format, such as 2.999381e-002". All coordinates were required to be positive numbers in the earlier STL versions, but this restriction is no longer enforced, and negative coordinates are now commonly encountered in STL files. STL files do not contain units (e.g., mm, inches), nor do they contain material information. The syntax "end-facet" indicates the end of the facet. Depending on the geometry's complexity, an STL file may contain more than one facet, usually thousands. If another facet exists, a new "facet" syntax can be found after the previous "endfacet" syntax. At the end of the file, the STL file uses "endsolid name" syntax.

5.6.3   Slicing 3D design into layers

Once the part's orientation and location on the substrate are determined, the part and the support structures, if any, are sliced into thin layers by the process planning software. The layer thickness for each material and machine can be specified by users. Even though the layers are very thin, they can lead to staircase effects in the part depending on the build orientation (Figure 5.6). The

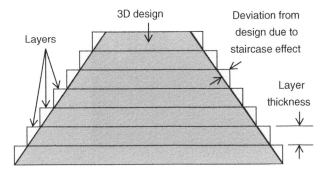

Figure 5.6 Slicing a 3D design into thin layers may cause staircase effects. *Source:* T. Mukherjee and T. DebRoy.

staircase effect indicates the deviation of "as built" dimensions from "as designed" dimensions and impacts the dimensional integrity of parts. The staircase effect often results in a poor surface finish and necessitates post-processing to achieve the specified dimensional tolerances.

Figure 5.7 shows the variations in the average slicing time per layer when parts with different geometric complexities such as the human heart, the Eifel Tower, and the dragon are sliced using different layer thicknesses. The consistency of the slicing time for each slicing layer thickness is noticeable. As a result, it can be represented as average slicing time, as illustrated in Figure 5.7. Each geometry has a different facet number, which indicates the variation in the complexity of the slicing process. Figure 5.7 depicts data for only 100 slices. However, depending on the resolution of the AM machine, this can be adjusted to any value. How long computer time is needed to finish the slicing calculation is explained in worked out example 5.5.

Worked out example 5.5

A polymeric cuboid block is designed to be printed using SLA process. Each side of the cube is 5 cm. A slicing program is executed on a computer with 2.30 GHz CPU and 6 Gb RAM for slicing the 3D design of the block into layers of 50 microns thick. The average slicing time per layer is 20 milliseconds. How long computer time will it take to finish the slicing computation for the entire block?

Solution:

The cuboid block of 5 cm dimension needs to be sliced into 50 microns thick layers. Therefore, total number of layers will be $(5 \times 10{,}000)$ microns/50 microns $= 1000$ layers.

Since the average slicing time per layer is 20 milliseconds, the total slicing time needed for 1000 layers will be $(1000 \times 20) = 20{,}000$ milliseconds $= 20$ seconds.

5.6.4 Selection of feedstock material

The selection of appropriate materials is critical in the design requirements for AM. Materials selection for AM emphasizes the ability to produce feedstock in a form amenable to the specific AM process, suitable material processing by AM, the ability to be acceptably post-processed to improve geometry and properties, and the achievement of necessary performance characteristics in service. Specific classes of material have become associated with specific AM processes as the technology has matured. Therefore, material selection has become an essential task in design for AM as a wide range of compatible materials are available for AM.

In AM, materials selection is done by optimizing the criteria mentioned in Table 5.4 such as application, aesthetics, mechanical properties, thermal properties, printability, feedstock availability, and cost by using various selection methodologies. For example, Figure 5.8 shows the use

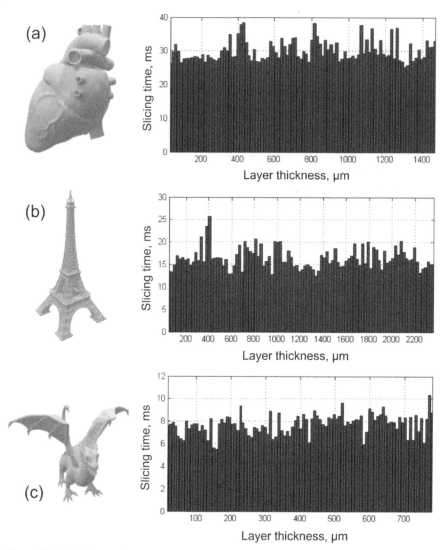

Figure 5.7 Variations in the average slicing time per layer when parts with different geometric complexities such as (a) the human heart, (b) the Eifel Tower, and (c) the dragon are sliced using different layer thicknesses. The slicing program was executed on LENOVO IdeaPad with i3-2350M CPU at 2.30GHz, 6GB RAM running on Windows 7 SP1 Home Premium (64-bit) operating system [5]. The figure is taken from an open-access article [5] under the terms and conditions of the Creative Commons Attribution (CC BY) license.

of an Ashby's materials selection chart showing ultimate tensile strength versus elongation for various AM materials. Appropriate material selection is necessary not only for improving product attributes but also for cleaner production and sustainable development.

5.6.5 Process and machine selection

An AM process and machine are selected based on the desired geometric properties of parts such as surface finish, minimum feature size, dimensional accuracy, and complexity, functional properties of parts such as material type and mechanical, chemical, and thermal properties, and manufacturing attributes such as cost, production time, batch size, and need for post-processing. Several commonly used AM processes and machines for metals and alloys, polymers, and ceramics

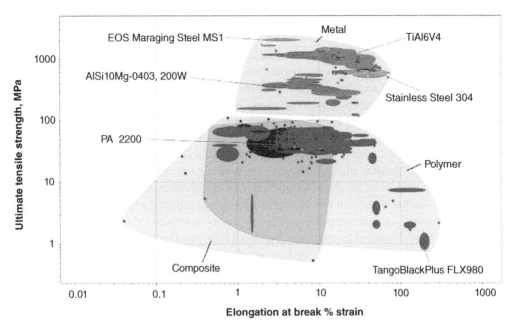

Figure 5.8 Ashby's materials selection chart shows ultimate tensile strength versus elongation for various AM materials [6]. The plot is based on property data for AM materials from the Senvol Database (refer to Table 5.5). The figure is taken from [6] with permission from Elsevier.

Table 5.8 Comparison of several AM processes for polymers based on the important process selection factors. Adapted from [7].

Process selection factors	Stereolithography	Selective laser sintering	Selective laser melting	Fused deposition modeling	Inkjet printing
Variety of materials	Small	Large	Medium	Medium	Small
Surface quality	Average	Good	Excellent	Average	Good
Dimensional accuracy	Excellent	Good	Excellent	Average	Good
Resistance to impact	Average	Good	Excellent	Good	Low
Flexural strength	Low	Excellent	Excellent	Excellent	Low
Cost of product	High	High	High	Low	Medium
Need for post-processing	Yes	Yes	No	No	No

are discussed in Chapters 2, 3, and 4 of this book, respectively. Readers can use that information and select an AM process and machine for their purpose. For example, Table 5.8 compares several AM processes for polymers based on important process selection factors.

5.6.6 Process planning and computer-aided manufacturing (CAM)

Process planning and computer-aided manufacturing (CAM) in AM rely on build plan specification, tool path and scanning strategy planning, and setting up the AM machine. Specifying the build plan considers important factors such as build time, dimensional integrity, material usage, power consumption, thermal history, and stress accumulation. Build time is driven primarily by the number of layers. A taller part requires more layers, which usually means longer build time. Build time is also affected by deposition rates and other process parameters. Dimensional integrity is controlled by slicing effects such as staircase effect, build orientation, size of unsupported features and overhanging geometries, and deviation due to the tessellation effects from STL file conversion. Material usage and power consumption are controlled largely by the size of the part. However, powder bed fusion, binder jetting, sheet lamination, and vat photo-polymerization processes require additional material to fill up the entire build volume that encloses the part during the AM process. Thermal history indicating the heating and cooling of material during the layer-by-layer printing drives the stress accumulation, microstructure evolution, and mechanical and chemical properties of the part that are significant considerations during AM processes.

The build planning software creates a toolpath or scanning pattern [8] that tells the machine where and how to add material in each layer. For fused deposition modeling, the nozzle temperature, speed, and deposition rate are specified to control the volume of materials deposited and how they are bonded together to achieve the desired part density. For powder-bed systems, the toolpath specifies the scanning pattern and hatch spacing for the heat source to follow. There are several scanning patterns commonly used in AM (Figure 5.9).

A zig-zag pattern (Figure 5.9 (a)) fills an area with parallel line segments that are evenly spaced. The spacing between these segments determines the density of the fill, and the direction of the parallel lines changes every two slices, allowing mechanical bias to accumulate. However, this fill contains many sharp corners, making complex shapes or hollow structures difficult and adding unnecessary de-acceleration and acceleration times for an extruder tool head in the fused deposition modeling process. Worked out example 5.6 illustrates a methodology to calculate the time required for printing a part.

In the contour-parallel pattern (Figure 5.9 (b)), concentric closed curves offset from the outer boundary with equal step sizes define tool paths. The concentric closed curves, on the other hand, are disconnected from each other, which can result in gaps. A combination of a few contour curves and then zig-zag fills is sometimes used, but gapping structures must be avoided. Contour-parallel paths often indicate widely varying degrees of complexity in terms of convexity/concavity of boundaries.

Spiral patterns (Figure 5.9 (c)) also use space-filling curves that are designed to avoid bending. However, even at the infinite resolution, it is not guaranteed to cover an entire 2D space. Although spirals are often similar to contour-parallel paths, they are advantageous because they can have arbitrary start and end points as well as continuous connections, allowing a curve to spiral both in and out.

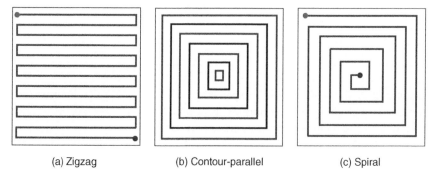

| (a) Zigzag | (b) Contour-parallel | (c) Spiral |

Figure 5.9 Commonly used scanning patterns in AM. *Source:* T. Mukherjee and T. DebRoy.

Worked out example 5.6

A polymeric cube of 5 mm in each side needs to be printed using a zig-zag scanning pattern (Figure 5.9 (a)) with a hatch spacing of 50 microns. If there is an idle time of 5 seconds between two layers for spreading the new layer of liquid, how long time will it take to print the entire block? Assume a laser scanning speed of 1000 mm/s. Assume that the part is printed in 1000 layers.

Solution:

Since the hatch spacing is 50 microns, total number of hatches per layer will be
   (5 × 10,000) microns/50 microns = 1000 hatches per layer.
The track length for each hatch is 5 cm. Therefore, printing time for each hatch will be
   (5 × 10) mm/1000 mm/s = 0.05 s.
Therefore, the total time required to print all 1000 hatches per layer will be
   (1000 × 0.05) = 50 s.
Therefore, the total time required to print all 1000 layers will be (50 × 1000) = 50,000 s.
It takes 5 seconds to spread liquid for each layer. Therefore, the total idle time for spreading liquid will be (5 × 1000) = 5000 s.
Therefore, the total time required for printing the block will be
   (50,000 + 5000) = 55,000 s = 15 hours 16 minutes and 40 seconds.

Consider the 3D model of a stool with four curvilinear legs as shown in Figure 5.10 (a). Four-layer objects are contained within each sliced layer. Figure 5.10 (b) depicts the layer-by-layer tool-path generated for AM. A thin layer of material is deposited during fabrication, and then the printhead moves horizontally to fabricate another thin layer of material at a different location in the same plane. Frequent travel between four legs may result in a significant waste of time during nonproductive travel. In such cases, care must be taken to optimize the tool path during AM.

The toolpath planning is a component of the overall process plan that tells the machine how to build the AM part. Other process parameter settings may be required to heat the build plate or build chamber, control the inert gas flow in a powder bed fusion system or directed energy deposition system, or switch materials when a machine can print support structures from an alternative material. Different machines require different settings for different process parameters, and default settings are built into the machine's build planning software.

(a)   (b)

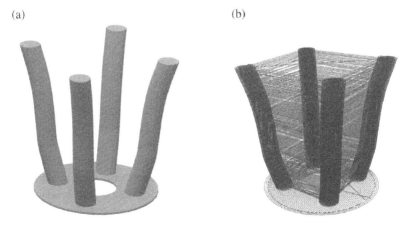

Figure 5.10   A CAD model is used for toolpath generation. (b) Toolpath that has many non-productive transition travels (indicated by lines) [9]. The figure is taken from [9] with permission from Elsevier.

5.6.7 Specifying sensing and control

In-situ process monitoring, sensing, and control of variables are important for controlling geometry, composition, defects, and part-to-part reproducibility during AM. The availability of high-quality and affordable optical and infrared cameras for monitoring hot surfaces has enabled reliable noncontact sensing of temperature fields and real-time defect identification during printing processes. Robust algorithms for processing sensing data have resulted in advancements in the monitoring of peak temperature, geometry, and defect evolution. Such monitoring is the first step toward developing effective real-time, closed-loop printing process control.

Developing in-situ measurements is critical, but interpreting the measured results requires insightful, well-tested process models based on scientific principles. To reduce defects and improve microstructure and properties, control algorithms must be based on predictive mechanistic models rather than heuristic algorithms based on empirical or intuitive correlations. By monitoring feedstock delivery rates, fusion zone geometry, and temperature field definitively, printing processes can be controlled to reduce defects, improve part density and mechanical properties, and save money.

5.7 Sustainability consideration in the design for additive manufacturing

The design for additive manufacturing (DFAM) affects material and energy usage, waste, and recycling. These factors affect sustainability. Since DFAM affects these key factors, there are opportunities to promote sustainability with appropriate DFAM by reducing energy and material usage, enhancing recycling, and reducing wastage.

Improving energy consumption can help to reduce greenhouse gas emissions. Optimization of design to minimize the amount of material used and the time required to print it can reduce energy consumption and enhance sustainability. However, an important challenge in determining energy usage is that different AM machines and processes use different amounts of energy. In addition, the production of feedstock materials such as powders and wires for AM is energy intensive and varies with the manufacturing method and alloy composition. Thus, it is a challenging task to accurately estimate the amount of energy used to make the feedstocks. Although lightweight part designs that meet the performance criteria are desirable, manufacturing parts with complex designs that require extensive support structures takes more time, energy, and material for the support structure. As a result, the complexity of parts made by AM adversely affects sustainability. However, for low-volume production, additive manufacturing may be more energy efficient than conventional manufacturing.

Recycling of material is another opportunity to improve sustainability in AM. Polymeric materials are widely used in additive manufacturing because of their high strength-to-weight ratio, stiffness, ductility and durability. Since a significant fraction of polymers end up in landfills and in the sea, their recycling is desirable. For example, Poly Lactic Acid (PLA), a thermoplastic made from renewable sources such as corn starch or sugar cane, can be recycled to produce 3D printing filament. However, some mechanical properties of recycled PLA such as the tensile strength degrade by a small amount owing to recycling which need to be considered in the design. However, virgin PLA may be mixed with the recycled PLA and used for the production of filament. In addition, the use of biodegradable polymers such as PLA can reduce adverse environmental impact. Among the metallic materials, steels and aluminum alloys are among the most recycled engineering materials. However, the value-added intricate metal parts made by additive manufacturing are small in volume compared with the total scrap recycling. As additive manufacturing is more widely adapted, the recycling rate will also increase. Biodegradable polymers and recyclable metals are examples of sustainable feedstocks. Product design with recyclable materials will also promote sustainability.

Since additive manufacturing can be used to produce parts on demand, there is no need to maintain a large inventory. This can help to conserve both material and energy. Efficient design for additive manufacturing will reduce the amount of support material and print failure, thus reducing wastage. Also, additive manufacturing can significantly reduce waste by using the amount of material needed to make the part. Use of recycled materials, such as recycled metal powders, can help to reduce the demand for virgin materials and reduce waste. By optimizing energy efficiency, sustainable materials, and reducing waste, additive manufacturing can promote sustainability. Sustainability can also be improved by enhancing the longevity of the product. The worked out example 5.7 illustrates the design considerations for load bearing ability for ensuring safety and sustainability. While engineers seek to design high-quality parts, errors in design are not uncommon. The common design errors are discussed in worked out example 5.8.

Worked out example 5.7

An additively manufactured frame (Figure E5.1) is subjected to a load of 6000 N acting at 45° to its horizontal axis. The frame has a rectangular section whose depth (b) is twice the thickness (t). Find the cross-sectional dimensions of the frame if the maximum permissible stress in the material of the frame is 60 MPa. Assume that the section modulus of the frame is $tb^2/6$.

Solution:

There are three contributors to the stress on the frame (1) bending stress due to horizontal component (P_H) of load, (2) bending stress due to vertical component (P_V) of load, and (c) direct stress due to vertical component (P_V) of load. Please note that the bending stresses are tensile on the top surface and compressive on the bottom surface of the frame. The three stresses are calculated as follows:

(1) Bending stress due to horizontal component (P_H) = Bending moment due to horizontal component/section modulus of the frame
   = $(P_H \times 75)/(tb^2/6) = (6000 \cos45° \times 75)/(t(2t)^2/6) = 477{,}225/t^3$ N/mm^2
(2) Bending stress due to vertical component (P_V) = Bending moment due to vertical component/section modulus of the frame
   = $(P_V \times 130)/(tb^2/6) = (6000 \sin 45° \times 75)/(t(2t)^2/6) = 827{,}190/t^3$ N/mm^2
(3) Direct stress due to vertical component (P_V) = $P_V/(bt) = (6000 \sin45°) / (2t^2) = 2121/t^2$ N/mm^2

For limiting case, $477{,}225/t^3 + 827{,}190/t^3 + 2121/t^2 = 60$
By iteratively solving, t = 28.4 mm and b = 56.8 mm.

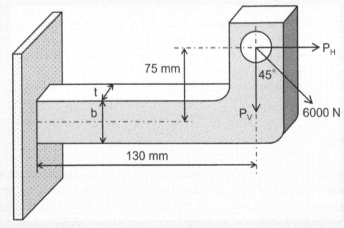

Figure E5.1 Schematic diagram of an additively manufactured frame.

Worked out example 5.8

Briefly discuss some of the common design errors that should be avoided when designing for additive manufacturing.

Solution:

Parts are often redesigned after a prototype is build and tested because optimizing the design of a complex part involves consideration of many variables and constraints. The design freedom that enables unprecedented functionality is a valuable gift but it does not represent a promise of compliance with the performance specifications and the needed serviceability and durability. However, the designers can avoid several common errors in design for additive manufacturing. The following is an incomplete list of the common difficulties.

Support structure: For complex parts, both too little and too much support structure has to be avoided. Too little support structure may distort some features such as overhangs. Too much support structure requires a large amount of feedstock material, more energy consumption, intensive post-processing to remove the support structure, and high cost. Also, the geometry of the part and the support structure should be such that there is enough free space to remove the support structure without damaging the part.

Wall thickness: They need to be thick enough to support the loads, but not too thick to save weight and money. On the other hand, if a part contains very thin wall, it may be susceptible to warping or in extreme cases, result in print failure. Commonly, wall thicknesses much below a millimeter is difficult to print.

Avoiding sharp corners: The very first jet aircrafts had square windows. Their fatigue failure taught us to avoid designs that are stress concentrators. This valuable lesson also applies to additive manufacturing. Rounding off corners and fillet joints is not only a good practice, it is mandatory.

Feature size: The minimum feature size depends on the additive manufacturing process, feedstock material, and machine. It is important to know the limitations of how small a feature can be accurately printed because it depends on many factors. For example, in PBF-L, the powder size, the laser beam radius, power, and the machine affect the feature size.

Faulty STL files: STL files are created from CAD files and a correct STL file contains closed and connected triangles. However, sometimes the following errors are present in the files. Missing triangles result in gaps or holes. Every triangle has a normal vector that determines which way to add material. A flipped normal vector is another type of error. Two triangles should not overlap or intersect but sometimes they do. Sometimes an edge is shared by more than two triangles and this error is called a non-manifold error. Another type of error is called noise shells that contain outside surfaces that are too small to build. The errors in the STL files may be repaired using an appropriate repair software.

5.8 Summary

In additive manufacturing, the design of a part considers its purpose, function, economics, sustainability, and aesthetics. It takes advantage of the unique capabilities of additive manufacturing, reduces materials waste, ensures functionality, optimizes shape, and complies with the desired specifications of properties and constraints. The design for additive manufacturing starts with choosing a manufacturing process and a feedstock material based on the principles of material and process selections. A 3D CAD model that considers the functionality and constraints and optimizes topology is created and the file is converted to an STL file. The design is divided into layers and a computer-aided manufacturing process planning (CAM) is created in preparation for the manufacturing process. The design also promotes sustainability by lowering energy usage, selecting eco-friendly feedstocks, and reducing waste.

Takeaways

Uniqueness of products

- Additive manufacturing allows designers freedom to create complex parts that would be difficult to produce using traditional methods.
- Many parts that used to be made by assembling multiple components can now be made in one piece by additive manufacturing.
- It is possible to build high performance lightweight structures by carefully controlling the size, shape, and orientation of each layer.
- Additive manufacturing enables construction of compositionally graded joints of dissimilar alloys, which avoids abrupt changes in mechanical properties and improves reliability.
- Because additive manufacturing produces less waste than traditional manufacturing, it is both environmentally friendly and cost competitive.
- Because it can produce only a few pieces quickly, additive manufacturing is sensitive to customer needs and can reduce lead time.
- Sensors and electronics may be integrated into a part, reducing the need for additional components while improving performance and functionality.

Constraints in the design

- The availability and properties of feedstock materials influence product design.
- Complex products may require redesign to reduce support structures and save time and money.
- Layer thicknesses affects the mechanical properties, surface finish, and dimensional accuracy.
- The orientation of a part influences its strength, surface finish, and support requirements. It is optimized to reduce machine time as well as feedstock and post-processing costs.
- The cost of the chosen material, machine time, and post-processing all have an impact on production rate and cost-competitiveness, which are important factors in design.
- Post-processing is required to remove support structure, improve surface finish and microstructure, and achieve high density. The design should minimize post-processing to save time and money.

Materials selection criteria

- Durability, functionality, reliability, and serviceability are among the key requirements. Some applications require biocompatibility, sterilization capabilities, toxicity certifications, chemical resistance, and other attributes.
- Tensile strength, yield strength, elastic modulus, fracture toughness, and fatigue life are important for materials selection. Post-processing can improve some properties.
- Printability is an important factor to consider when deciding on an appropriate process-material combination, taking into account susceptibility to common defects as well as metallurgical and mechanical properties.

(Continued)

Takeaways (Continued)

- Many printable materials are unavailable as powder or wire feedstock. A strong business case is required to pay for testing and qualification of a new material.
- The feedstock material's cost is an important consideration in material selection.
- Certain components require unique aesthetics, such as color, surface polish, and appearance. Even after post-processing, some materials offer superior aesthetics to others.

Materials selection methodologies

- Design engineers can benefit from materials selection charts or diagrams that compare the physical and mechanical qualities of various materials.
- Materials selection indices are numerical values used to compare and evaluate the properties of various materials. They include the weighted property index, cost per unit property index, and printability index.

Materials database for additive manufacturing

- The National Institute of Standards and Technology's open-access Additive Manufacturing Materials Database contains vendor information, machine information, and information on part design, pre-process, in-process, and post-process information, as well as testing.
- Massachusetts Institute of Technology's open-access database contains mechanical properties of additively manufactured parts of polymers, metals, and alloys.
- The Additive Manufacturing Standards Database of the Society of Manufacturing Engineers contains data on design, testing, and materials, powder bed fusion, binder jetting, and directed energy deposition processes, and metal, polymers, and ceramics in powder and liquid forms.
- The National Aeronautics and Space Administration Material database includes test data and results on Inconel 718 AM processes and properties. It is available to registered users.
- InssTek's AM materials database book contains data for AM of tool steels, stainless steels, Ti-6Al-4V, Inconel alloys, Co-Cr alloys, and Al-Bronze. This is a public database.
- Senvol's open-access database of Additive Manufacturing Machines and Materials contains two modules, Machine Search and Materials Search containing data for different feedstock materials and various AM machines.

Topology optimization

- Topology optimization enables designers to create complex and intricate structures with internal voids and complex features that would otherwise be impossible to make. It also contributes to waste reduction and sustainability.
- It takes into account part performance constraints, geometric constraints, the specific AM process, build chamber dimensions, as well as material properties, manufacturing process, and cost.
- The topology optimization involves identifying the performance criteria and the part specifications, optimizing the material distribution, modeling the mechanical behavior of the structure to determine the part performance, and making the AM build plan. The support structure is then removed and the part is tested to ensure that it satisfies the performance specifications.
- Optimization of topology to create efficient and superior parts involve many complex steps. There are a number of topology optimization software available to achieve this task.

Takeaways (Continued)

Process selection for design for additive manufacturing

- The design for additive manufacturing starts with a computer-aided design (CAD) file. There are many software available to make this file. However, the CAD file is not sent to the AM machine. It is converted to a stereolithography (STL) file which is made up of a triangulated mesh that approximates the 3D solid model's boundaries and surfaces.
- Once the part's orientation and location on the substrate are determined, the part and the support structures, if any, are sliced into thin layers by the process planning software and an appropriate feedstock material is selected.
- An AM process and machine are selected based on the desired attributes of the part and the manufacturing constraints such as cost, production time, batch size, and need for post-processing.
- A process planning and computer-aided manufacturing (CAM) file containing the build plan specification, tool path and scanning strategy, and information required for setting up the AM machine is then made. It considers factors such as build time, dimensional integrity, material usage, power consumption, thermal history, and stress accumulation.
- Finally, the part is built aided by in-situ process monitoring, sensing and control of variables to controlling geometry, composition, defects, and part-to-part reproducibility during AM.

Sustainability in the design for additive manufacturing

- With appropriate design for additive manufacturing, there are opportunities to promote sustainability by reducing energy and material usage, improving recycling, and reducing waste.

Appendix – Meanings of a selection of important technical terms

Build volume: This is the maximum size of the object that can be printed in a machine. Build volume is limited by the chamber size in a powder bed fusion machine.

Build orientation: It is the geometric placement of a part with respect to the build plate that affects the surface finish and the amount of support structure needed.

Computer-aided design (CAD) file: It is a computer file used to create digital 3D models of a part and visualize how a model will look when printed.

Stereolithography: It is a technique to make a prototype using layer-by-layer deposition where a laser beam heats a photosensitive liquid to locally polymerize the liquid to form a solid. The process is continued to make a three-dimensional component.

Stereolithographic (STL) file: STL is an acronym for stereolithography, also referred to as Standard Triangle Language or Standard Tessellation Language. The STL file format is commonly used in additive manufacturing. The file is created from a CAD file to a series of triangles that shows the volume and the surface of a part.

Slicing: A process of dividing a three-dimensional model into thin layers.

Support structure: Complex parts in additive manufacturing often contain parts that require support to ensure compliance with specified part geometry. The support structure is removed after the part is manufactured.

Topology: It refers to the geometry of the faces, edges, and vertices of a 3D model that can affect the strength and surface smoothness of a part made by additive manufacturing.

Practice problems

1) Explain with the help of examples how the properties of materials affect product design in additive manufacturing.
2) What are some of the important considerations for designing a product using additive manufacturing?
3) How can sustainability be considered in the design of a product using additive manufacturing?
4) What factors are important in determining the cost of a design using additive manufacturing and how can the cost be minimized?
5) Discuss the unique features of additive manufacturing that need to be considered in the product design.
6) Indicate and explain the main constraints of additive manufacturing that need to be considered in the design for additive manufacturing.
7) What are the important factors that need to be considered for selecting a process in the design for additive manufacturing?
8) How is the design for additive manufacturing tested and verified?
9) What is topology optimization and what is its purpose?
10) How is the topology of the products optimized in the design for additive manufacturing?
11) How is sustainability considered in the design for additive manufacturing?

References

1 Zhou, Y., Lu, H., Ren, Q. and Li, Y., 2019. Generation of a tree-like support structure for fused deposition modelling based on the L-system and an octree. *Graphical Models*, 101, pp.8–16.

2 DebRoy, T. and Bhadeshia, H.K.D.H., 2021. *Innovations in Everyday Engineering Materials*. Springer, Switzerland.

3 Ashby, M.E., 2017. *Material Selection in Mechanical Design*, 5th Edition. Butterworth-Heinemann.

4 Zhu, J., Zhou, H., Wang, C., Zhou, L., Yuan, S. and Zhang, W., 2021. A review of topology optimization for additive manufacturing: status and challenges. *Chinese Journal of Aeronautics*, 34(1), pp.91–110.

5 Adnan, F.A., Romlay, F.R.M. and Shafiq, M., 2018. Real-time slicing algorithm for Stereolithography (STL) CAD model applied in additive manufacturing industry. In *IOP Conference Series: Materials Science and Engineering*. 342(1), article no.012016.

6 Bourell, D., Kruth, J.P., Leu, M., Levy, G., Rosen, D., Beese, A.M. and Clare, A., 2017. Materials for additive manufacturing. *CIRP Annals*, 66(2), pp.659–681.

7 Mançanares, C.G., de, S., Zancul, E., Cavalcante da Silva, J. and Cauchick Miguel, P.A., 2015. Additive manufacturing process selection based on parts' selection criteria. *The International Journal of Advanced Manufacturing Technology*, 80, pp.1007–1014.

8 Zhao, H., Gu, F., Huang, Q.X., Garcia, J., Chen, Y., Tu, C., Benes, B., Zhang, H., Cohen-Or, D. and Chen, B., 2016. Connected fermat spirals for layered fabrication. *ACM Transactions on Graphics (TOG)*, 35(4), pp.1–10.

9 Liu, W., Chen, L., Mai, G., and Song, L., 2020. Toolpath planning for additive manufacturing using sliced model decomposition and metaheuristic algorithms. *Advances in Engineering Software*, 149, article no.102906.

6

Sensing, Control, and Qualifications

Learning objectives

After reading this chapter the reader should be able to do the following:

1) Know several applications of sensing and control in additive manufacturing.
2) Understand the working principles of commonly used sensors for additive manufacturing.
3) Select an appropriate sensor for a desired application.
4) Recognize the volume of data generated and collected during additive manufacturing processes and the storage devices needed to store them.
5) Understand the different types of controls models and their applications in additive manufacturing.
6) Appreciate the roles of sensing and control for improving the standardization and part qualification in additive manufacturing.

CONTENTS

Theory and Practice of Additive Manufacturing, First Edition. Tuhin Mukherjee and Tarasankar DebRoy.
© 2024 John Wiley & Sons, Inc. Published 2024 by John Wiley & Sons, Inc.

6.1 Introduction

The diversity of microstructure, properties, and performance of 3D printed components results from the wide variations of controllable variables such as the heat source power, scanning speed, power density, geometry, composition, the delivery rate of the feedstock, and many other variables. Avoiding defects and controlling part dimensions and properties are important for their qualification. In-situ monitoring and control of variables are important for controlling geometry, composition, defects, and part-to-part reproducibility.

The availability of improved optical and infrared cameras for monitoring hot surfaces at an affordable cost has enabled reliable noncontact sensing of temperature fields and the identification of defects in real-time during printing processes. Robust algorithms for the processing of sensing data have resulted in progress in the monitoring of peak temperature, geometry, and evolution of defects. Such monitoring is needed to develop effective real-time, closed-loop control of printing processes.

Developing in-situ measurements are an important requirement but interpretation of the measured results needs insightful well-tested process models based on scientific principles. Control algorithms need to have a basis on predictive mechanistic models and not be guided by heuristic algorithms based on empirical or intuitive correlations to reduce defects and improve microstructure and properties. Printing processes can then be controlled to reduce defects, improve part density and mechanical properties, and save cost by monitoring feedstock delivery rates, fusion zone geometry, and temperature field definitively.

6.2 Sensors

A variety of sensing equipment, such as optical and infrared image sensors, thermocouples to monitor temperatures in the base plate, acoustic emission monitors, sensors that can detect displacements and other devices have been used for the monitoring of additive manufacturing processes and parts. Most commonly used electro-optic sensors convert radiations from visible, infrared, and ultraviolet emissions to electrical signals. Pyrometers and photodiodes typically provide one spatially integrated output signal from their field of view, while both optical and infrared cameras provide spatially resolved signals. All sensors need to have short response times and cameras need to have a good spatial resolution. The number of pixels of a camera determines the resolution of the images. More pixels provide better resolution for determining the melt pool surface area and the temperature field from the signals generated by each pixel. They collect a much larger volume of data than single-channel photodiodes. However, the high volume of data acquisition rates of high-resolution cameras presents a significant challenge in data storage and analysis.

6.2.1 Optical cameras

When electromagnetic radiation impinges on a surface, the energy is either absorbed, reflected, or transmitted. The property that determines the fraction absorbed is called absorptivity, the fraction that is transmitted is termed transmissivity and the portion that is reflected is known as reflectivity. The values of these properties change with the temperature of the surface, the nature of the material, and the wavelength of the radiation. Each detector is capable of sensing a range of wavelengths of radiation. For example, a sensor for a home camera needs to detect the visible wavelength, i.e., between 400 and 700 nm. On the other hand, a detector for an infrared camera needs the capability to detect longer wavelengths in the range of 700 nm to 1 mm. Pyrometers determine the temperature from the intensity and wavelength of the emitted radiation.

Charge-coupled devices (CCD) that are tiny capacitors can detect radiation intensity. The signals from each capacitor are processed by an electrical circuit and converted to an image. Complementary metal oxide semiconductor (CMOS) detectors are now increasingly used for capturing images. The quality of the images is affected by the small size of the molten pool, the interference of the light by the falling powder particles in DED, the ejected spatter particles in PBF systems, and radiation from the heat source. It is important to use an appropriate filter to select the right wavelength of radiation and use additional illumination when needed to obtain high-quality images.

Several attributes need to be considered in selecting cameras. They include the frame rate, the number of pixels, and the dynamic range which is the ratio of the maximum to minimum light intensity. High dynamic range cameras provide better resistance to interference, more pixels provide better resolution and higher frame rates deliver improved temporal resolution of images. As a practical matter, the relatively low frame rate does not affect sensing significantly since the processing of the images in a computer also takes time.

Image quality improvements require good illumination of the specimen using added light sources such as a light emitting diode (LED) or a laser. Light sources can be made coaxial with the heat source such as a laser beam or installed at an angle with the heat source. Optical filters are commonly used and in 3D cameras, they are usually a part of the camera. Illumination using light of radiation in the range of 405–520 nm is often used to improve image quality.

6.2.2 Infrared cameras

Charge coupled device (CCD) and complementary metal oxide semiconductor (CMOS) devices can also detect infrared radiations and are used to monitor objects at high temperatures such as the molten pool and the surrounding areas. They are also used for noncontact monitoring of temperatures. Emissions with short (0.9–1.7 micrometer), medium (3–5 micrometer), and long (8–14 micrometer) wavelengths are captured depending on the hardware used. Appropriate filters and wavelength ranges are useful to avoid interference. For example, when a laser beam of 1.06 micrometer wavelength is used, the infrared signal from the build area can interfere with the radiation from the laser beam which needs to be avoided. Higher temperatures shift the emitted radiations to shorter wavelengths consistent with Wien's displacement law that relates the peak wavelength (λ_P) of the emitted radiation with temperature (T) as:

$$\lambda_P = \frac{b}{T} \tag{6.1}$$

where b is Wien's displacement constant and equals 2898 μmK where λ_P is in μm and T is in K. Therefore, at high temperatures, the selection of the filters and wavelength ranges to avoid interference with the laser beam needs to be carefully done. In addition, the emissivity which affects the ability of a surface to emit radiation depends on temperature and a sharp variation of emissivity with temperature can affect temperature measurements.

Figure 6.1 (a) shows an infrared image acquired by a CCD camera with a field of view coaxial with the Nd:YAG heat source. A filter was used to block the radiation from the Nd:YAG laser at 1.06 μm wavelength and an infrared filter was used to reduce the high intensity of the light from the molten pool by blocking radiation below 700 nm. The camera took gray images with a 128 × 128

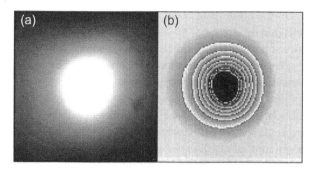

Figure 6.1 (a) An infrared image of a molten pool of H13 tool steel on a mild steel substrate during DED-L and (b) isotherms from gray level values of the infrared image [1]. The image is reprinted with permission from Elsevier.

resolution at 800 frames/s from the melt pool and nearby regions. The isotherms are shown in Figure 6.1 (b) on grayscale. The solidus isotherm was determined by simultaneously taking optical images of the region by a second camera. The edge of the molten pool from the optical camera was mapped to the infrared data to obtain the gray level corresponding to the solidus temperature which served as the melt pool boundary. Worked out example 6.1 explains how the filter of an infrared camera is selected for measuring temperature during additive manufacturing.

Worked out example 6.1

An infrared camera is used to track the molten pool boundary during DED-L where the molten pool boundary is defined by the solidus isotherm of the alloy. A wavelength filter is used to filter the signals of lower wavelengths. Specify the wavelength of a filter that can be used to track pool boundary for aluminum alloy AlSi10Mg. Can the same filter be used for tracking the pool boundary for titanium alloy Ti-6Al-4V? The solidus temperature of AlSi10Mg and Ti-6Al-4V are 831 K and 1889 K, respectively.

Solution:

To track the molten pool boundary for AlSi10Mg the camera needs to detect the solidus isotherm of 831 K. From Eq. 6.1, the peak wavelength corresponding to 831 K is 2898 µmK/831 K = 3.5 µm. Therefore, a filter for 3.5 µm wavelength is needed for the measurement.

The solidus temperature of Ti-6Al-4V (1898 K) is significantly higher than that of AlSi10Mg. Therefore, the same filter will not work for Ti-6Al-4V. To detect the molten pool boundary for Ti-6Al-4V, a filter for (2898 µmK/1898 K) = 1.5 µm wavelength is needed.

6.2.3 Photodiodes and pyrometers

Photodiodes and pyrometers provide a single output such as a voltage or temperature rapidly from their field of view. They are easy to use, inexpensive, and do not require contact with the object. Their detectors absorb light to generate electrical energy. The type of detector determines the wavelength of radiation that can be detected. Silicon, germanium, gallium arsenide, and indium gallium arsenide detectors are commonly used. The type of detector affects the wavelength range, sensitivity to input energy, response speed, and cost. Silicon (Si) photodiodes are most sensitive at the 800–900 nm range and can provide high data acquisition speed with low noise. Germanium provides good sensitivity at wavelengths of about 1400 nm and is cheaper than Indium Gallium Arsenide (InGaAs) that provides good sensitivity in the 1000–1600 nm infrared range.

6.2.4 Thermocouples

Infrared pyrometers provide noncontact, indirect measurements. However, the measurements need knowledge of emissivity which requires calibration with a standard. Thermocouples embedded in the base plate provide a more direct and accurate measure of the local temperature as a function of time. Thermocouples have also been used to measure temperatures in monitoring locations within the deposited layers. Thermocouples have hot and cold junctions. The hot junction is placed where temperature needs to be measured. The cold junction maintains a reference temperature, generally taken as the room temperature. The voltage generated by a thermocouple is a function of the temperature difference between the two junctions. The thermocouple voltage is measured with a high impedance voltmeter. The relationship between thermocouple voltage and temperature is nonlinear. Manufacturers supply a conversion table to determine the temperature from the measured voltage (see worked out example 6.2). Worked out example 6.3 explains how to calculate temperature from the measured voltage.

Although the use of thermocouples is the cheapest way of measuring temperature, their exact placement within a metal layer before the start of the deposition process is not straightforward and requires special care. Temperatures reach well above the solidus temperature in the layers and care needs to be taken for the selection of the thermocouples and the location where they can measure temperature safely. The response time for the thermocouples can be shortened by reducing the diameters of the thermocouple wires, but very thin wires may break easily. Temperature versus time data from thermocouples embedded within the base plate or the build is often used to validate calculations of temperature using mechanistic heat transfer models.

Worked out example 6.2

While measuring the temperature on the substrate far from the deposit during DED-L, a thermocouple voltage of 3699 µV is recorded. The cold junction of the thermocouple is kept at 30 °C. What is the temperature at the hot junction of the thermocouple? The manufacturer of the thermocouple supplied the following Table E6.1 to convert voltage to the temperature for the cold junction temperature of 0 °C.

Table E6.1 Data from the manufacturer to convert voltage to the temperature.

Temperature °C	0	30	60	90	120	150	180	210	240	270	300
Voltage (µV)	0	397	798	1203	1612	2023	2436	2851	3267	3682	4096

Solution:

From the above table, the voltage of the cold junction at 30 °C is 397 µV. Since the cold junction is kept at 30 °C, a correction voltage of 397 µV should be added to the voltage measured at the hot junction. Therefore, the corrected voltage at the hot junction is (3699 + 397) = 4096 µV. From the table above, the temperature at the hot junction corresponding to 4096 µV voltage is 300 °C.

Worked out example 6.3

For a thermocouple used to measure temperature during additive manufacturing, the relationship between the voltage (V) in µV and the temperature difference between two junctions (ΔT) in °C is approximately described by an empirical relation $V = \Delta T^2$. If the thermocouple voltage reading is 23,000 µV and the temperature at the cold junction is 25 °C, calculate the temperature at the hot junction.

(Continued)

Worked out example 6.3 (Continued)

Solution:

For a thermocouple whose hot junction is at 25 °C and the cold junction is at 0 °C, the voltage is equal to $(25 - 0)^2 = 625$ μV. Therefore, a correction voltage of 625 μV should be added to the voltage measured when the cold junction is kept at 25 °C. Therefore, the corrected voltage is $(23,000 + 625) = 23,625$ μV. Therefore, the temperature at the hot junction equals $\sqrt{23,625} = 153.7$ °C. Note that empirical correlations are unreliable and must be validated with experimental data.

6.2.5 Optical emission spectroscope

When a laser beam interacts with metal vapors, a plasma consisting of ionized species, excited atoms, and electrons may form. The emissions from a plasma result from the change in the energy states of electrons from higher to lower energy states. The difference in energy equals the Planks constant times the frequency of emission. The wavelengths of emissions are signatures of energy changes of electrons that can be detected by an optical emission spectroscope. It detects the intensities of emissions from a plasma plume as a function of wavelengths by collecting light from the plasma plume using a fiber optic cable. Figure 6.2 shows a typical intensity versus wavelength plot during the DED-L of Ti-6Al-4V alloy. Emissions characteristic of titanium and aluminum are observed.

The spectra give valuable information about the electron densities and the kinetic energy of the electrons commonly expressed as "electron temperature." The wavelengths of emissions also identify the elements present in the plasma. A metal vapor-dominated plasma that forms during metal printing consists of volatile alloying elements vaporized mostly from the hottest region of the fusion zone located directly below the heat source. The procedures for the calculations of electron temperatures and electron densities are available in textbooks [2]. However, the properties of the plasma are not uniform throughout the three-dimensional plume. When a fiber optic cable collects emissions from its line of sight, the emissions come from all locations within its view. If the plasma plume has a nearly circular cross-section centered at the axis of the heat source, the intensity of the emission is the summation of all the emissions at various radial locations. As a result, the measured raw data do not provide plasma properties at any specific location and the data needs to be processed using an appropriate algorithm to extract spatial resolution of the plasma properties.

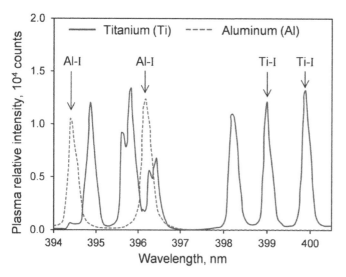

Figure 6.2 Intensity versus wavelength plot obtained from a plasma plume above the melt pool during DED-L of Ti-6Al-4V alloy showing characteristic peaks of Al and Ti. The figure is plotted by T. Mukherjee and T. DebRoy based on the data reported in [3].

Spectroscopy of plasma during laser welding has been investigated for the last several decades and electron temperatures and densities have been studied during DED. Spectroscopy is a complex process and no commonly available commercial additive manufacturing machines are currently available with spectroscopic facilities.

6.2.6 Acoustic emission sensors

Stress waves generate sounds that can indicate important events in real-time in a nondestructive manner during a process such as additive manufacturing or laser welding. Electrical signals may be related to part quality by placing sensors below the build plate. Acoustic emission sensors have been investigated because of their potential benefits. For example, acoustic emissions from powder flow have been investigated to monitor flow rate and uncover possible correlations between acoustic signals and defects in parts. In welding, the frequency response of the acoustic signals has been correlated with several attributes of welds such as keyhole formation and crack propagation. Commonly available commercial directed energy deposition and powder bed fusion equipment do not currently come with any acoustic emission sensors and considerable work is needed before an integrated acoustic emission system may be developed for routine use.

Currently, there is no ASTM standard for the use of acoustic emission sensors for additive manufacturing. However, for laser welding, the ASTM E749 indicates the specification of the acoustic emission sensors. An associated ASTM standard E650 specifies the location of the sensors for the most effective use. They may serve as valuable guides for the development of standards applicable for AM.

6.2.7 Displacement sensors

Displacement sensors are used in metal printing for a variety of purposes, such as layer thickness measurement in directed energy deposition systems, tracking the location of the deposition head, directing the trajectory of the deposition track, and determining the location of the build platform in powder bed fusion. The displacement sensor often consists of a low-power blue laser and a sensor to detect the reflected laser energy. The time taken for the laser to impinge on a surface and return to the sensor is a measure of the distance between the position of the laser and the object. Since the blue laser does not significantly penetrate the surface of metallic materials, it provides good measurement accuracy. In addition, there are other types of displacement sensors such as an eddy current sensor for applications that require high precision.

Surface imperfections and deformations during DED can be detected by laser profile scanners (Figure 6.3). A laser beam scans the part surface and collects the coordinates of the surface along the length of the deposit. The coordinate data is processed to obtain the surface profile of the component in 3D. By comparing the surface profile both before and after the deposition, the thickness of the deposited

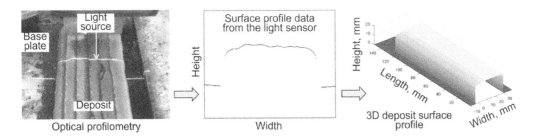

Figure 6.3 Actual component and the light source; surface profile constructed from the light sensor and 3D view of the simulated virtual component [4]. The figure is reprinted with permission from Elsevier.

layer can be determined. The evolution of the part geometry can be tracked by comparing the uniformity and thickness of the layer to that in the CAD file. The distance between the print head and the depositing surface is an important parameter in DED. The displacement sensor can help in the accurate positioning of the print head during the entire process. In powder bed systems, the build platform is adjusted after each layer by a very small distance, sometimes smaller than a human hair. The eddy current-based sensors are commonly used for this purpose because of their precision and speed.

6.2.8   X-ray and other tools

X-rays are commonly used for the determination of crystal structures and flaws. Most X-ray applications in additive manufacturing have focused on the determination of porosity and other defects ex-situ. X-ray tomography can provide high-resolution pictures of internal features that are larger than 1 μm. It is widely used for detecting voids, lack of fusion defects, and pores as small as 10 μm after a part is built. Figure 6.4 shows dynamic X-ray images [5] of keyhole mode laser powder bed

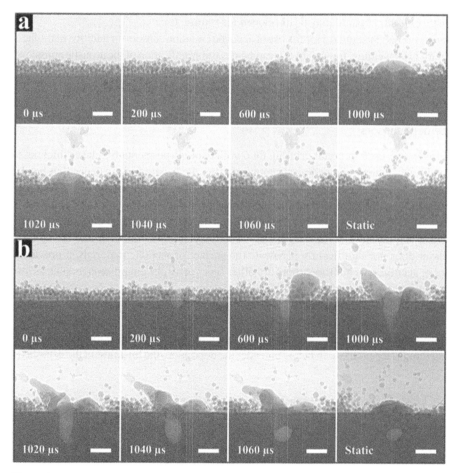

Figure 6.4   Dynamic X-ray images [5] of keyhole mode laser powder bed fusion processes of Ti-6Al-4V. Figures (a) and (b) are for 340 W and 520 W, respectively. The images were taken at different times with a frame rate of 50 kHz. All the scale bars are 200 μm. The voids are attributed to keyhole instability. The figure is taken from an open-access article [5] under the terms and conditions of the Creative Commons Attribution (CC BY) license.

fusion processes of Ti-6Al-4V. The voids observed are larger than 25 μm. In-situ determination of flaws of porosity and defects has been limited to research laboratories. Very powerful X-rays available from synchrotrons have been used for real-time determination of crystal structure change and other scientific investigations during welding, but they remain out of reach for most practicing engineers. X-rays and neutron diffraction have been used for the detection of residual stresses. In addition, hole drilling has been used for the determination of residual stresses.

Laser-ultrasonic testing used in the aerospace and metallurgical industry is being adapted for use in additive manufacturing. A pulsed laser generates ultrasonic waves within an object and these waves are detected by a separate device. The waves are altered in the presence of discontinuities and surface roughness. Their detection using a second laser forms a basis for flaw detection. The procedure is usable in hot objects and can be applied to detect defects in each layer. Worked out example 6.4 discusses how sensors can be selected for different applications during AM.

Worked out example 6.4

Which sensors should be selected for the applications below?

a) To monitor the temperature on the top surface of the molten pool during DED-GMA.
b) To track molten pool boundary during PBF-L.
c) To detect porosity during the PBF-L process.
d) Dimensional accuracy of the deposited track consistent with the part design.

Solution:

a) Temperatures on the top surface of the molten pool are very high, higher than the liquidus temperature of the alloy. Therefore, a noncontact temperature measurement device such as an infrared camera should be used.
b) Molten pool boundary can be monitored by tracking the solidus isotherm of the alloy. A noncontact temperature sensor such as an infrared camera can be used. In PBF-L, the molten pool moves rapidly because of the high scanning speed. Therefore, the sensor should have very high spatial and temporal resolutions.
c) Gas pores originate inside the deposited track. Therefore, in-situ X-ray tomography or acoustic emission sensors can be used for this purpose.
d) Dimensional accuracy is checked by monitoring the deflection of the part during deposition. A contact-based sensing using strain gauges or noncontact displacement sensors can be used.

6.3 Applications

Different types of sensors discussed in the previous section are used for controlling layer thickness and geometry, tracking molten pool boundary, monitoring the powder flow, and reducing distortion as described below.

6.3.1 Layer thickness control

Layer thickness during additive manufacturing affects the time necessary to build parts, surface finish, and cost. Parts printed using thinner layers require more time to manufacture and increase the total production cost. If a part has curved surfaces, thinner layers can result in smoother surfaces and avoid large steps near the curved surface owing to the staircase effect. In addition, layer thickness also determines the feature size since thinner layers are needed to build parts with finer features. Layer thickness also affects cooling rates. Thin layers exhibit rapid heat transfer from the molten pool to the substrate or previously deposited layers and increase cooling rates. Since thinner layers cool faster, finer microstructures are obtained. Control of layer thickness is usually based on optical measurements as discussed below.

The optical sensors for layer thickness measurements collect images from multiple locations at different angles. Charge-coupled devices collect the images and calculate the layer height from the data. In DED systems, the controller can adjust the layer thickness when needed by adjusting the powder feed rate and the scanning speed. The difference between the heights of the powder and the solidified layers depends on the attributes of the powder particles, particularly the size range. An appropriate algorithm is used to adjust the layer thickness. For systems with wire feeding, the wire feed speed and scanning speed are important parameters for determining the metal deposition rate that affects the layer thickness.

Camera image from one camera can provide data about the two-dimensional surface of the build. An additional camera that can capture a view along the build direction is needed to obtain the image for the elevation. The type of filters used depends on the heat source. A 3D scanning system based on a low-power laser, 5 mW at 512 nm wavelength has been used to measure layer height. Displacement sensors have also been used for layer height measurements.

6.3.2   Melt pool and seam tracking

Monitoring and control of temperature and melt pool size are important for ensuring part quality and repeatability. Since they are affected by the process parameters such as the heat source power and scanning speed, control models can adjust process parameters to ensure the desired fusion zone geometry and peak temperature. For example, images acquired by the cameras can be used in a closed-loop control system for the effective control of both track width and temperature by controlling variables such as heat source power and speed.

Optical and infrared cameras and photodiodes have been used for seam tracking and melt pool imaging. The cameras need to be calibrated using an appropriate black body source for accurate measurements of temperatures. For this purpose, the previous works on welding and cladding have served as a basis for these applications. Since the fusion zone is often smaller than 1 mm in dimensions, the resolution of the cameras needs to be sufficient for accurate measurements of fusion zone geometry and temperature field. Positioning of pyrometers and infrared cameras are important factors for the measurements of temperature. Both sensing along the axis of the heat source and off-axis monitoring are practiced. Viewing the location and positioning of sensors are important because of the need to avoid interference caused by objects such as feedstock, limited space, and interference with other radiations such as a laser beam heat source. The position of the camera is important because feedstocks such as wire or powder in DED may block its view of the fusion zone resulting in errors in the measurement of the pool size and temperature field. Erroneous sensing may affect the functioning of the control model and result in defective parts. When the cameras are at fixed locations and do not move with the deposition head during DED, the output of the camera may be affected by the tool path. Careful selection of wavelengths of the illuminations is also important. To acquire signals from the fusion zone, an appropriate filter is required to remove signals from outside the melt pool, i.e., regions below the solidus temperature.

The measurement of the area of the fusion zone is commonly accomplished by adding the areas of pixels enclosed by the solidus temperature contour. Figure 6.5 on the right shows infrared camera

Figure 6.5 Infrared image of the molten pool. The left images are captured with an optical camera and the right images are the pool boundaries obtained from an infrared camera data [6]. The figure is taken from an open-access article [6] under the terms and conditions of the Creative Commons Attribution (CC BY) license.

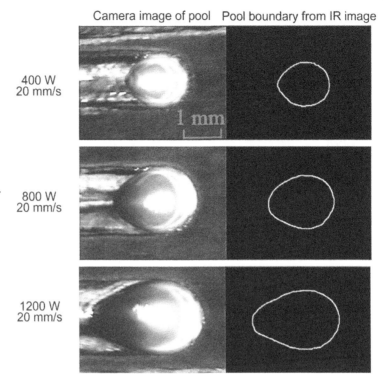

Camera image of pool   Pool boundary from IR image

400 W
20 mm/s

800 W
20 mm/s

1200 W
20 mm/s

images of melt pools captured with a resolution of 320 × 240 pixels for various laser powers [6]. The imaging setup was installed at the laser head and an infrared filter (>695 nm) was used. The boundary of melt pools was determined using an optical filter for various laser powers and shown in the figures on the right. The separation of the molten pool and the surrounding area required data from both an infrared camera and an optical camera to correctly identify the molten pool boundary.

The track width after the deposition can be compared with that determined from the real-time sensing. The ability to determine the sensitivity of the variables such as heat source power, scanning speed, and the supply rate of the feedstock can serve as a basis for using the data in heuristic control models. For example, the experimentally determined values of fusion zone width, length, and area as a function of heat source power and speed have been used in the control model. The speed of detection is important to accommodate the timely implementation of the parameter changes suggested by the control model in response to the observed deviation of the track width or temperature from the anticipated values.

Figure 6.6 shows an example of real-time closed-loop melt pool size control through laser power modulation [7]. An in-axis infrared camera was used with filtering to receive emission below the laser wavelength during DED-L with Ti-6Al-4V wire as a feedstock. The melt pool size could be controlled to a predetermined value when the print speed was changed in a stepwise manner.

6.3.3 Monitoring of powder flow

Powder flow control is important because the variation of the flow rate results in an uneven deposition. In extreme conditions, interruptions in the flow of a powder stream can result in the disruption of the printing process. Control of powder delivery rate is particularly important for the

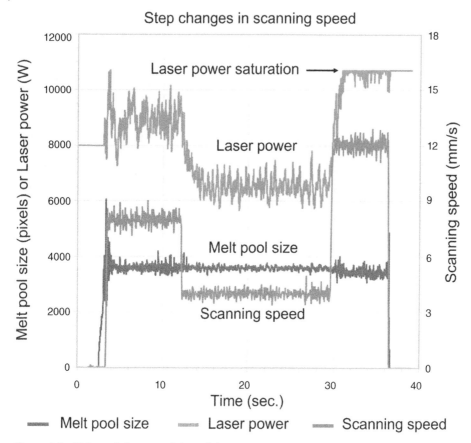

Figure 6.6 Melt pool size control through laser power modulation during a step-change in the printing speed. Melt pool size monitoring was done through an in-axis infrared camera [7]. The figure is reprinted with permission from Elsevier.

deposition of functionally graded alloys where the local composition is controlled by supplying different powders at predetermined flow rates from multiple powder feeders. Sensing and control of powder supply rate are important for accurate control of composition and fusion zone attributes such as layer thickness.

Let us consider the DED systems with powder or wire feedstocks. The rate of powder delivery or the speed of wire feed, and the location where the feedstock is delivered with respect to the heat source, all must be precisely controlled to achieve a good part. The changes in the pressure of the gas tank containing the carrier gas, the difference in the build chamber pressure, and the clogging of one of the delivery nozzles can cause difficulties. Powder flow monitoring and control address these difficulties for powder-blown systems. Automated control of the wire feed system is necessary for the wire-arc or electron beam-wire systems.

Two methods have been employed to measure the flow rates of powder during additive manufacturing. The first one is based on the measurement of the weights of the powder hoppers continuously. The powder supply rate can be calculated from the rate of decrease in the weight of the hopper. A screw feeder controls the supply rate of the powder from the hopper. The rate can be controlled by adjusting the rotational speed of the motor that controls the powder delivery rate. However, the steady powder flow rate is not attained instantaneously. There is a short delay between the start of the screw feeder and the attainment of a steady powder flow at the delivery nozzle because of the time it takes for the powder to reach from the hopper to the nozzle. Therefore, when an adjustment of the flow rate is made, there is also a short delay in the realization of the new steady flow rate of the powder at the nozzle exit. This delay depends on the length of the tubes through which the powder must travel to the discharge nozzle. Reduction of such delays by appropriate design of the hardware and a provision for the temporal adjustment of the flow rate by the control models are needed for the smooth operation of the powder delivery system.

The powder delivery rate has also been measured by an optical method. A laser beam of a wavelength of 600–710 nm and a maximum power of 500 mW from a laser diode was passed through a flowing stream of powder inside a glass tube (Figure 6.7). Since the beam was partially absorbed and partially reflected by the powder particles, the laser beam was attenuated after passing through the tube. The relation between the intensity of the original laser beam (I_0) and that of the attenuated laser beam (I) follows the Beer Lambert law [8]:

$$\frac{I}{I_0} = \exp(-\varepsilon CL) \tag{6.2}$$

where ε is the attenuation coefficient, C is the concentration, and L represents the optical path length. The intensity of the attenuated laser beam was measured using a photodiode which provided a voltage as a measure of the energy. For known values of ε and L, the concentration (C) can be calculated using Eq. 6.2 from which the mass flow rate of the powder can be estimated. Worked out example 6.5 explains the procedure for calculating the mass flow rate. The attenuation was more pronounced when more powder passed through the tube and a calibration curve was set up by correlating the powder flow rate with the output voltage of the photodiode. Depending on the signal from the photodiode, a control model was able to adjust the powder feed rate to a predetermined value enabling a closed-loop control of the powder delivery rate. The typical delivery rate of 3–22 gm/min could be achieved at a sampling rate of 10 Hz using a commercial data acquisition card installed on a personal computer.

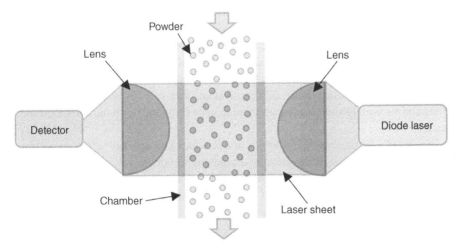

Figure 6.7 Powder flow rate sensor based on the absorption of laser energy by the flowing powder [6]. The figure is taken from an open-access article [6] under the terms and conditions of the Creative Commons Attribution (CC BY) license.

Worked out example 6.5

A powder flow rate sensor (Figure 6.7) is used to measure the flow rate of stainless steel 304 powders used in a DED-L process. The laser beam loses 30% of its intensity after being attenuated by the flowing powders through a cylindrical chamber of 20 mm diameter. A flow of argon gas at a rate of 40 mm^3/s aids the powder flow. The value of the attenuation coefficient is known to be 0.1 cm^2/g [9]. What is the mass flow rate of the powders?

Solution:

In Eq. 6.2, the intensity of the attenuated laser beam (I) is 70% of that of the original laser beam (I_0). The optical path length can be approximated as the diameter of the cylindrical chamber. The value of ϵ is taken as 0.1×10^{-4} m^2/g and L is 20×10^{-3} m. Therefore, from Eq. 6.2, 0.7 = exp (-0.1×10^{-4} × C × 20×10^{-3}). The concentration (C) is calculated as 0.2×10^7 g/m^3. This is the concentration of the powders in the gas. Therefore, the mass flow rate of the powders (g/s) = concentration (g/m^3) × gas volume rate (m^3/s). Therefore, the calculated powder flow rate is 0.08 g/s or 4.8 g/min.

6.3.4 Distortion control

Dimensions of the components made by additive manufacturing often deviate from those in the design. The control of track geometry, melt pool size, and layer thickness does not automatically guarantee a component free of residual stresses and deformation. Nonuniform heating and cooling, solidification, and solid-state phase transformations may result in significant irreversible deformation of components, warping, delamination, and in extreme cases, part rejection. A more complete explanation of the origins of distortion and the common methods of reducing distortion including adjustment of process parameters, scanning strategies, support structure, and alloy composition is available in Chapter 11.

The measurement of distortion of the build plate can be undertaken by commonly used strain gauges. In-situ measurements of part distortion entail the use of laser-based displacement sensors, which can optically monitor the displacement of multiple layers as previously described, and digital image correlation (DIC), which relies on noncontact optical measurements to determine distortion. An area of interest can be divided into uniform grids and data from each grid location are examined. Individual pixels within marked areas are tracked by two cameras and intensities from a series of images are correlated to their locations. Strains are calculated from the displacements. Unlike strain gauges, it can take data from an area and does not require physical contact with the specimen. After the parts are built, a 3D optical scanner can be used to measure the dimensions of the parts accurately and the measurements can be compared with the corresponding values in the design.

The sensing of images at various stages of building parts provides a comparison of the actual part with that of the CAD data. However, if there is any disagreement between the two, the changes needed in process parameters cannot be easily determined because many process variables affect distortion. In addition, there is no straightforward functional relation between each process variable and distortion. Efforts are continuing to develop a methodology for real-time control of distortion.

Thermomechanical simulations are often undertaken because they provide an estimate of how much distortion to expect before building a component. These calculations use commercial codes that are computationally very intensive. They calculate temperature fields, residual stresses, and distortion as outputs for a given set of thermophysical properties and process variables as input. The calculations of heat transfer are often simplified based on the solution of the heat conduction equation that ignores convection to make the calculations tractable. The calculations of residual stresses and distortion can typically take several days except for very small specimens. These codes typically have high license fees. However, the calculations allow exploration of the simulated effects of alternative sets of process variables. The computed results for several sets of process variables can be compared to select a set where the distortion is lower. The process does not guarantee an optimum selection of process variables because only a few cases can be explored because of the computationally intensive nature of the calculations and high cost.

6.3.5 Commercially available equipment

Real-time nondestructive estimation of the process signatures such as the melt pool geometry, temperature distribution, lack of fusion, porosity, and track geometry is important for both directed energy deposition and powder bed fusion processes. The sensing data are the basis for the control models for quality assurance of parts. The standards for sensing equipment and the control models are in their initial stages of development. However, there are several types of equipment available from various vendors as add-ons to commercial printing machines. A selection of the available equipment and its functions are presented in Table 6.1.

Table 6.1 A selection of commercial powder bed fusion and directed energy deposition monitoring equipment.

Make/model	Equipment	Function
B6 Sigma/ PrintRite3D	Coaxial and off-axis camera and software	Determines part quality based on layer by layer images and their comparison with the CAD data during PBF-L
Concept laser/QM melt pool	Coaxial camera with a CMOS detector	Melt pool imaging to determine area and emission intensity for multiple parts during PBF-L
EOS-Plasma Industrietechnic/ EOSTATE Meltpool	Camera and photodiode monitoring	Melt pool imaging to determine light emission intensity during PBF-L
Arcam/LayerQam	Camera	Captures images of every layer to determine defects during PBF-EB
Stratonics/ThermaViz	Two wavelength pyrometers	Melt pool temperature measurement during DED-L
Promotec/PD2000, PM7000	Camera with CMOS detector, Photodetector	Melt pool and layer height monitoring during DED-L
DM3D Technology/DMD-closed-loop	Dual-color pyrometer and three CCD cameras	Melt pool and layer height monitoring during DED-L
DEMCON/LLC100	Camera	Melt pool monitoring during DED-L

6.4 Processing and storage of the sensing data

The sensors produce a large volume of data. The data need to be processed quickly during the deposition process for real-time process control. In addition, the data need to be archived to make the best use of them. A few AM machines can generate about petabytes of data per year [10]. One would need about 500 units of external hard drives, 2TB each, to store the data costing hundreds of thousands of dollars just to buy the storage devices. Worked out example 6.6 explains the volume of data generated during AM. High-capacity cloud storage devices such as Redundant Array of Inexpensive Disks (RAID) and Network Attached Storage (NAS) are becoming available. However, retrieving and processing voluminous sensing data in AM remains a major challenge. The volume of data must be significantly reduced to make both the processing and storage of data tractable. One solution is the generation of a reduced set of data using appropriate statistical metrics. The reductions could be based on a better understanding of which data are most important for the purpose for which the data are being collected. Such reductions can result in an about 100-fold reduction of data volume and the entire raw data set is not stored.

Worked out example 6.6

Melt pool sensing data are collected during PBF-L where a part is made with a geometric tolerance of 10 microns. The scanning speed of the laser beam is 500 mm/s. The system collects data on five variables, temperature, surface elevation, and three coordinates. If the data are stored in single precision that occupies 32 bits, how much data are collected per second?

Solution:

Since the part is made with a geometric tolerance of 10 microns, data collection at a spatial resolution of 10 microns would be necessary which means one data point for every 10 micrometers. Since the laser beam scans 500 mm every second, the number of data points collected per second is 500 mm/10 µm = 50,000. Since the system collects data on 5 variables, the total number of data collected per second = (5 × 50,000) = 250,000. Since each data occupies 32 bits or 4 bytes, the total data collection rate is (4 × 250,000) = 1,000,000 bytes per second or 1 megabyte per second.

Since the additive manufacturing processes are relatively recent, not all machine manufacturers have built-in hardware for the sensing and control of various parameters discussed here. Both the hardware and the software for the sensing and control are evolving and if the advancement of digital technology in welding is any indication of what to expect, the development of standards in this area would require considerable work. 3D printing hardware integrated with rigorous sensing and control would facilitate wider adaptation of metal printing.

6.5 Control models

Process control is used for real-time regulation of one or more process variables to achieve the desired part attribute. The sensing data during the AM process are analyzed using a control model and the model makes decisions based on which control signals are sent to the machine for adjusting the input process variables. For example, Figure 6.8 shows that the sensing data on molten pool temperature measured by a thermal imaging device or substrate distortion captured by a strain

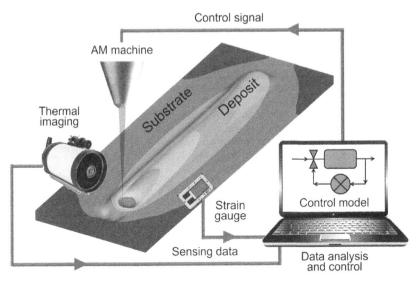

Figure 6.8 A schematic representation of AM process control using control models. *Source:* T. Mukherjee and T. DebRoy.

gauge can be fed to a computer. The computer uses a control model to analyze the sensing data, make a decision, and send a control signal to the AM machine for corrective action if one is needed.

The control models deal with four types of variables. First, the input process variables to AM machine include heat source power, scanning speed, layer thickness, powder flow rate, and energy distribution. Second, variables that cause undesirable changes in the process such as the process instability or change in part geometry due to thermal distortion. Third, sensing variables monitored by different sensors include the deposit dimensions, molten pool temperature, and surface features. Finally, control variables are the output of the control models which are sent to the AM machine. Controls models used in AM can be broadly classified into four categories, statistical regression, operations research-based techniques, fuzzy logic, and machine learning methods. The salient features of these categories and examples of their applications in AM are summarized in Table 6.2.

Table 6.2 Commonly used control models, their salient features, and selected applications in AM.

Control models	Features	Applications in AM
Statistical regression models	• Values of a part attribute obtained from multiple experiments are fitted with the corresponding process parameters. • During the process, the sensing data on the part attribute is compared with the fitting. • If a deviation from the fitted value is observed, the computer sends a control signal to the machine to adjust process variables.	Control of track width in DED-L [11], defect in PBF-L [12], molten pool temperature in DED-L [13], molten pool size in DED-L [6]

(Continued)

Table 6.2 (Continued)

Control models	Features	Applications in AM
Operations research techniques for decision making and control	• This method relies on a set of alternative values of the sensing variable and a decision-making criterion. • The algorithm selects the best value of the sensing variable and the computer sends a control signal to the machine to achieve that. • Commonly used algorithms are the Markov chain [12], multi-attribute decision making (MADM) model [14], and proportional integral derivative (PID) model [14].	Control of defect in PBF-L [14], molten pool temperature in DED-L [15], surface roughness [16]
Fuzzy logic-based methods	• This method relies on a set of values of the sensing variable corresponding to different sets of process parameters obtained from previous measurements and a decision-making criterion. • The logic makes a decision and the computer sends a control signal to the machine to satisfy that.	Control of deposit dimensions in DED-GMA [4,17], deposit height in DED-L [18]
Machine learning techniques	• Machine learning algorithms are trained using the data obtained from previous experiments. • Trained algorithms are used for the new sensing data based on which an output control variable or a decision is generated. • Commonly used machine learning methods are the neural network, decision tree, random forest, and support vector machines.	Control of track width in DED-L [11], defects in PBF-L [19,20] and DED-L [21], powder flow [22]

When the input-output relationship of an AM process can be modeled or determined precisely from the sensing data by varying the input process variables, the controller can be designed using control models based on statistical regression or operation research-based techniques. However, when the process is complex and only a limited number of input and output data are available, fuzzy logic-based controls are best suited. In contrast, machine learning-based approaches can perform well both for large datasets and limited data.

6.6 Sensing and control in part qualification

Additive manufacturing is a relatively new technology. The protocols necessary for ensuring the reproducibility of components, reducing defects, attaining desirable microstructure, properties, and serviceability in a cost-competitive manner are still being developed. Attaining these goals will require a well-tested protocol that will produce reliable parts even when parts are manufactured from the same design in different machines. The standard procedures and specifications will enable the certification of parts and will reduce part costs. Both the manufacturers and the end-users will benefit from such standardization. The sensing and control of additive manufacturing processes and products will play a central role in the standardization process. While individual sensing solutions have been discussed separately in this chapter, several sensing techniques may be deployed simultaneously. Based on the sensing data, AM processes can be controlled, and sensing data will help in the part qualification.

The standardization will involve the unified specification of feedstock, process parameters and equipment, and validation of parts based on sensing and control. Characterization of microstructure and internal defects such as lack of fusion and porosity may also be required as a part of the standardization process. Sensing and control will reduce defects and ensure compliance of printed part geometry with its design. Well-tested simulation of AM processes will avoid problems and establish appropriate process variables for producing high-quality parts. Calculations of temperature fields, melt pool geometry, solidification structure, microstructural features, and susceptibility to certain defects can be simulated before building parts. Predicting mechanical properties remains a major challenge. However, emerging digital tools such as mechanistic models and machine learning based on available data can uncover the effects of process variables on attributes that affect part quality. Once the process variables have been determined and specifications have been established for feedstock and equipment, sensing and control can support making and qualifying reliable parts in a cost-competitive manner.

6.7 Summary

The effective use of sensing data for the mitigation of defects, achieving process repeatability, and effective process control to attain good microstructure and properties now remains mostly a field of active research with low technological readiness levels. Improved integration of sensing and control with the printing hardware will lead to better parts and more widespread use of additive manufacturing. Features such as the geometry of the fusion zone and the deposit track, temperature field, and layer thickness can be monitored using available equipment in laboratories. Reliable in-situ detection of defects, real-time data processing, and interfacing with control models remain active fields of research with limited integration with commercial machines. Additive manufacturing machines can generate a large volume of data that are difficult to store, retrieve, and process. Appropriate methodologies for reducing the volume of data and their processing to achieve better control of the manufacturing process are needed. Sensing and control can play an important part in part qualification and standardization.

Takeaways

Sensors

- Optical and infrared image sensors, thermocouples, acoustic emission monitors, optical emission spectroscope, and displacement sensors have been used in additive manufacturing.
- Thermocouples are the cheapest sensors for measuring temperatures. However, their usage during layer by layer manufacturing often poses a challenge. Infrared cameras have been used for that purpose.
- Acoustic emission sensors and X-ray-based sensing have been used for the detection of defects such as cracks, pores, and lack of fusion.

Applications

- Layer thickness that affects cooling rates and surface roughness can be measured using optical and infrared cameras.
- Molten pool boundary during additive manufacturing has been tracked using optical and infrared cameras and optical emission spectroscope.
- Powder flow rate during directed energy deposition can be monitored using powder flow sensors where flow rate is estimated based on the attenuation of a laser beam through powders.
- Distortion of additively manufactured parts may be determined by measuring the displacement using strain gauges or noncontact displacement sensors.

(Continued)

Takeaways (Continued)

Processing and storage of the sensing data

- Additive manufacturing machines can generate a large volume of data per second that needs post-processing and storage.
- Both the hardware and the software for the sensing and control are evolving and development of standards in this area will require considerable work.

Control models

- Control models used in additive manufacturing include statistical regression models, operations research tools, fuzzy logic-based methods, and machine learning techniques.
- The sensing data are analyzed using a control model to adjust the process variables during additive manufacturing.

Sensing and control in part qualification

- Sensing and control of part dimensions, temperature, and defect formation are needed because these factors affect part quality.
- Sensing and control can help in part qualification by providing the ability to accurately measure and adjust various process and product attributes during additive manufacturing.

Appendix – Meanings of a selection of technical terms

CAD: The acronym of CAD stands for computer-aided design. CAD represents a three-dimensional design of a part.

Infrared: Wavelength of electromagnetic radiation greater than that of the red end of the visible light spectrum but less than that of microwaves. Infrared radiation has a wavelength from about 800 nm to 1 mm.

Laser beam: The acronym laser stands for "light amplification by stimulated emission of radiation." A focused beam of laser is often used in additive manufacturing as a heat source.

Light-emitting diode: A light-emitting diode (LED) is a semiconductor light source that emits light when electric current flows through it.

Operations research: An analytical method of problem-solving and decision-making commonly used in management.

Optical fiber: A flexible, transparent fiber made from glass or plastic. Light travels through an optical fiber by bouncing off the walls of the cable repeatedly.

Plasma: An electrically neutral gas consisting of electrons, excited atoms, and positively charged ions. A plasma arc is often used as a heat source in additive manufacturing.

Semiconductor: Materials with electrical conductivity between conductors and insulators. Semiconductors can be pure elements, such as silicon or germanium, or compounds such as gallium arsenide or cadmium selenide.

Strain gauge: A sensor to measure the strain in a component. Its resistance varies with the applied force.

Welding: A manufacturing process that uses a heat source to melt two or more parts to form a sound joint after solidification.

Practice problems

1) How can the peak temperature during powder bed fusion be monitored?
2) Which sensors can be used to monitor part dimensions during additive manufacturing?
3) When is it appropriate to use infrared cameras to monitor additive manufacturing? Can photodiodes be used instead of infrared cameras?
4) Can optical emission spectroscopy be used to monitor additive manufacturing? Discuss the capabilities of this technique.
5) What in-situ technique can you recommend to measure layer thickness during additive manufacturing?
6) Crack formation is one of the major concerns in the additive manufacturing of metallic materials. Detection of cracks during the deposition process is often desirable so that processing conditions can be adjusted to minimize them. What type of sensors can be used for this purpose?
7) A powder flow rate sensor is used to measure the flow rate of the powders used in a DED-L process. If the laser beam loses 20% of its intensity after being attenuated by the flowing powders at 5 g/min, how much loss of intensity is expected if the powder flow rate is doubled keeping the same gas flow velocity?
8) How can the powder flow rate be measured during directed energy deposition?
9) What types of control models are appropriate for additive manufacturing?
10) Discuss the roles of sensing and control in qualifying metallic parts made by additive manufacturing.

References

1 Hu, D. and Kovacevic, R., 2003. Sensing, modeling and control for laser-based additive manufacturing. *International Journal of Machine Tools and Manufacture*, 43(1), pp.51–60.

2 Griem, H.S., 1997. *Principles of Plasma Spectroscopy*, 1st Edition. Cambridge University Press, USA.

3 Song, L., Huang, W., Han, X. and Mazumder, J., 2016. Real-time composition monitoring using support vector regression of laser-induced plasma for laser additive manufacturing. *IEEE Transactions on Industrial Electronics*, 64(1), pp.633–642.

4 Li, Y., Li, X., Zhang, G., Horváth, I. and Han, Q., 2021. Interlayer closed-loop control of forming geometries for wire and arc additive manufacturing based on fuzzy-logic inference. *Journal of Manufacturing Processes*, 63, pp.35–47.

5 Zhao, C., Fezzaa, K., Cunningham, R.W., Wen, H., De Carlo, F., Chen, L., Rollett, A.D. and Sun, T., 2017. Real-time monitoring of laser powder bed fusion process using high-speed X-ray imaging and diffraction. *Scientific Reports*, 7(1), article no.3602.

6 Ding, Y., Warton, J. and Kovacevic, R., 2016. Development of sensing and control system for robotized laser-based direct metal addition system. *Additive Manufacturing*, 10, pp.24–35.

7 Gibson, B.T., Bandari, Y.K., Richardson, B.S., Henry, W.C., Vetland, E.J., Sundermann, T.W. and Love, L.J., 2020. Melt pool size control through multiple closed-loop modalities in laser-wire directed energy deposition of Ti-6Al-4V. *Additive Manufacturing*, 32, article no.100993.

8 Hulst, H.C. and van de Hulst, H.C., 1981. *Light Scattering by Small Particles*. Courier Corporation.

9 Singh, V.P. and Badiger, N.M., 2013. Study of mass attenuation coefficients, effective atomic numbers and electron densities of carbon steel and stainless steels. *Radioprotection*, 48(3), pp.431–443.

10 Spears, T.G. and Gold, S.A., 2016. In-process sensing in selective laser melting (SLM) additive manufacturing. *Integrating Materials and Manufacturing Innovation*, 5(1), pp.16–40.

11 Vandone, A., Baraldo, S. and Valente, A., 2018. Multisensor data fusion for additive manufacturing process control. *IEEE Robotics and Automation Letters*, 3(4), pp.3279–3284.

12 Khanzadeh, M., Tian, W., Yadollahi, A., Doude, H.R., Tschopp, M.A. and Bian, L., 2018. Dual process monitoring of metal-based additive manufacturing using tensor decomposition of thermal image streams. *Additive Manufacturing*, 23, pp.443–456.

13 Lee, J. and Prabhu, V., 2016. Simulation modeling for optimal control of additive manufacturing processes. *Additive Manufacturing*, 12, pp.197–203.

14 Yao, B., Imani, F., and Yang, H., 2018. Markov decision process for image-guided additive manufacturing. *IEEE Robotics and Automation Letters*, 3(4), pp.2792–2798.

15 Song, L. and Mazumder, J., 2010. Feedback control of melt pool temperature during laser cladding process. *IEEE Transactions on Control Systems Technology*, 19(6), pp.1349–1356.

16 Qie, L., Jing, S., Lian, R., Chen, Y., and Liu, J., 2018. Quantitative suggestions for build orientation selection. *The International Journal of Advanced Manufacturing Technology*, 98(5), pp.1831–1845.

17 Xiong, J., Shi, M., Liu, Y., and Yin, Z., 2020. Virtual binocular vision sensing and control of molten pool width for gas metal arc additive manufactured thin-walled components. *Additive Manufacturing*, 33, article no.101121.

18 Hua, Y. and Choi, J., 2005. Adaptive direct metal/material deposition process using a fuzzy logic-based controller. *Journal of Laser Applications*, 17(4), pp.200–210.

19 Imani, F., Chen, R., Diewald, E., Reutzel, E. and Yang, H., 2019. Deep learning of variant geometry in layerwise imaging profiles for additive manufacturing quality control. *Journal of Manufacturing Science and Engineering*, 141(11), article no.111001.

20 Amini, M. and Chang, S.I., 2018. MLCPM: A process monitoring framework for 3D metal printing in industrial scale. *Computers & Industrial Engineering*, 124, pp.322–330.

21 Mazumder, J., 2015. Design for metallic additive manufacturing machine with capability for "Certify as You Build". *Procedia CIRP*, 36, pp.187–192.

22 DeCost, B.L., Jain, H., Rollett, A.D. and Holm, E.A., 2017. Computer vision and machine learning for autonomous characterization of am powder feedstocks. *JOM*, 69(3), pp.456–465.

7

Heat Transfer in Additive Manufacturing

Learning objectives

After reading this chapter the reader should be able to do the following:

1) Understand the differences in heat absorption mechanisms in powder bed fusion and directed energy deposition processes.
2) Appreciate the fundamental differences in heat transfer mechanisms during directed energy deposition and powder bed fusion processes and how they are affected by the thermophysical properties of alloys.
3) Calculate approximate values of peak temperatures, thermal cycles, molten pool dimensions, and cooling rates considering heat conduction during powder bed fusion and directed energy deposition processes.
4) Compare the relative magnitudes of Marangoni, buoyancy, and the electromagnetic forces that drive the flow of liquid metal in the fusion zone.
5) Calculate the maximum velocities of the liquid metal inside the molten pool for different driving forces.
6) Calculate the important dimensionless numbers such as Fourier, Peclet, and Marangoni numbers for the heat transfer and fluid flow and appreciate their significance.
7) Compare the effects of different alloy properties and process conditions on peak temperature, pool dimensions, and cooling rates.

CONTENTS

Theory and Practice of Additive Manufacturing, First Edition. Tuhin Mukherjee and Tarasankar DebRoy.
© 2024 John Wiley & Sons, Inc. Published 2024 by John Wiley & Sons, Inc.

7.1 Introduction

In additive manufacturing, a laser beam, or an electron beam, or an electric arc is used as a heat source to melt a powder or wire feedstock layer-upon-layer to construct a three-dimensional part. Heat transfer from the heat source to the part controls the temperature fields, cooling rates, molten pool geometry, solidification parameters, microstructure, properties, defects, and performance of parts made by additive manufacturing (AM) [1]. Figure 7.1 shows schematically the interaction between the feedstock and the heat source that leads to the rapid heating, melting, and vigorous circulation of liquid metal in the molten pool.

The heat transfer and fluid flow affect the transient temperature distribution in the solid deposit and substrate, the shape and size of the fusion zone, and the solidification behavior. The flow of liquid metal is important since convective heat transfer is often the main mechanism of heat transfer within the

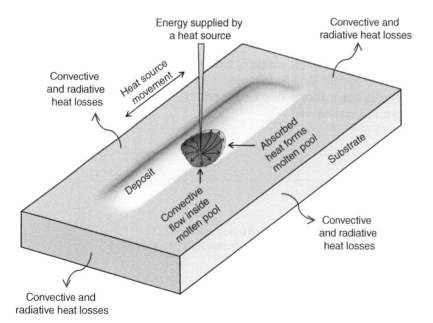

Figure 7.1 A schematic representation of heat transfer in directed energy deposition (DED) process. Heat energy from a heat source is partially absorbed by the feedstock. Molten metal vigorously recirculates within the molten pool. Heat is transported away from the molten pool through the solid deposit and substrate and eventually is lost to the environment by convection and radiation. (*Source:* T. Mukherjee and T. DebRoy).

molten pool. The circulation of liquid metal is driven by one or more of the following: the spatial gradient of surface tension, buoyancy, electromagnetic force, and powder or liquid metal droplet impingement. The driving forces depend on the specific AM process variant. The variation of temperature with time, often referred to as the thermal cycle, affects microstructure, properties, the evolution of residual stresses, and distortion of the part. On the surface of the molten pool, the temperature distribution affects the loss of alloying elements by evaporation, which in turn can alter the final composition of the component from that of the feedstock. Therefore, the control of the temperature fields, fusion zone geometry, cooling rates, temperature gradients, solidification growth rates, and many other parameters that are affected by heat transfer influence the part attributes. Figure 7.2 shows a partial selection of many attributes including deposit geometry, spatial variation of microstructure and properties, certain types of defects, residual stresses, and distortion that are affected by heat transfer.

While the temperature field during AM is required to understand the microstructure and properties of parts, it cannot be measured during a highly transient process like AM by thermocouples. Thermal imaging techniques can only measure surface temperatures along the line of sight. Therefore, quantitative calculations are necessary to gain insight into the heat transfer phenomena during AM. In addition, various AM processes exhibit significantly different shapes and sizes of deposits depending on the process conditions used [2]. These differences originate from the

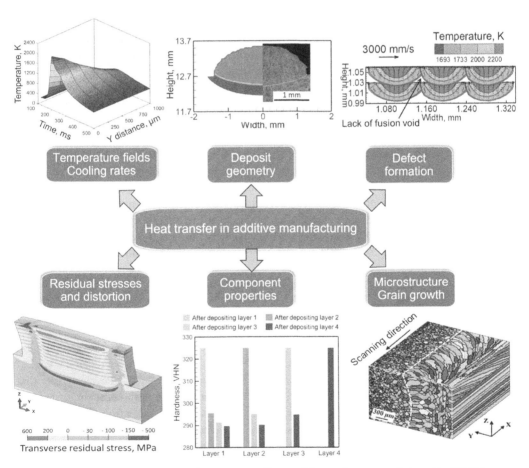

Figure 7.2 Heat transfer affects many important attributes of the products manufactured by AM. This figure shows a selection of these attributes. (*Source:* T. Mukherjee and T. DebRoy).

materials, process parameters, and the resulting heat transfer during the AM processes. A better understanding of heat transfer during AM processes is necessary to explain the differences.

In this chapter, the fundamentals of heat transfer in powder bed fusion (PBF) and directed energy deposition (DED) processes are examined. The discussion is focused on laser, electron beam, and arc heat sources with powder and wire feedstocks. The heat absorption mechanisms for PBF and DED processes and the factors important for heat transfer are explained here. Several heat conduction calculations for both PBF and DED processes to estimate thermal cycles, fusion zone size, and cooling rates for different alloys and process conditions are presented. In the molten pool, heat is transported by both conduction and convection. Several driving forces and their effects on convective flow are explained. This chapter includes several calculations to predict the maximum velocities inside the molten pool, which provide important insight into convective heat transfer. With the improvements in computer hardware and computational algorithms, more realistic heat transfer and fluid flow calculations are now performed numerically. These calculations are also introduced in this chapter.

7.2 Heat sources

In additive manufacturing, a heat source supplies energy to melt the powder or wire feedstock which upon solidification forms the deposit. Commonly used heat sources are laser beams, electron beams, and electric arcs. The important characteristics of a heat source are its power, power density distribution, effective radius, and for a laser beam, its wavelength. Power density distribution or how the input heat energy is distributed at the plane perpendicular to the heat source axis affects the shape and size of the fusion zone. Figure 7.3 (a) shows how the cross-section of the laser beam changes along the heat source axis. Often the beam converges at a certain distance from the source to form a waist (Figure 7.3 (a)) and then diverges. However, at any transverse plane (XY) perpendicular to the heat source axis (Z-direction) the power density follows a certain distribution. Generally, the beam has circular cross-sections and the power density distribution is axisymmetric which has been verified by measurements. For example, a Faraday Cup [3] was used to measure the power density distribution of an electron beam heat source. The split anode technique [4] was used to measure the power density distribution of an electric arc. These measurements confirm that at any transverse plane (XY plane) the power density distribution (P_d) of commonly used heat sources generally follow an axisymmetric Gaussian distribution, mathematically expressed as [1]:

$$P_d = \frac{2P}{\pi \left(r_b(Z)\right)^2} exp\left(-2\frac{r^2}{\left(r_b(Z)\right)^2}\right) \tag{7.1}$$

where P is the total power of the heat source, r_b is the radius of the beam which varies as a function of Z, the distance along the heat source axis (Figure 7.3 (a)). The symbol r is the radial distance of any point from the axis of the heat source which is also equal to $\sqrt{X^2 + Y^2}$. Power density that follows a Gaussian distribution has the maximum value near the axis of the heat source ($r = 0$). The power density decreases exponentially with the radial distance (r) away from the heat source axis. For example, Figure 7.3 (b) shows the distribution of the measured values of the power density of an electron beam heat source. The measurement was performed using a Faraday cup which provided the power density distribution as a function of radial distance from the heat source axis ($r = 0$ mm). The power density distributions can be fitted in a Gaussian distribution as a function of radial distance from the heat source axis.

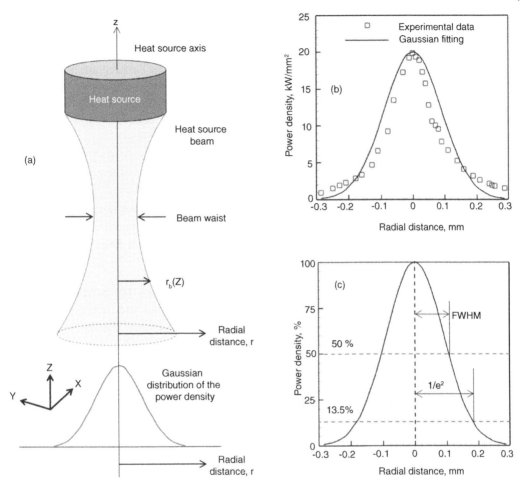

Figure 7.3 (a) A schematic representation of a heat source. (b) Experimentally measured variation of power density as a function of radial distance from the axis of a circular, axisymmetric electron beam heat source. The experimentally measured data is best fitted with a Gaussian distribution (Eq. 7.1) where the peak power density is 20 kW/mm^2. The experimental data are taken from [3]. (c) FWHM (full width at half-maximum) and 1/e^2 beam radius were obtained from the Gaussian distribution of the power density of the heat source. (*Source:* T. Mukherjee and T. DebRoy).

The effective radius of the beam can be obtained from the power density distribution. There are three commonly accepted conventions for determining the beam radius from the power density distribution, as explained in Figure 7.3 (c). First, the FWHM (full width at half-maximum) is measured from the power density distribution curve corresponding to the distance between the two points closest to the peak that has 50% of the maximum power density, as shown in Figure 7.3 (c). Second, the 1/e^2 radius is commonly used for laser and electron beam heat sources. Eq. 7.1 provides the maximum power density at the heat source axis ($r = 0$) as $\dfrac{2P}{\pi\left(r_b(Z)\right)^2}$. For $r = r_b$, the power density is $\dfrac{2P}{\pi\left(r_b(Z)\right)^2} exp(-2)$ or 1/e^2 times the maximum power density since $exp(-2)$ and 1/e^2 are both 0.1353. Therefore, the 1/e^2 radius is the radius of the beam corresponding to the 1/e^2 or 13.5%

of the maximum power density, as shown in Figure 7.3 (c). The third definition of the beam radius is called D4σ. In this method, the beam diameter (equals to twice the radius) is equal to four times the standard deviation (σ) of the Gaussian distribution concerning the heat source axis ($r = 0$). All three aforementioned methods are used to indicate laser beam radius. An accurate estimation of the beam radius is important because it defines the area where most of the heat energy is distributed. A smaller radius can concentrate more energy and as a result, produces a higher temperature. Worked out example 7.1 illustrates the relation between the effective radius and power density of a laser beam.

Worked out example 7.1

A 2000 W Gaussian laser beam is used in DED-L of Ti-6Al-4V. The effective radius of the beam on the molten pool is 1 mm. If the power is reduced to 1000 W, what should be the effective radius of the beam to maintain the same maximum power density on the fusion zone?

Answer:

The power density of the laser beam can be calculated using Eq. 7.1. The maximum power density is achieved at the center of the beam ($r = 0$). For a 1 mm radius and 2000 W power, the maximum power density is $(2000 \times 2)/[\pi\ (0.001)^2]$. To achieve this power density for 1000 W power, if the beam radius needed is r, then,

$$(2000\times2)/\left[\pi\left(0.001\right)^2\right]=(1000\times2)/\left[\pi\left(r\right)^2\right]$$

Solving the above equation, the radius of the beam (r) is found to be 0.00071 m or 0.71 mm.

The actual power density distribution often deviates from the ideal Gaussian distribution. For those heat sources, Eq. 7.1 is often rewritten as [1]:

$$P_d = \frac{fP}{\pi\left(r_b\left(Z\right)\right)^2}\exp\left(-f\,\frac{r^2}{\left(r_b\left(Z\right)\right)^2}\right) \tag{7.2}$$

where f is the power density distribution factor whose value can vary between 1 and 3. The power distribution factor $f = 2$ represents the ideal Gaussian distribution (Eq. 7.1). Figure 7.4 shows the power density distribution of a 500 W beam having a circular cross-section of 1 mm radius for different values of the distribution factor (f). The incident energy becomes more focused on higher values of the distribution factor which results in high peak temperatures underneath the heat source. The distribution factor significantly affects the temperature distribution in the part and therefore the shape and size of the fusion zone.

In practice, a power density distribution different from the Gaussian distribution is sometimes used for specific advantages. For example, a donut-shaped power distribution profile (Figure 7.5) has been used in PBF-L [5]. The power density drops near the middle of the beam (Figure 7.5 (a)). The location of the maximum power density is shifted radially away from the heat source axis. A donut-shaped beam can distribute a given power at a larger area which increases the molten pool width and makes the pool shallower (Figure 7.5 (b)). The net heat energy supplied by the heat source is absorbed by the previously deposited layer and feedstock to form the molten pool as discussed in the next section.

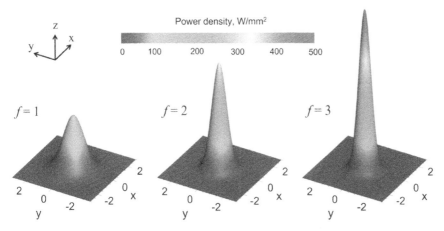

Figure 7.4 Power density distribution of a 500 W laser heat source of circular cross-section of 1 mm radius for different distribution factors (a) f = 1, (b) f = 2, and (c) f = 3. (*Source:* T. Mukherjee and T. DebRoy).

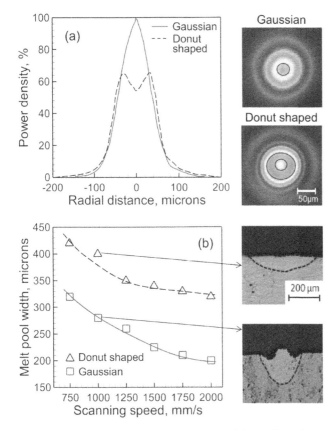

Figure 7.5 (a) Comparison of the Gaussian and donut-shaped power density distributions of a heat source. (b) Comparison of molten pool width during PBF-L of AlSi10Mg at 800 W laser power and different scanning speeds for Gaussian and donut-shaped laser beams. The two plots on the left were made by T. Mukherjee and T. DebRoy using the data reported in [5]. The four images on the right are reproduced from [5] with the permission from The Laser Institute of America.

7.3 Heat absorption by the feedstock

Absorptivity, or the ability of feedstock to absorb heat from an incident heat source, depends on the material, temperature, and wavelength of the beam. Absorptivity (a) is related to the wavelength and the electrical resistivity as [6]:

$$a \propto \left(\frac{R}{\lambda}\right)^{1/2} \tag{7.3}$$

where λ is the wavelength and R is the electrical resistivity of the material, which varies significantly with temperature. A beam with a large wavelength cannot easily penetrate through the interatomic space of the material resulting in diminished absorptivity. Therefore, the absorptivity is inversely proportional to the wavelength as indicated in Eq. 7.3. At high temperatures, the atoms vibrate rapidly, which transfers energy efficiently and increases absorptivity. Therefore, absorptivity increases with temperature. However, the faster atomic vibrations at high temperatures hinder the movement of free electrons and increase electrical resistivity. Therefore, at high temperatures both the electrical resistivity and absorptivity increase.

The aforementioned behavior of metallic materials can be justified experimentally. Figure 7.6 (a) shows the effect of laser wavelength on the absorptivity of different metallic materials at room temperature. For all four materials, absorptivity decreases with an increase in the wavelength of the incident beam. The absorptivity of a 1 μm wavelength laser beam by a flat plate of copper at room temperature is about 2.5%, which indicates that about 97.5% of the energy is reflected and not used to heat the material. In contrast, about 30% of the incident energy from a 1 μm wavelength laser beam can be absorbed by a flat plate of iron at room temperature. The poor absorptivity of silver and copper can be attributed to their low electrical resistivity as both metals are excellent conductors of electricity.

For a beam of a given wavelength, its absorptivity by a metallic material increases with temperature (Figure 7.6 (b)). The figure also shows that at a given temperature, the absorptivity increases

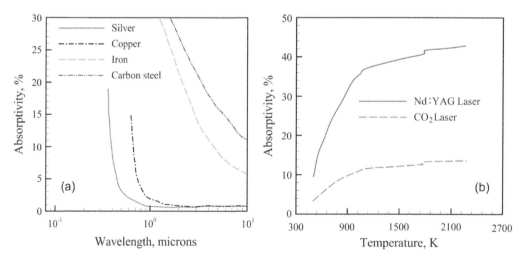

Figure 7.6 (a) Dependence of the absorptivity of different metallic materials on the laser wavelength at room temperature. (b) Variation of absorptivity of Nd:YAG (wavelength =1.06 μm) and CO_2 laser (wavelength =10.6 μm) by pure iron at different temperatures. Figures (a) and (b) are plotted using the data reported in [7] and [6], respectively. (*Source:* T. Mukherjee and T. DebRoy).

with a decrease in wavelength. The absorptivity is higher for the Nd:YAG laser which has a wavelength of 1.06 μm than for the CO_2 laser having a wavelength of 10.6 μm. When the power density is higher than a threshold value a metal may vaporize intensely and a deep and narrow vapor cavity called a keyhole may form. The laser beam may undergo multiple reflections within the cavity and as a result, more energy can be absorbed.

In additive manufacturing, the laser, electron beam, and electric arc are the three common heat sources used. The feedstock in the form of either powder or wire absorbs a portion of the energy supplied by the heat source. The absorbed energy affects the establishment of the time-dependent temperature and velocity fields, the formation of the molten pool, and the shape and size of the deposit. The physical phenomena that influence the energy absorption by the feedstock material are unique for each AM process variant. The absorption of heat by the feedstock depends on the type of heat source and the temperature of the solid and liquid regions. Also, the powder and the wire feedstocks absorb heat differently. The mechanisms of heat absorption during different PBF and DED processes for various types of feedstocks are discussed below.

7.3.1 Heat absorption mechanisms in powder bed fusion processes

In the PBF, part of the laser or electron beam energy is absorbed by the powder particles and the rest is reflected as shown schematically in Figure 7.7. The process continues until the beam emerges out of the powder bed or its intensity becomes negligible [8]. The heat absorbed by the powder particles in the PBF processes depends on the particle size, powder packing density, and material properties. Since the net energy absorbed increases with each reflection and subsequent absorption, the absorptivity can be significantly higher than that of a flat surface.

Apart from the powder characteristics, the power density of the laser beam plays an important role in heat absorption. At low power density, the maximum temperature inside the molten pool is well below the boiling point of the material and the molten pool has a low depth-to-width ratio and this mode of interaction is called the conduction mode (Figure 7.8 (a)). The beam energy is absorbed only once and the remaining energy is reflected from the surface. Consequently, the

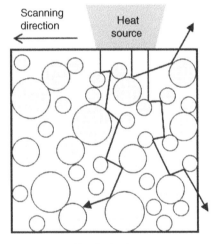

Figure 7.7 Heat absorption by the interparticle reflection of the incident beam in PBF. (*Source:* T. Mukherjee and T. DebRoy).

energy is mainly transported from the surface into the interior of the material through heat conduction and convection in the melt pool. In contrast, the high power density results in the intense vaporization of metals and the formation of a vapor-filled cavity known as the keyhole (Figure 7.8 (a)). The peak temperature reaches the boiling point of the material and the fusion zone has a high depth-to-width ratio (Figure 7.8 (b)). The heat absorption mechanism in a keyhole mode PBF process involves multiple reflections of the beam within the keyhole. In each location, a portion of the incident beam is absorbed, and the remaining portion is reflected. Due to the multiple reflections of the beam inside the keyhole, a high value of absorptivity often exceeding 80% is observed. In some practices, the powder bed is preheated to reduce the cooling rate and avoid defects.

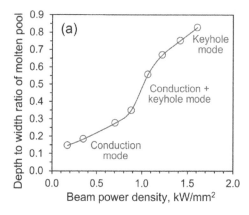
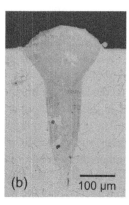

Figure 7.8 (a) Identification of the three modes in laser welding depending on the laser beam power density. The plot is made by T. Mukherjee and T. DebRoy using the data reported in [9]. (b) Transverse section of the fusion zone in a keyhole mode PBF-L process showing high depth-to-width ratio [10]. The image is reproduced from [10] with the permission from Elsevier.

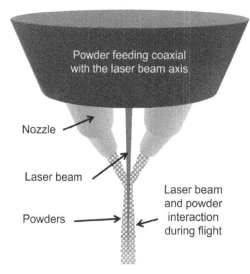

Figure 7.9 Heat absorption by powders during flight from nozzle to the substrate in DED-L. (*Source:* T. Mukherjee and T. DebRoy).

7.3.2 Heat absorption mechanisms in directed energy deposition processes

In powder-feeding DED processes, the powders absorb heat in two stages. First, a fraction of the available heat is absorbed as the powders emerge from the nozzle and travel through the beam as shown in Figure 7.9. This in-flight heating of the particles depends on the type, shape, size, and speed of the powders, beam characteristics, flight duration through the beam, and shielding gas type and velocity. The powder particles are generally heated to a high temperature below their solidus temperature. Second, after the flight, the hot powder particles impinge on the molten pool and depositing surface, and the powder continues to absorb heat. The extent of absorbed heat and melting depends on the beam characteristics, the absorptivity of the alloy powders for the energy source used, deposit geometry, and the shielding gas.

In the wire feeding DED process, which typically uses an electric arc or plasma arc or in some cases laser or electron beam as the heat source, heat absorption by the wire depends on the heat source characteristics, efficiency, molten pool size, wire type, wire diameter, and feeding rate, shielding gas or vacuum environment. The process is similar to multilayer gas-metal welding where the electrode melting and the droplet transfer mode are important features of the process. Due to the higher surface-to-volume ratio of the powder, the powder-based AM processes have a higher rate of heat absorption than the wire-based processes. Sometimes, the wire is preheated by resistive or inductive heating or using a secondary heat source to increase its heat content.

7.4 Heat conduction through the deposit and substrate

Part of the absorbed heat is used to melt the previously deposited layer and feedstock materials to form a molten pool. A portion of the remaining heat is conducted away from the molten pool, resulting in a high-temperature solid region called the heat-affected zone. The rate of heat conduction affects the deposit shape and size, evolution of the microstructure, and properties, susceptibility to defects, and the development of the thermally induced stresses. Since heat transfer significantly affects the formation of the fusion zone and the surrounding heat-affected zone, it is important to understand how the heat is distributed from the heat source through the workpiece. The transport of heat from the fusion zone through the solid deposit and substrate is discussed here. The effects of the thermophysical properties of alloys on heat transfer are explained. In addition, approximate heat conduction calculations of temperature distribution, peak temperature, fusion zone size, and cooling rates are presented.

7.4.1 Heat conduction

Heat transfer mechanisms vary significantly in different AM processes depending on the types of feedstock and heat source. For a multitrack build with many layers and hatches, the primary direction of heat transfer from the molten pool changes continuously as different hatches and layers are built. However, the primary direction of heat conduction is from the hot molten pool to the relatively cooler substrate.

In PBF, the properties of the powder bed significantly affect the rate and pattern of heat conduction. The maximum amount of heat is conducted from the molten pool through the substrate since the thermal conductivity of the solid substrate is significantly higher than that of the powder bed. The thermal conductivity of the powder bed is lower than that of a solid material because the interparticle space in the powder bed is filled with shielding gas. A bed containing large powder particles has low thermal conductivity because of smaller contact areas between particles [9]. Also, the bed with large powder particles may have higher porosity depending on the packing and particle morphology. Figure 7.10 shows that the thermal conductivity of the powder bed decreases significantly as the particle size increases.

As the deposition of multiple tracks progresses, the maximum heat conduction occurs from the molten pool to both the substrate and the already deposited tracks because the thermal conductivity of the solidified build is higher than the powder bed [11]. One side of the molten pool is already solidified and the other side has the powders (Figure 7.11). Therefore, the rate of heat transfer from the molten pool to the two sides is unequal because of the high thermal conductivity of the already-built hatch on one side and powder on the other side. This difference in heat transfer may affect the molten pool shape. The molten pool is narrower on the side of the already deposited solid due to relatively faster heat transfer (Figure 7.11). Since the distance between the molten pool and the substrate increases as more layers are deposited, the rate of heat conduction away from the molten pool decreases for the upper layers. However, the maximum heat conduction occurs from the molten pool to the substrate in the direction opposite to the build direction.

In DED, the maximum amount of heat is conducted from the molten pool through the substrate, since the temperature of the substrate is significantly lower than that of the molten pool resulting in a high-temperature gradient that drives the heat conduction. As the deposition of multiple tracks progresses, heat conduction occurs from the molten pool to both the substrate and the already deposited tracks. Since the distance between the molten pool and the substrate increases as more layers are deposited, the rate of heat conduction decreases for the upper layers, often

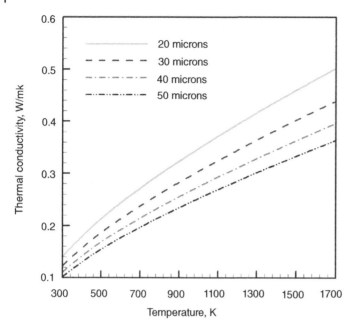

Figure 7.10 Variation of thermal conductivity of a powder bed for different sizes of the stainless steel 316 powder particles (in microns). The size represents the average size of powders in a powder bed. (*Source:* T. Mukherjee and T. DebRoy).

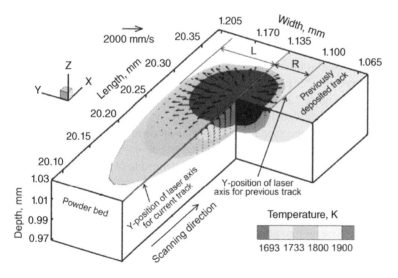

Figure 7.11 Asymmetry of the molten pool size due to the difference in heat conduction rate on both sides of the molten pool during the PBF process. This example is for PBF-L of SS 316 using 60 W laser power, 1000 mm/s scanning speed, and 30 μm layer thickness. The colored contours represent the isotherm indicated in the contour legend. The previously deposited track is shown in a gray color. The scanning direction is along the positive X-axis. "R" represents the distance of the right side (side of the previous track) of the molten pool from the laser beam axis (y = 1.135 mm). "L" represents the distance of the left side (side of the powder layer) of the molten pool from the laser beam axis. L > R indicates the difference in molten pool shape due to the difference in the rate of heat transfer. The result is obtained from the numerical model of additive manufacturing developed by T. Mukherjee and T. DebRoy.

resulting in higher peak temperatures and a bigger molten pool for the upper layers. In both PBF and DED processes, both the heat conduction pattern and its rate can significantly vary depending on the thermophysical properties of alloys and the scanning pattern of the heat source.

7.4.2 Alloy properties affect heat conduction

Heat transfer during additive manufacturing processes depends on the thermophysical properties of alloys such as thermal conductivity, specific heat, density, latent heat, and liquidus and solidus temperatures. Table 7.1 summarizes important thermophysical properties of several commonly used alloys in AM such as stainless and tool steels, titanium alloys, nickel-based superalloys, and aluminum alloys. A comparison of the rates of heat transfer and resultant temperature fields, molten pool sizes, and cooling rates requires the thermophysical properties of alloys. For example, the same heat input can melt more material for the alloy with lower density resulting in a bigger molten pool. An alloy with high thermal conductivity dissipates heat rapidly resulting in a low peak temperature. Also, alloys with a large difference between the solidus and liquidus temperatures often exhibit a large mushy zone.

The effective thermophysical properties that drive the heat transfer vary for each variant of the AM process. For example, in PBF, the density, thermal conductivity, and specific heat of the packed powder bed depend on the shielding gas entrapped among the powder particles and the packing efficiency of the powder bed. Also, the properties responsible for heat transport such as thermal conductivity and specific heat vary significantly with temperature. At the beginning of the AM process, the substrate is at room temperature or preheat temperature. During the process, the properties of the substrate and the already deposited tracks vary based on the spatially variable, transient temperature field. At the end of the deposition process, when both the deposited component and the substrate cool down, the properties also vary depending on the temperature field.

7.4.3 Heat conduction calculations

Thermal cycle, peak temperature, fusion zone size, and cooling rates can be approximately estimated using simplified, analytical heat conduction calculations. The readers should appreciate that these calculations are approximate because they ignore heat transport within the molten zone by convection, and the mixing of the hot and cold fluids is not considered. As a result, the calculated peak temperatures and cooling rates are overpredicted and the molten zone dimensions are unreliable. However, they are useful to compare AM of two alloys or evaluate trends of values for different process variables. These calculations were originally derived for fusion welding [12] and can be applied in additive manufacturing for approximate estimations and trends of the computed variables.

Thermal cycle:
The thermal cycle, which is the variations of temperature with time at different locations, can be approximately estimated by an analytical solution of a three-dimensional equation of heat conduction [13].

$$\frac{d^2T}{d\xi^2} + \frac{d^2T}{dy^2} + \frac{d^2T}{dz^2} = -\frac{\rho C_p}{k_s}\left(V\frac{dT}{d\xi}\right) \tag{7.4}$$

where T = Temperature, K
y = Distance from the heat source axis in the width direction, m

Table 7.1 Thermophysical properties of alloys commonly used in AM. "T" denotes the temperature in K which varies between room and solidus temperatures [1]. Density values are given at room temperature. Viscosity, surface tension, and $d\gamma/dT$ are given at the liquidus temperatures of alloys. SS 316: Stainless steel 316; Ti-6Al-4V: Titanium alloy; IN718: Nickel alloy Inconel 718; H13: H13 tool steel; 800H: Nickel alloy 800H; AlSi10Mg: Aluminum alloy.

Properties	SS316	Ti-6Al-4V	IN718	H13	800H	AlSi10Mg
Liquidus temperature (K)	1733	1928	1609	1725	1675	867
Solidus temperature (K)	1693	1878	1533	1585	1608	831
Thermal conductivity (W/m K)	$11.82 + 0.0106\,T$	$1.57 + 1.6 \times 10^{-2}\,T$ $-\,1 \times 10^{-6}\,T^2$	$0.56 + 2.9 \times 10^{-2}\,T$ $-\,7 \times 10^{-6}\,T^2$	$18.29 + 7.5 \times 10^{-3}\,T$	$0.51 +$ $2.0 \times 10^{-2}\,T$ $-\,6 \times 10^{-6}\,T^2$	$113 +$ $1.06 \times 10^{-5}\,T$
Specific heat (J/kg K)	$330.9 + 0.563\,T -$ $4.015 \times 10^{-4}\,T^2$	$492.4 + 0.025\,T$ $-\,4.18 \times 10^{-6}\,T^2$	$360.4 + 0.026\,T$ $-\,4 \times 10^{-6}\,T^2$	$341.9 + 0.601\,T$ $-\,4.04 \times 10^{-4}\,T^2$	$352.3 + 0.028\,T$ $-\,3.7 \times 10^{-6}\,T^2$	$536.2 + 0.035\,T$
Density (kg/m³)	7800	4430	8100	7900	7270	2670
Viscosity (kg/m s)	7×10^{-2}	4×10^{-2}	5×10^{-2}	7×10^{-2}	7.5×10^{-2}	1.3×10^{-2}
Surface tension (N/m)	1.50	1.52	1.82	1.90	1.82	0.82
$d\gamma/dT$ (N/m K)	-0.40×10^{-3}	-0.26×10^{-3}	-0.37×10^{-3}	-0.43×10^{-3}	-0.40×10^{-3}	-0.35×10^{-3}

z = Distance from the heat source axis in the depth direction, m
$\xi = x - Vt$ where x is the distance from the heat source axis in the scanning direction, m
V = Scanning speed, m/s
t = Time, s
k_s = Thermal conductivity of solid alloy, W/m K
C_p = Specific heat of solid alloy, J/kg K
ρ = Density of solid alloy, kg/m^3

In this equation, the coordinate along the scanning direction (x-direction) is transformed into "ξ" to include the effect of scanning speed (V) where the scanning direction is along the positive x-axis. The analytical solution of Eq. 7.4 is derived based on the following assumptions [13].

• The substrate is assumed to be a thick plate (significantly thicker than molten pool depth) and has infinite width.
• The heat source energy is assumed to be concentrated at a point on the top surface of the work plate.
• The heat source moves at a constant speed on a straight path relative to the substrate.
• Constant values (temperature-independent) of the thermophysical properties are used.
• Heat losses from the surface through convection and radiation are neglected.

The thermal cycle is obtained from the analytical solution of Eq. 7.4 for the following three initial and boundary conditions:

Initial condition:
$T = T_0$ at $t = 0$ for all values of x, y, and z.

Boundary conditions:
$T = T_0$ for all values of t for $x = \infty$, $y = \infty$, and $z = \infty$ and

$$-k_s A \frac{dT}{dz} = P \text{ at } x = 0, y = 0, \text{ and } z = 0 \text{ for all values of } t, \text{ where } A \text{ is an area}$$

The solution is given by the following expression:

$$T = T_0 + \frac{\eta P / V}{2\pi k_s t} e^{-\frac{y^2 + z^2}{4\alpha_s t}} \tag{7.5}$$

T_0 = Initial temperature, K, which can be equal to the ambient or preheat temperature.
P = Heat source power, W
η = Fraction of the energy absorbed
α_s = Thermal diffusivity of solid alloy, m^2/s, which is equal to $k_s / (\rho C_p)$.

A sample calculation is provided in Table 7.2 for different combinations of heat source power and scanning speed during AM of stainless steel 316. The thermal cycles estimated using Eq. 7.5 are shown in Figure 7.12. It can be noted that the temperature rises rapidly, attains the highest temperature at about 100 ms, and gradually cools down to room temperature. The highest temperature decreases with the depth from the top surface owing to an increase in distance from the top surface of the deposit where the heat source is located. The highest temperature attained at the deposit centerline on the top surface just beneath the heat source axis is the peak temperature inside the molten pool. Table 7.2 reports the peak temperatures for different combinations of heat source power and speed. The peak temperature increases with an increase in power or decrease in scanning speed.

Table 7.2 Example calculations of peak temperature, fusion zone depth, and cooling rate on the top surface at the deposit centerline at various temperatures during AM of stainless steel 316 at different combinations of heat source power and scanning speed using Eqs. 7.5 and 7.7. The thermophysical properties of stainless steel 316 are taken as the following constant values: k_s =28 W/m K, ρ =7200 kg/m^3, C_P= 611 J/kg K, α_s=6.36×10^{-6} m^2/s. In addition, the initial temperature (T_0) and the fraction of the energy absorbed (η) are taken as 298 K and 0.9, respectively, where the symbols have their usual meanings as indicated before.

Cases	Peak temperature	Fusion zone depth	Fusion zone width	T_C in Eq. 7.7	Cooling rate on the top surface at the deposit centerline
Power = 500 W	2857 K	1.24 mm	2.48 mm	800 K	985 K/s
Speed = 10 mm/s				1000 K	1926 K/s
Power = 300 W	1833 K	0.50 mm	1.00 mm	800 K	1641 K/s
Speed = 10 mm/s				1000 K	3209 K/s
Power = 500 W	2004 K	0.72 mm	1.44 mm	800 K	1477 K/s
Speed = 15 mm/s				1000 K	2888 K/s

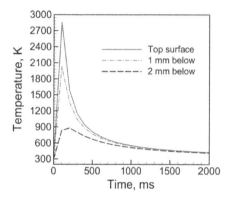

Figure 7.12 Thermal cycles calculated at the deposit centerline (y = 0 m) and three depths, the top surface (z = 0 m), 1 mm from the top surface (z = 0.001 m), and 2 mm from the top surface (z = 0.002 m) using Eq. 7.5 for 500 W power and 10 mm/s scanning speed for stainless steel 316. Other conditions are the same as those provided in Table 7.2. (*Source:* T. Mukherjee and T. DebRoy).

Fusion zone geometry:

Fusion zone depth and width can be estimated from the calculated temperature distribution. The maximum depth is observed at the deposit centerline below the heat source axis. Figure 7.12 shows that at the deposit centerline, for all depths the maximum temperature is reached at about 100 ms. Therefore, fusion zone depth is equal to the "z" value for which the peak temperature is equal to the solidus temperature of the alloy (1693 K for stainless steel 316) at 100 ms. It can be calculated by putting t = 100 ms and T = 1693 K in Eq. 7.5 for the condition mentioned in Table 7.2. The calculated fusion zone depths and widths for different powers and speeds are provided in Table 7.2. The depth increases with the increase in power or a decrease in scanning speed both owing to an enhancement in the energy input from the heat source per unit length of deposit. The fusion zone width is equal to twice the fusion zone depth because the calculations assume the formation of a semicircular molten pool considering uniform heat transfer both to the width and depth directions.

Cooling rate:

Heat energy applied to the feedstock is distributed by conduction in the solidified deposit and substrate. The resulting thermal cycle exhibits a sharp temperature rise and then gradual cooling by conducting heat away primarily through the substrate. The cooling rate is the slope of the thermal cycle during cooling which varies with temperature. Therefore, at a particular temperature (T_c), the cooling rate can be calculated by differentiating Eq. 7.5 (thermal cycle) for different values of time. For example, on the top surface at the deposit centerline (y = 0 and z = 0), Eq. 7.5 can be written as:

$$T = T_0 + \frac{\eta P / V}{2\pi k_s t} \qquad (7.6)$$

Therefore, the cooling rate on the top surface at the deposit centerline at a given temperature (T_c) is given by:

$$R_c = \left(\frac{dT}{dt}\right)_{T=T_c} = -\frac{\dfrac{\eta P}{V}}{2\pi k_s t^2} = -\frac{2\pi k_s}{\eta P / V}\left(T_c - T_0\right)^2 \qquad (7.7)$$

R_c = Cooling rate on the top surface at the deposit centerline, K/s. A negative value indicates decreasing slope of temperature with time signifying cooling.

P = Heat source power, W

V = Scanning speed, m/s

η = Fraction of the energy absorbed

k_s = Thermal conductivity of solid alloy, W/m K

T_c = Temperature at which cooling rate is calculated, K

T_0 = Initial temperature, K. Initial temperature can be equal to the ambient or preheat temperature.

Table 7.2 shows the cooling rates at different temperatures for different combinations of heat source power and speed. At higher temperatures, the slope of the thermal cycle is more as shown in Figure 7.12. Therefore, the calculated cooling rates increase with temperature as observed in Table 7.2. Both high power and low scanning speed result in a high peak temperature and a large molten pool that cools slowly resulting in a lower cooling rate. Therefore, the cooling rate decreases with an increase in power or a reduction in scanning speed. The relation between cooling rate and scanning speed is explained in worked out example 7.2. Alloys with low density result in a large molten pool. Also, alloys with high thermal diffusivity result in rapid heat conduction and fast cooling rates.

Worked out example 7.2

It is stipulated that the cooling rate at solidus temperature (1693 K) of a stainless steel 316 part must not exceed 5000 K/s during DED-L. In addition, a minimum of 1.5 mm pool depth is required to ensure proper fusional bonding with the previously deposited layer and avoid the lack of fusion defects. The laser beam power is 500 W. What is the maximum scanning speed that can be used? Data: thermal conductivity (k_s): 28 W/m K, density (ρ): 7200 kg/m^3, specific heat (C_p): 611 J/kg K, thermal diffusivity (α_s): 6.36 × 10^{-6} m^2/s, the fraction of the energy absorbed (η): 0.9, and the initial temperature (T_0): 298 K.

Answer:

Eq. 7.7 can be used to calculate the maximum cooling rate on the top surface at the trailing edge of the molten pool. Since the maximum allowable cooling rate at 1693 K is 5000 K/s, we get a scanning speed of 6.58 mm/s from the following equation.

$$5000 = \frac{2\pi \times 28}{0.9 \times 500 / V}\left(1693 - 298\right)^2$$

At this scanning speed, the fusion zone depth can be calculated as 1.62 mm (see Table 7.2) which is higher than the minimum depth indicated (1.5 mm). Therefore, a scanning speed of 6.58 mm/s satisfies the stipulated requirements. A higher scanning speed would increase the cooling rate and reduce the fusion zone depth.

Fourier number:

In AM, a portion of the absorbed heat energy is stored in the part and the substrate, and the rest are lost to the surroundings. The relative measure of heat dissipation rate to heat storage rate is indicated by the Fourier number (F_o):

$$F_o = \frac{\alpha}{Vd} \tag{7.8}$$

where α, V, and d refer to the thermal diffusivity of the alloy, scanning speed of the heat source, and fusion zone depth, respectively. A high Fourier number characterizes an AM process that exhibits rapid heat dissipation which tends to lower the peak temperature. Fourier number varies typically in the range of 10^{-3} to 1 depending on the AM process, process variables, and the alloy used. Alloys with high thermal diffusivity exhibit a rapid rate of heat dissipation resulting in a high Fourier number. For PBF, a fast scanning speed reduces the Fourier number according to Eq. 7.8. Therefore, Fourier numbers encountered during PBF processes are smaller than those for DED-L and DED-GMA, as shown in Figure 7.13. An AM process with a high Fourier number experiences rapid heat dissipation that often results in a fast cooling rate. This phenomenon is illustrated in worked out example 7.3. It is well-known that a high cooling rate often refines grains to achieve superior mechanical properties of a part. Therefore, in an AM process, process parameters can be controlled to achieve a high Fourier number which is desirable for achieving finer grains.

The calculations presented here are useful to compare AM of two alloys or compare trends of values for different process variables and may not be relied upon to compute values of a single variable. The temperature distributions and fusion zone size depend on the transport of heat by both heat conduction and convection owing to the flow of liquid metal inside the molten pool (Section 7.5). In addition, the heat transfer and the rate of cooling depend on the convective and radiative heat losses (Section 7.6) from the deposit to its surroundings. Furthermore, in AM, the heat source energy often follows a Gaussian distribution (Section 7.2) that cannot be accurately represented as a point heat source. Estimations of accurate temperature fields, cooling rates, and fusion zone

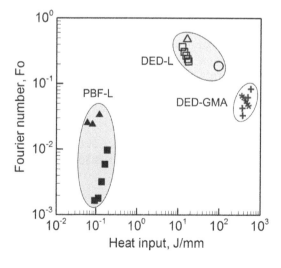

Figure 7.13 Fourier numbers for different AM processes. The plot is for stainless steel 316. Here, heat input represents heat source power/scanning speed. The plot is adapted from [9] with the permission from Elsevier.

Worked out example 7.3

Two stainless steel 316 deposits are made using DED-L with 500 W and 300 W laser powers and 10 mm/s scanning speed. Assume that the depths of fusion zone were 1.24 mm and 0.5 mm for 500 W and 300 W power, respectively. All other processing conditions are kept constant. Which of the two deposits will experience a faster cooling rate? Suggest a dimensionless number that could explain the difference in the rate of cooling and explain the significance of the difference. The ambient temperature is 298 K and the thermal diffusivity of stainless steel is 6.36×10^{-6} m^2/s.

Answer:

Fourier number, which is the ratio of the heat dissipation rate to the heat storage rate, can be used for this purpose. A high Fourier number indicates rapid heat dissipation which results in rapid cooling. Therefore, the computed values of the Fourier number (Eq. 7.8) for the two cases can reveal which of the two deposits will exhibit faster cooling. The results are shown in Table E7.1 below.

Table E7.1   Computed results showing the Fourier number values for two cases.

Cases	Parameters in Eq. 7.8		Fourier number
Power = 500 W	α	6.36×10^{-6} m^2/s	0.5
	V	10 mm/s	
	d	1.24 mm	
Power = 300 W	α	6.36×10^{-6} m^2/s	1.3
	V	10 mm/s	
	d	0.5 mm	

The results show that the deposit made with a lower laser power of 300 W exhibits a higher Fourier number and thus experiences faster cooling. A lower laser power results in a small pool that cools faster.

---

dimensions require the consideration of heat transfer by both conduction and convection in the solution of the energy conservation equation. Numerical models (Section 7.8) of AM are emerging for this purpose, but they are computationally intensive and time-consuming.

7.5   Convective heat transfer within the molten pool

During AM, the interaction between the feedstock and the heat source leads to the rapid heating and melting of the feedstock to form the molten pool. In the molten pool, the liquid metal undergoes circulatory motion (Figure 7.14) driven by the surface tension gradient, buoyancy, electromagnetic forces for electron beam and arc-based AM processes, and powder or droplet impingement. The magnitude of velocities of the liquid metal can vary significantly from a few centimeters per second to a few meters per second depending on the AM process, alloy, and process parameters used. The maximum velocities of the convective flow inside the molten pool can be comparable with the maximum speed of a swimmer in the Olympic games which is about 2.5 m/s.

Figure 7.14 The computed flow of liquid metal during gas-tungsten arc welding of low carbon steel, using 160 A, 18 V, and a welding speed of 2.7 mm/sec. The colors represent temperatures in Kelvin and the dotted lines show the flow of liquid metal. The two loops near the surface are caused by the Marangoni effect and those below the surface originate from the electromagnetic force. (*Source:* G.G. Roy and T. DebRoy).

The mixing of hot and cold liquids due to convective flow reduces the temperature gradient inside the molten pool and affects the peak temperature, cooling rate, temperature gradient, and solidification growth rate. Also, convective heat transfer is often very important in determining the shape and size of the fusion zone, which in turn affects the macro- and microstructure of the deposit. In the following subsection, different driving forces and their contributions to convective flow are discussed along with several calculations for estimating the maximum velocities in the molten pool. The effect of convective flow on the shape and size of the molten pool is also examined.

7.5.1 What drives the flow of liquid metal?

There are several driving forces for the convective flow of liquid metal in AM. The most important force for fluid motion is often the Marangoni force which arises because of the spatial variation of interfacial tension on the surface of the molten pool. Additionally, when an electric arc is used as the heat source in DED-GMA, the momentum from the droplet transfer becomes an important driver of the fluid flow. The interaction between the current and the magnetic fields induced by the current in PBF-EB, DED-EB, and DED-GMA results in an electromagnetic force field that also contributes to the convection of the liquid metal. The vaporization of alloying elements from the fusion zone results in a downward force acting upon the molten pool surface, known as the recoil force, which also contributes to the convective flow. The spatial variation of density results in a buoyancy force which in most cases is much weaker than the other forces.

7.5.1.1 Marangoni force

The spatial gradient of surface tension on the liquid surface is stress, known as Marangoni stress that results in a convective flow of the liquid. The liquid flows from a region with low surface tension to other regions with high surface tension. Let us consider the following experiment to show the flow of water whose surface is sprayed with black pepper (Figure 7.15). A cotton bud with a little soap is dripped in the middle of the surface. The surface tension drops locally because of the addition of soap and the water flows from the middle to the periphery as observed by the motion of the black pepper powders on the surface of the water.

Step 1   Step 2   Step 3

Black pepper powders floating on the water surface   Soap on the cotton bud   Soap reduces the surface tension at the center and high surface tension near the periphery drags the pepper

Figure 7.15 An experiment to demonstrate surface tension gradient-driven convective flow. (*Source:* T. Mukherjee and T. DebRoy).

During AM, the spatial variation of the surface tension at the molten pool surface may arise owing to variations in both temperature and composition. Convection results from the spatial gradient of surface tension at the molten pool surface. In most cases, the difference in the surface tension is due to the temperature variation at the molten pool surface. For such a situation, the Marangoni stress (τ) can be expressed as:

$$\tau = -\frac{d\gamma}{dT}\frac{dT}{dY} \tag{7.9}$$

where, $\dfrac{d\gamma}{dT}$ is the gradient of surface tension (γ) with the temperature (T) and $\dfrac{dT}{dY}$ is the spatial gradient of temperature along the Y-direction. In Figure 7.16 (a), velocity along the Y-direction depends on the Marangoni stress along the Y-direction which is affected by $\dfrac{dT}{dY}$. Assuming the molten metal as a Newtonian fluid, the stress on the top surface of the molten pool, Marangoni stress, can also be expressed as:

$$\tau = \mu \frac{dU}{dZ} \tag{7.10}$$

where μ is the viscosity of molten liquid and $\dfrac{dU}{dZ}$ is the gradient of velocity U along the depth direction, Z, as shown in Figure 7.16 (b). Maximum velocity of the liquid metal (U_m) is observed on the top surface of the molten pool which originates from the highest temperature difference on the top surface of the molten pool, $T_m - T_S$, where T_m is the peak temperature and T_S is the solidus temperature. The solidus isotherm indicates the boundary of the molten pool as shown in Figure 7.16. The temperature gradient in the width direction is $\dfrac{T_m - T_S}{W/2}$ where W is the pool width as indicated in Figure 7.16 (a). In the depth direction (Z-direction), the molten metal velocity changes its direction depending on the shape of the fusion zone near the half-depth, as schematically shown

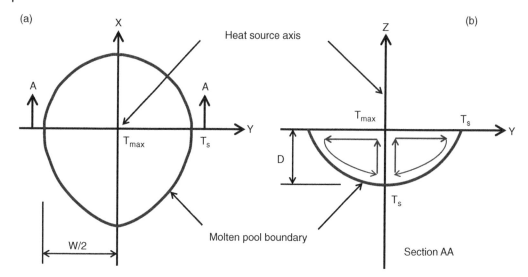

Figure 7.16 Schematic representations of a molten pool showing (a) top view on XY plane and (b) transverse view (on YZ plane) at a section AA shown in (a) which is at the same location as the heat source axis. "X" is the scanning direction and "Z" is the vertical direction. Therefore, the heat source axis is along the "Z" axis and perpendicular to the plane of the paper for figure (a). (*Source:* T. Mukherjee and T. DebRoy).

in Figure 7.16 (b). Therefore, for the maximum velocity, $\dfrac{dU}{dZ} = \dfrac{U_m}{D/2}$ where D is the depth of the molten pool. Therefore, the maximum velocity is obtained by rearranging Eqs. 7.9 and 7.10:

$$\mu \frac{U_m}{D/2} = -\frac{d\gamma}{dT}\frac{T_m - T_S}{W/2} \tag{7.11}$$

This equation can provide an approximate value of the maximum velocity of liquid metal inside the molten pool if the peak temperature and pool dimensions are known. For commonly used engineering alloys such as stainless steel 316, the value of $\dfrac{d\gamma}{dT}$ is negative, since its surface tension decreases with an increase in temperature, i.e., lower surface tension at the periphery and higher surface tension at the center of the molten pool. Since molten metal flows from the low to high surface tension, the liquid metal flows from the center to the periphery of the molten pool (Figure 7.17 (a)). However, accurate calculations of the three components of velocities at all locations require advanced numerical simulation.

The dominance of the surface tension force (Marangoni force) on the fluid flow is evaluated by comparing the magnitude of the surface tension force with the viscous force. The ratio of the surface tension force to viscous force is expressed by the surface tension Reynolds number as [15]:

$$M = \frac{\rho L_c \Delta T}{\mu^2}\left|\frac{d\gamma}{dT}\right| \tag{7.12}$$

where ρ is the density of liquid alloy, L_c is the characteristic length which may be taken as the half-width of the fusion zone, μ is the viscosity of the alloy, ΔT is the difference between the peak temperature and the solidus temperature of the alloy, and $d\gamma/dT$ is the temperature coefficient of surface tension gradient. A high value of the surface tension Reynolds number indicates the

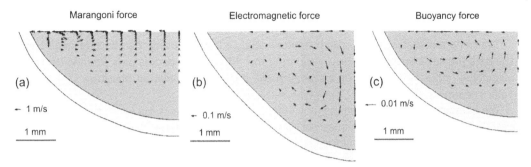

Figure 7.17 Calculated velocity fields due to (a) Marangoni (b) electromagnetic and (c) buoyancy forces during gas tungsten arc welding of aluminum 6061 shown at the transverse section of the pool where the direction of heat source movement is perpendicular to the plane of the paper. The heat source power is 2250 W. Half of the molten pool is shown due to symmetry. The magnitude of velocity due to the Marangoni force is the highest followed by velocities due to electromagnetic force and buoyancy force. Here, the electromagnetic force results in a radially inward flow, unlike a radially outward flow due to Marangoni and buoyancy forces. The plots are made based on the results reported in [14]. (*Source:* T. Mukherjee and T. DebRoy).

vigorous convective flow of the molten metal driven by the surface tension force. In addition to the surface tension force, fluid flow inside the molten pool can also be driven by buoyancy force as discussed in the next section. The calculation of velocities owing to Marangoni convection is explained in worked out example 7.4.

Worked out example 7.4

a) Derive an approximate analytical expression for the maximum velocity of liquid metal owing to the spatial gradient of surface tension. Assume that the stress in a boundary layer can be given by the following expression: $\tau = -0.332\,\rho^{1/2}\,\mu^{1/2}\,u^{3/2}/y^{1/2}$, where ρ is the density of the liquid metal, μ is the viscosity, u is the velocity, and y is the distance from the location of the axis of the heat source. State all assumptions you make.

b) Using the formula you derive, calculate the maximum velocity for the following conditions: the width of the liquid pool (W), density, viscosity, temperature coefficient of surface tension $\left(\dfrac{d\gamma}{dT}\right)$, and the spatial gradient of temperature along the width (ΔT) are 5 mm, 7200 kg/m³, 6×10^{-2} kg/m s, -0.5×10^{-3} N/m K, and 60,000 K/m, respectively. Discuss how the validity of the computed velocity is checked.

Answer:

Assume that the maximum velocity occurs on the top surface of the molten pool halfway between the beam axis and the periphery of the melt pool along the width direction perpendicular to the scanning direction. The maximum velocity occurs at a distance W/4 from the beam axis where W is the width of the deposition track. By combining the above equation with Eq. 7.9, and assigning y = W/4, one gets the expression for the maximum velocity as:

$$U_{max}^{3/2} = \frac{\Delta T}{0.664\left(\rho\mu W\right)^{1/2}}\left|\frac{d\gamma}{dT}\right| \tag{E7.1}$$

(Continued)

Worked out example 7.4   (Continued)

The computed maximum velocity is approximately 620 mm/s using the above formula and the data provided in the problem. Considering that this molten pool is only 5 mm wide, this velocity is rather large. However, a search of the numerical simulation literature [14] gives values of similar magnitudes.

The computed results are commonly validated by measurements. However, the commonly used velocity measuring techniques such as laser Doppler and particle image velocimetry, Schlieren optics, and interferometric techniques are difficult to apply in AM because of high temperatures, small melt pool size, rapidly moving melt pool, and limited access to the melt pool under the conditions of AM. It is not surprising that no commercial systems are available for measuring velocities in the liquid pool in additive manufacturing.

7.5.1.2   Buoyancy force

Buoyancy force originates from the density differences in a molten pool primarily due to the variations in local temperatures. Liquids are most dense at temperatures near the solidus temperature. As temperature increases above the solidus temperature, the density of the liquid decreases. Therefore, hot regions of liquid metal in a molten pool are of lower density than cold regions. Cooler, more dense molten metal sinks under the force of gravity, and hotter, less dense molten metal rises. The maximum velocity (U_m) of the molten metal due to buoyancy force can be estimated based on scaling analysis [15]:

$$U_m = \sqrt{g\beta D(T_m - T_S)} \tag{7.13}$$

where g is the acceleration due to gravity, β is the volumetric expansion coefficient of the alloy, D is the pool depth, T_m is the maximum temperature inside the pool, and T_S is the solidus temperature of the alloy. The buoyancy force is much weaker than the Marangoni force as evident from the lower velocities in the buoyancy-driven flow than that for the surface tension-driven flow (Figures 7.17 (a) and (c)). The liquid metal sinks near the edge of the pool because of its relatively low temperature and high density. The hotter and lighter liquid metal near the middle of the molten pool rises. The resulting circulation or convection pattern is radially outward similar to that observed for the Marangoni force (Figure 7.17 (a) and (c)). The role of the buoyancy force on the fluid flow is evaluated from the Grashof number which is the ratio of the buoyancy force to viscous force [15].

$$Gr = \frac{g\beta\rho^2 L_c^3 \Delta T}{\mu^2} \tag{7.14}$$

where g is the acceleration due to gravity, β is the coefficient of volumetric expansion, ρ is the density, μ is the viscosity of the alloy, ΔT is the difference between the peak temperature and the solidus temperature, and L_c is the characteristic length which may be taken as the depth of the fusion zone. The magnitude of the velocity owing to buoyancy force is calculated in worked out example 7.5. The relative magnitude of buoyancy force and Marangoni force is computed in worked out example 7.6. In addition to the surface tension and buoyancy forces, the flow inside the molten pool can also be driven by the electromagnetic force for the arc and electron beam-based processes as discussed in the next section.

7.5.1.3 Lorentz or electromagnetic force

The interaction between the current density and the induced magnetic flux results in the electromagnetic or Lorentz force field as [15]:

$$F_e = \vec{j} \times \vec{B} \tag{7.15}$$

where F_e is the electromagnetic force, $\vec{j}$ is the current density vector, and $\vec{B}$ is the magnetic flux vector. This force can cause convective flow in electron beam and arc-based processes such as PBF-EB, DED-EB, and DED-GMA. The maximum characteristic velocity (U_m) of molten metal due to the electromagnetic forces can be calculated from dimensional (scale) analysis starting with the equation of conservation of momentum [15,16]:

$$U_m = \frac{DI}{\pi r_b^2} \sqrt{\frac{3\mu_m r_b}{2\rho L_S}} \tag{7.16}$$

where D is the pool depth, I is the arc current, μ_m is the magnetic permeability of the material used, r_b is the effective radius of an arc or electron beam, ρ is the density of the material, and L_S is a length scale that is taken as substrate thickness. The result shows that the velocity is proportional to both the current and the depth of the molten pool. The magnitude of velocities due to electromagnetic force is generally lower than that from the Marangoni force but higher than that owing to the buoyancy force as shown in Figure 7.17. The relative magnitudes of the velocities that originate from Marangoni, buoyancy, and electromagnetic forces are calculated in worked out example 7.7.

Worked out example 7.5

Calculate the approximate value of the maximum velocity owing to buoyancy force for the following data. Volumetric thermal expansion coefficient: 1×10^{-5}/K, acceleration due to gravity: 9.81 m/s^2, molten pool depth: 1.1 mm, peak temperature: 2000K, solidus temperature: 1693 K, and liquid metal viscosity: 6.3×10^{-2} kg/(m s). The depth of the molten pool is affected by many factors such as the concentration of surface-active elements in steels. How would the depth of the fusion zone affect the contribution of the velocity component driven by the buoyancy force?

Answer:

Using Eq. 7.13, the computed velocity is 17.1 mm/s. As expected, this velocity is much smaller than that attained when the convection is driven by the surface tension gradient.

The computed maximum velocity is proportional to the square root of the depth of the liquid pool (Eq. 7.13). If the depth of penetration of the liquid pool changes, the velocity resulting from the buoyancy force will be affected.

7.5.1.4 Other driving forces

For DED-L and DED-GMA, where a feedstock material is deposited on the molten pool, there is a force generated by the impingement of the liquid droplets or powder particles impacting the molten pool. Because of the high temperature during AM, significant vaporization of alloying elements takes place. The vaporization results in a downward force acting upon the molten pool surface, known as the recoil force. This recoil force often contributes to the convective flow of molten metal. Both the impact force and recoil force are perpendicular to the top surface of the molten pool, causing a radially inward convective flow. Maximum velocities due to these forces cannot be estimated analytically because of the complex nature of these forces. Numerical calculations show that the maximum velocities due to these forces are significantly lower than those due to the Marangoni force for the most commonly used AM process conditions. Although the Marangoni force is the primary driving force for the convective flow of liquid metal, all forces act simultaneously, and the resulting convective flow affects the temperature distribution and fusion zone geometry.

Worked out example 7.6

Evaluate the relative magnitude of surface tension force to buoyancy force for the following data. Density: 7800 kg/m^3, fusion zone depth: 3.0 mm, fusion zone half-width: 4.0 mm, peak temperature: 2000K, solidus temperature: 1693 K, viscosity: 6.3×10^{-2} kg/m s, temperature coefficient of surface tension gradient: -0.5×10^{-3} N/m K, volumetric thermal expansion coefficient: 1×10^{-5}/K, and acceleration due to gravity: 9.81 m/s^2.

Answer:

The relative magnitude of the surface tension force to buoyancy force can be obtained from the ratio of the surface tension Reynolds number to the Grashof number. Using the data provided, we first compute the surface tension Reynolds number using Eq. 7.12 (= 1210). In addition, the computed value of the Grashof number is 12.5 which implies that the buoyancy force is much higher than the viscous force. The relative magnitude of surface tension force (Marangoni) to buoyancy force is given by 1210/12.5 = 96.8. The large value of the ratio indicates the dominance of the Marangoni force over the buoyancy force (Figure 7.17).

Worked out example 7.7

Compare the maximum velocities of molten metal driven by surface tension (Marangoni force), buoyancy, and electromagnetic forces when each of these forces is considered to act individually from the following data. Power: 500W, speed: 10 mm/s, thermal conductivity (k_s): 8 W/m K, density (ρ): 7200 kg/m^3, specific heat (C_p): 611 J/kg K, thermal diffusivity (α_s): 6.36×10^{-6} m^2/s, the fraction of the energy absorbed (η): 0.9, initial temperature (T_0): 298 K, solidus temperature (T_S): 1693 K, acceleration due to gravity (g): 9.81 m/s^2, viscosity (μ): 7×10^{-2} kg/m s, volumetric expansion coefficient (β): 3.5×10^{-5}/K, temperature coefficient of surface tension $\left(\dfrac{d\gamma}{dT}\right)$: -0.3×10^{-3} N/m K, arc current (I): 100 A, magnetic permeability (μ_m): 1.256×10^{-6} N/A^2, effective radius of the arc (r_b): 2 mm, and substrate thickness (L_S): 10 mm. The fusion zone depth and width can be found in Table 7.2. as 1.24 mm and 2.48 mm, respectively.

Answer:

The maximum velocity of the molten metal driven by the surface tension (Marangoni force), buoyancy force, and electromagnetic force can be calculated using Eqs. 7.11, 7.13, and 7.16, respectively. For the calculations, maximum temperature, pool depth, and width are needed. Eq. 7.5 can be used to calculate the thermal cycles at the deposit centerline ($y = 0$ m) on the top surface ($z = 0$ m). The maximum temperature (T_m) of the cycle is the peak temperature which is equal to 2857 K. Molten pool depth is equal to the "z" value in Eq. 7.5 for which the maximum temperature at the deposit centerline ($y = 0$ m) is equal to the solidus temperature of the alloy (1693 K for stainless steel 316). The maximum velocity of molten metal driven by surface tension force is calculated using Eq. 7.11 as 2.49 m/s. The maximum velocity driven by buoyancy force is calculated using Eq. 7.13 as 0.022 m/s. The corresponding velocity driven by the electromagnetic force is calculated using Eq. 7.16 as 0.071 m/s. Therefore, the surface tension force (Marangoni) results in the largest velocity, followed by the electromagnetic force, and buoyancy force, as is also evident from Figure 7.17.

7.5.2 Heat conduction vs. convection in the molten pool – Peclet number

The Peclet number indicates the relative importance of heat transfer by convection and conduction. Peclet number (Pe) is given by:

$$Pe = \frac{\rho C_p U L}{k} \qquad (7.17)$$

ρ = Density of liquid metal, kg/m^3
C_p = Specific heat of liquid metal, J/kg K
k = Thermal conductivity of liquid metal, W/m K
U = Characteristic velocity, m/s
L = Characteristic length, often taken as molten pool half-width, m

Convective heat transfer dominates if the Peclet number is much greater than unity. When the Peclet number is less than unity, the convec-

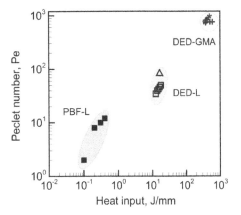

Figure 7.18 Peclet numbers for different AM processes. The plot is for stainless steel 316. Here, heat input represents heat source power/scanning speed. The plot is adapted from [9] with the permission from Elsevier.

tive flow does not affect the molten pool geometry in a major way. In most AM processes, the Peclet number is higher than unity, as shown in Figure 7.18. Therefore, in all AM processes, convection is the dominant heat transfer mechanism inside the molten pool and it significantly affects the deposit geometry. However, the conduction of heat in the solid region is affected by the thermal conductivity of the alloy. Also, at a very high scanning speed, the molten pool solidifies rapidly before the convective flow can mix the hot and cold liquids. In those conditions, heat conduction has more influence on heat transfer. The relative importance of heat transfer by conduction and convection inside the molten pool is estimated in worked out example 7.8.

Worked out example 7.8

Evaluate the relative importance of heat transfer by convection and conduction inside the molten pools for the DED-L of stainless steel 316 deposited using two process conditions. While for depositing track "A," the maximum velocity of the liquid and the half-width of the molten pool are 100 mm/s and 2 mm, respectively, the corresponding values for track "B" are 200 mm/s and 3 mm, respectively. Data: thermal conductivity (k): 28 W/m K, density (ρ): 7200 kg/m^3, and specific heat (C_p): 611 J/kg K.

Answer:

The relative importance of the heat transfer by convection and conduction inside the molten pool can be quantified by the Peclet number (Eq. 7.17). Using Eq. 7.17 and the data provided here, the Peclet number for track "A" can be calculated as 31.4. Similarly, the Peclet number for track "B" is 94.2. Since Pe $\gg$ 1, convection is the dominant mechanism of heat transfer inside the molten pool. For the two tracks, Pe_A for deposit "A" and Pe_B for deposit "B" can be compared for the deposition of stainless steel 316.

$$\frac{Pe_A}{Pe_B} = \frac{31.4}{94.2} = \frac{1}{3}$$

The Peclet number for deposit "B" is higher than that for deposit "A." Therefore, the effect of convective heat transfer inside the molten pool is more pronounced for deposit "B."

7.5.3 Effects of liquid metal convection on fusion zone geometry

Fusion zone geometry is often largely affected by the direction and magnitude of the convective flow which is primarily driven by the Marangoni force. The direction of convective flow inside the molten pool depends on the temperature coefficient of surface tension $\left(\dfrac{d\gamma}{dT}\right)$. Negative values of $\dfrac{d\gamma}{dT}$ mean higher surface tension at lower temperatures and vice versa. In this case, the liquid metal on the surface of the molten pool flows from the middle to the periphery. The direction of flow is reversed for a positive value of $\dfrac{d\gamma}{dT}$. For most of the commonly used engineering alloys, a negative value of $\dfrac{d\gamma}{dT}$ results in a radially outward flow on the top surface from directly under the heat source to the periphery of the molten pool (Figure 7.17 (a)). However, for steels containing surface-active elements such as sulfur, oxygen, selenium, or tellurium, $\dfrac{d\gamma}{dT}$ can be positive depending on the temperature and concentration of the elements. This is because, for steels with surface-active elements, the surface tension (γ) varies significantly depending on the temperature and concentration of the elements as [17]:

$$\gamma = \gamma_m + A\left(T - T_m\right) - RT\Gamma_s \log\left(1 + KC\right) \tag{7.18}$$

where γ_m is the surface tension of the material without surface active element at melting point (T_m) which is often taken as the solidus temperature. A is the negative of the temperature coefficient of surface tension when the concentration (in weight percent) of surface-active element (C) is zero, T is the local temperature of the molten metal, and K is the adsorption coefficient represented by [17]:

$$K = k_1 exp\left(\dfrac{-\Delta H^0}{RT}\right) \tag{7.19}$$

The other variables in Eqs. 7.18 and 7.19 are explained in Table 7.3. As an example, for stainless steel 316 with sulfur as a surface-active element, the values used in Eqs. 7.18 and 7.19 are provided in Table 7.3. The temperature coefficient of surface tension $\left(\dfrac{d\gamma}{dT}\right)$ can be obtained by differentiating Eq. 7.18 with temperature as [17]:

$$\dfrac{d\gamma}{dT} = A - R\Gamma_s \log\left(1 + KC\right) - \dfrac{KC}{1 + KC}\dfrac{\Gamma_s}{T}\Delta H^0 \tag{7.20}$$

Table 7.3 Data to calculate the temperature coefficient of surface tension from Eqs. 7.18 and 7.19 [17].

Variables	Value
Surface tension of material without sulfur at solidus temperature, γ_m (N m^{-1})	2.0
dγ/dT of material without sulfur, A (N m^{-1} K^{-1})	-3.0×10^{-4}
Surface excess of sulfur at saturation, Γ_s (mol m^{-2})	1.3×10^{-5}
Enthalpy of segregation, ΔH° (J kg^{-1} mol^{-1})	-1.66×10^{5}
Entropy factor, k_1	3.18×10^{-3}
Gas constant, R (J K^{-1} mol^{-1})	8.314

The temperature coefficient of surface tension is negative for stainless steel 316 without any sulfur resulting in a radially outward flow. However, the temperature coefficient of surface tension of stainless steel 316 containing sulfur can be positive depending on temperature and weight percent of sulfur as shown in Figure 7.19. In that case, the liquid metal flows from the periphery to the middle of the molten pool. The calculation of temperature coefficient of surface tension is presented in worked out example 7.9.

The presence of surface-active elements affects the flow pattern of the liquid metal in the molten pool and the shape and size of the molten pool. Radially outward flow results in a wide and shallow pool while a narrow and deep molten pool is formed for a radially inward flow. Numerical simulation [17] can uncover the flow pattern. The cross-sections of laser-welded stainless steel samples with 20 ppm and 150 ppm sulfur using 1900 W and 5200 W of laser powers are shown in Figure 7.20. When the samples are welded using a laser power of 5200 W, the weld containing 150 ppm sulfur exhibits a radially inward flow resulting in a deeper penetration (Figure 7.20 (b)) than that containing 20 ppm sulfur that shows radially outward flow (Figure 7.20 (a)). However, at a laser power of 1900 W, the pool geometries in the two cases are similar, although the flow directions are different (Figure 7.20 (d)). The concentration of sulfur may or may not affect the molten pool geometry depending on process conditions. The reasons for this behavior are explained below.

The role of sulfur concentration on deposit geometry can be explained based on the Peclet number calculations (Eq. 7.17). At a laser power of 1900 W, the maximum values of Peclet number for the steels with 20 and 150 ppm sulfur are 0.18 and 0.91, respectively. These low Peclet numbers (<1) indicate that heat transfer by conduction is more important than by convection. As a result, the direction of fluid flow is not important in determining the fusion zone shape and size since most of the heat is transported by conduction. Therefore, there is no significant difference between the molten pool geometries for steels containing 20 and 150 ppm sulfur at a laser power of 1900 W. In contrast, at a laser power of 5200 W, the calculated Peclet numbers are large (>200), making convective heat transport the primary mechanism of heat transfer inside the molten pool. The convective flow field and its direction significantly affect the molten pool geometry. Only when the convective heat transfer is the main mechanism of heat transfer (Section 5.2) can the surface-active elements play an important role in controlling the fusion zone geometry. The role of laser power on the depth/width ratio of the molten pool of steel containing sulfur is explained in worked out example 7.10.

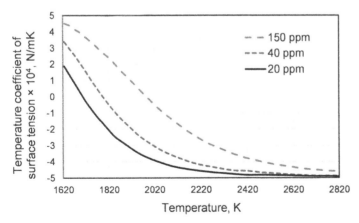

Figure 7.19 Variations in the temperature coefficient of surface tension of stainless steel 316 with sulfur as a surface-active element with temperature and sulfur concentration (in ppm) calculated using Eq. 7.20 using the data in Table 7.3. (*Source:* T. Mukherjee and T. DebRoy).

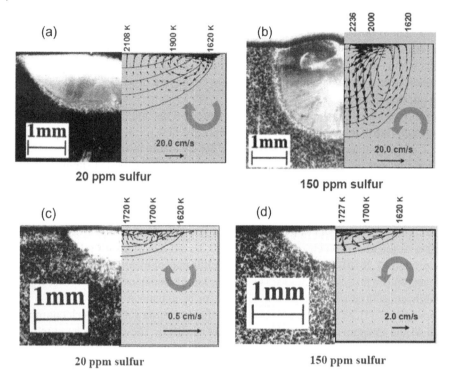

Figure 7.20 Effect of a surface-active element, sulfur, on the convective flow of molten metal and fusion zone shape and size during laser spot welding of stainless steel using 5200 W laser power and (a) 20 ppm and (b) 150 ppm of sulfur and 1900 W laser power and (c) 20 ppm and (d) 150 ppm of sulfur. Numerical calculation shows the convective flow pattern and predicted fusion zone shape and size agree with experiments for all four cases. These results were generated by W. Pitscheneder and T. DebRoy.

Apart from the surface-active elements, the orientation of the melt pool also plays an important role in the convective flow of liquid metal and the fusion zone geometry [18]. In AM, the molten pool does not always maintain a horizontal orientation. The effects of part orientation on fluid flow and fusion zone geometry are well-explained in the fusion welding literature. For example, Figure 7.21 shows fillet welding of two configurations, V-shaped and L-shaped joints. A numerical model of arc welding was used to calculate the temperature and velocity fields. It can be seen that the free surface of the molten pool deforms differently in the two cases. In addition, the fluid flow pattern varies significantly for the two joints resulting in a striking difference in molten pool shape and size. Therefore, a better understanding of the effects of part configuration on the fluid flow and molten pool geometry is needed to control the part geometry.

Worked out example 7.9

A part is made using DED-L of stainless steel 316 containing 150 ppm of sulfur using 500 W laser power and 15 mm/s scanning speed. Calculate the value of the temperature coefficient of surface tension corresponding to the peak temperature. Determine the approximate location where the maximum surface tension is attained. Data: thermal conductivity (k_s): 28 W/m K, density (ρ): 7200 kg/m³, specific heat (C_p): 611 J/kg K, and thermal diffusivity (α_s): 6.36 × 10⁻⁶ m²/s, solidus temperature: 1693 K, fraction of the energy absorbed (η): 0.9, and the initial temperature (T_0): 298 K.

Worked out example 7.9 (Continued)

Answer:

The peak temperature calculated in Table 7.2 for 500 W laser power and 15 mm/s scanning speed in a stainless steel 316 specimen is 2004 K. At this temperature, the temperature coefficient of surface tension can be calculated using Eq. 7.20 where the data needed for the calculations are available in Table 7.3. The computed value of the temperature coefficient of surface tension corresponding to the peak temperature is 2.1×10^{-4} N/mK.

The variation of surface tension with temperature for stainless steel 316 with 150 ppm sulfur can be calculated using Eq. 7.18 where the data needed for the calculations are available in Table 7.3. Figure E7.1 below shows the computed variation of surface tension with the temperature. Above the solidus temperature (1693 K), the surface tension increases up to a maximum value and then decreases with the temperature. Therefore, the maximum value of the surface tension is observed at around 1893 K. Between the solidus temperature and 1893 K, the gradient of surface tension with respect to the temperature is positive.

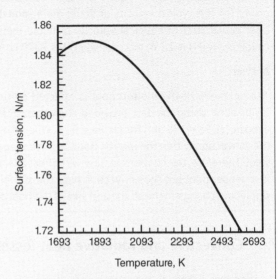

Figure E7.1 Surface tension vs. temperature data for stainless steel 316 with 150 ppm sulfur.

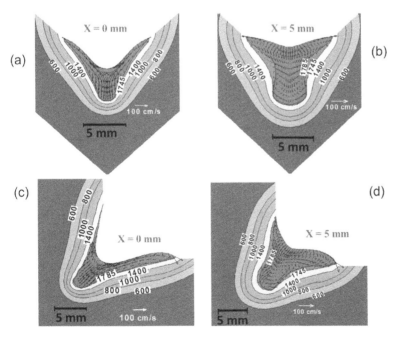

Figure 7.21 Numerically calculated temperature and velocity fields at different cross-sections perpendicular to the weld direction directly under the arc (X = 0 mm), and 5 mm behind the arc (X = 5 mm) are shown during gas-metal arc welding of (a) and (b) symmetrical V-shaped joints and (c) and (d) L-shaped joints. The temperatures are given in Kelvin. All figures are for welding stainless steel at a welding speed of 100 cm/s. These results were obtained from a numerical model of welding developed by A. Kumar and T. DebRoy.

Worked out example 7.10

It is known that the presence of sulfur does not always change the molten pool depth/width ratio (Figure 7.20). Determine which of the following two cases would 150 ppm of sulfur significantly change the molten pool shape as determined by the depth/width ratio. (a) 2000 W laser power, 0.4 m/s typical velocity of liquid metal, and 0.002 m molten pool half-width. (b) 400 W laser power, 0.01 m/s typical velocity of liquid metal and 0.0005 m pool half-width, thermal conductivity (k_s) is 28 W/m K, density (ρ) is 7200 kg/m^3, and specific heat (C_p) is 611 J/kg K.

Answer:

The geometry of the molten pool is affected by the mechanism of heat transfer within the fusion zone. Using the data provided, the values of Pe can be calculated using Eq. 7.17. For the deposit (a), Pe = 125 and for (b), Pe = 0.75. When the Pe $\gg$ 1, the heat is transported mainly by the convection within the molten pool, and the shape and size are affected by the flow of liquid steel. Therefore, the convective flow will affect the molten pool shape and size in (a) resulting in a deeper pool but the convective heat transfer is not pronounced in (b) and as a result, no significant change in the shape and size of the pool owing to the presence of sulfur is expected.

7.6 Convective and radiative heat losses from the part

Heat conducted through the deposit and substrate is ultimately lost to the surrounding atmosphere by convection and radiation from all surfaces of the deposit and substrate. The factors governing convective heat loss include the nature of the shielding gas, its flow rate, and the ambient temperature. Radiation losses are proportional to the fourth power of the absolute temperature of the metal surface. The heat loss by convection and radiation from the surfaces of the deposit and substrate is given by the following:

$$q = h_c\left(T - T_a\right) + \varepsilon\sigma\left(T^4 - T_a^4\right) \tag{7.21}$$

where
q = Heat flux normal to the deposit or substrate surface, W/m^2
h_c = Convective heat transfer coefficient, W/m^2 K
T = Local temperature at the surface of the deposit or substrate, K
T_a = Surrounding temperature, K
ε = Emissivity of the solid material
σ = Stefan-Boltzmann constant, 5.67×10^{-8} W m^{-2} K^{-4}

The first term of the right side of Eq. 7.21 represents heat transfer by convection, while the second term represents radiative heat loss. The temperature for the calculation of the radiative heat loss must be expressed in an absolute temperature scale. Significant variations in the heat loss rates can occur depending on the specific AM conditions. The surrounding temperature (T_a) can be different from the room temperature (around 298 K) in PBF-EB since this process is carried out in a closed chamber where the temperature can vary depending on the accumulation of metal vapor inside the chamber. In DED processes, the convective heat transfer coefficient (h_c) can vary significantly depending on the velocity of the shielding gas. At a low velocity of shielding gas, free convection occurs, while forced convection can be observed at a high velocity of gas where the velocity of the gas can change the value of h_c. For example, a convective heat transfer coefficient of 1.0×10^{-6} W/mm^2 K for free convection and 21.0×10^{-6} W/mm^2 K for forced convection was reported for DED-L of Ti-6Al-4V based on a combination of calculations and experiments. The role of shielding gas on the convective heat transfer is explained in worked out example 7.11.

Worked out example 7.11

A rectangular block of Ti-6Al-4V is to be made using DED-GMA. To maintain a surface temperature of 400 K, a constant rate of heat flux removal of 100 W/m^2 is required. An appropriate shielding gas needs to be selected among four available gases, A, B, C, and D. The convective heat transfer coefficients for natural convection are provided for the four gases in Table E7.2 below. Which shielding gas would comply with the requirement? Assume that the radiative heat loss is negligible and the ambient temperature is 300 K.

Answer:

The convective heat loss can be calculated using Eq. 7.21, which gives $100 = h_c (400 - 300)$. Therefore, the required convective heat transfer coefficient should be 1.0 W/m^{2}K. Therefore, gas "A" should be an appropriate shielding gas.

Table E7.2 The convective heat transfer coefficients for the four gases.

Shielding gas	A	B	C	D
Convective heat transfer coefficients for natural convection (W/m^{2}K)	1.0	2.0	0.5	1.5

7.7 Temperature and velocity fields, thermal cycles, and heating and cooling rates

In AM, temperature, and velocity fields are transient and spatially nonuniform. The peak temperature in the molten pool can be several hundred degrees above the liquidus temperature of the alloy, sometimes as high as the boiling point of the alloy for the keyhole mode. In AM, the measurement of the temperature field is difficult because the heat source moves rapidly, and the temperature field is highly transient. The most commonly used method of measuring temperature is by placing thermocouples at monitoring locations in the solid away from the molten region. However, the thermocouples need to be very thin to avoid significant errors in measurements due to thermocouple inertia and significant temporal and spatial variations of temperature. Also, it is difficult to get a complete temperature field even with multiple thermocouples. Infrared thermography is also used to measure the temperature distribution on the surface of the build. However, this method measures only the surface temperatures and is unable to provide a 3D transient temperature distribution in the interior of the specimen. Therefore, calculations are often used to estimate the temperature profiles and local cooling rates. Here we discuss the variations in temperature and velocity fields and cooling rates for different AM processes, processing conditions, and alloys.

Heat transfer patterns vary widely for different AM processes as described in Section 7.3. Also, the range of processing conditions such as heat input varies significantly for different processes. Therefore, there are significant variations in the resultant temperature and velocity fields, fusion zone geometry, and cooling rates for different AM processes. Numerical models are often used to explain those differences. These models solve the transient equations of conservation of mass, momentum, and energy in 3D. The equations of conservation of mass and momentum can be written as [1]:

$$\frac{\partial u_i}{\partial x_i} = 0 \tag{7.22}$$

$$\rho \frac{\partial u_j}{\partial t} + \rho \frac{\partial \left(u_i \, u_j \right)}{\partial x_i} = \frac{\partial}{\partial x_i} \left(\mu \frac{\partial u_j}{\partial x_i} \right) + S_j \tag{7.23}$$

where ρ and μ are the density and the viscosity of the alloy, respectively, u_i and u_j are the velocity components along the i and j directions, respectively, x_i is the distance along the i direction, and S_j is the source term for the jth momentum conservation equation [9]. The energy conservation equation is written as [1,9]:

$$\frac{\partial h}{\partial t} + \frac{\partial (u_i h)}{\partial x_i} = \frac{\partial}{\partial x_i}\left(\alpha \frac{\partial h}{\partial x_i}\right) - \frac{\partial \Delta H}{\partial t} - \frac{\partial (u_i \Delta H)}{\partial x_i} + S_v \tag{7.24}$$

where h denotes the enthalpy, t is the time, α and ΔH are the thermal diffusivity and the latent heat of fusion of the alloy, respectively, u_i is the velocity components along the i direction, and S_v is the source term of heat generation or supply. The solution domain consists of a substrate, deposited layers and hatches, and shielding gas. The heat source is applied on the top surface of the depositing layer. The heat source is either applied as a boundary condition to the energy conservation equation or considered as the source term in the energy conservation equation (Eq. 7.24). Convective and radiative heat losses are applied to all surfaces of the solution domain as boundary conditions to Eq. 7.24 as discussed in Section 6. Marangoni stress (Eq. 7.10) is applied on the top surface of the molten pool as a boundary condition to the momentum conservation equation (Eq. 7.23) to compute the velocity field of the liquid metal. The buoyancy force is generally applied as a source term in Eq. 7.23. Temperature-dependent thermophysical properties (Table 7.1) of the alloy feedstocks are used in these calculations. The conservation equations of mass, momentum, and energy are generally discretized in the 3D Cartesian coordinate. All discretized equations are solved simultaneously using a numerical method to obtain temperature and velocity fields. Generally, an error threshold value is specified to determine the convergence. The calculation procedure continues until all the hatches and layers are completed. The final computed results reveal important insights about the process.

Figure 7.22 (a-c) shows the experimentally determined and numerically computed size of the molten pool in three printing processes. The molten pool sizes vary widely depending on the heat sources and heat input. The linear heat input (power/speed) in PBF-L is of the order of 0.1 J/mm which results in a very small pool whose dimensions are in micrometers [2]. However, linear heat inputs in DED-L and DED-GMA are in the order of 10 J/mm and 100 J/mm, respectively [2]. Therefore, the molten pool width in DED-GMA is the largest followed by that in DED-L. The pool dimensions in PBF-L are around 10% and 30% of those for DED-GMA and DED-L respectively. Therefore, the molten pool size in DED-GMA is the largest followed by that in DED-L and PBF-L.

Figure 7.23 (a-c) shows the computed temperature and velocity fields in 3D during the printing of stainless steel 316 using PBF-L, DED-L, and DED-GMA, respectively. The shape and size of the molten pool are defined by the solidus temperature (1693 K) isotherm. The velocity vectors for the convective flow are shown by black arrows. The reference vector is used to estimate the magnitudes of velocities inside the pool. The shape of the molten pool varies significantly for the three printing processes depending on the unique features of the processes [19–21]. For example, in DED-GMA, the molten metal near the arc axis is depressed by the arc pressure and the impact force of the impinging droplets. The liquid metal is pushed to the rear part of the molten pool and forms a curved deposit as it solidifies. The impingement of the droplets also results in deep penetration in DED-GMA. However, in DED-L the curved pool surface is formed immediately under the laser beam axis due to the accumulation of powder particles. In contrast, the top surface of the molten pool in PBF-L is flat because of the addition of thin flat layers of powders while printing the component.

Cooling rates vary significantly depending on the process and processing conditions. Table 7.4 compares the cooling rates between the liquidus and solidus temperatures of common manufacturing

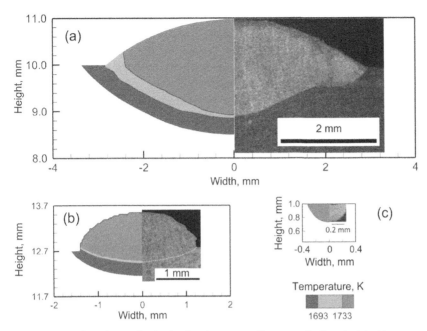

Figure 7.22 Experimentally obtained and corresponding numerically calculated transverse sections of stainless steel 316 deposit for (a) DED-GMA, (b) DED-L, and (c) PBF-L. For figure (a), arc current and arc voltages are 150 A and 14.2 V respectively, resulting in a heat source power of 2130 W. Scanning speed is 10 mm/s. For figure (b), laser power and scanning speed are 1500 W and 10.6 mm/s, respectively. For figure (c), laser power and scanning speed are 110 W and 100 mm/s respectively. These results were generated by T. Mukherjee and T. DebRoy.

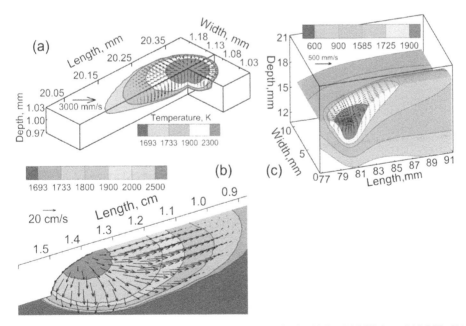

Figure 7.23 Differences in molten pool shape and size for (a) PBF-L (b) DED-L, and (c) DED-GMA of stainless steel 316. For figure (a), arc current and arc voltage are 150 A and 14.2 V respectively, resulting in a heat source power of 2130 W. Scanning speed is 10 mm/s. For figure (b), laser power and scanning speed are 1500 W and 10.6 mm/s respectively. For figure (c), laser power and scanning speed are 110 W and 100 mm/s respectively. These results were obtained from the numerical models of additive manufacturing developed by T. Mukherjee and T. DebRoy.

processes with that for AM. Typically, there are two useful temperature ranges for which cooling rates are most important. First, the cooling rates from the liquidus to solidus temperatures are useful for determining the scale of the solidification microstructure. Second, the thermal history affects the solid-state transformations in alloys. In steels, the cooling rate in the 800 to 500 °C temperature range affects the solid-state transformations and the microstructure.

Different process parameters such as heat source power and scanning speed significantly affect the fusion zone geometry, temperature distribution, and cooling rates. The effects of different process parameters on the peak temperature, fusion zone size, and cooling rate during solidification are summarized in Table 7.5. The relative rates of cooling of stainless steel parts made by DED-GMA, DED-L, and PBF-L are calculated in worked out example 7.12. In addition, their dependence on alloy properties is summarized in Table 7.6. A comparison of the fusion zone depth of parts made by stainless steel 316 and Ti-6Al-4V is presented in worked out example 7.13.

Table 7.4 Comparison of cooling rates of AM with other common manufacturing processes.

Process	Cooling rate [K/s]	Alloy	References
Additive manufacturing processes			
DED-L	$3 \times 10^3 - 7 \times 10^3$	SS 316 deposit	[22]
DED-L	$1 \times 10^3 - 4 \times 10^3$	SS 316 deposit	[23]
DED-L	$5 \times 10^3 - 3 \times 10^4$	IN 718 deposit	[24]
DED-GMA	$10^2 - 10^3$	Ti-6Al-4V	[25]
PBF-EB	5×10^4	IN 718 deposit	[26]
PBF-EB	5×10^4	Ti-6Al-4V	[27]
PBF-L	5×10^5	Ti-6Al-4V	[28]
PBF-L	$1 \times 10^6 - 6 \times 10^6$	Al alloy deposit	[29]
Common manufacturing processes			
Casting	$10^0 - 10^2$		[30]
Arc welding	$10^1 - 10^3$		
E-beam welding	$10^2 - 10^4$		
Laser welding	$10^2 - 10^6$		

Note: The cooling rates are either specified in the literature or estimated from the thermal cycles provided. All cooling rates specified here are measured or calculated between the liquidus and the solidus temperatures of the alloy.

Table 7.5 Effects of process parameters on peak temperature, fusion zone size, and cooling rate.

Variables	Peak temperature	Fusion zone size	The cooling rate during solidification
Heat source power	High power produces high peak temperature	High power increases fusion zone size	High power creates a big molten pool that cools slowly during solidification
Scanning speed	High scanning speed decreases peak temperature	High scanning speed results in a small fusion zone size	High scanning speed results in a small molten pool that cools rapidly
Spot size	Large spot size reduces power density; decreases peak temperature	Large spot reduces power density and decreases fusion zone size	A large spot reduces power density and creates a small molten pool that cools rapidly during solidification

Table 7.5 (Continued)

Variables	Peak temperature	Fusion zone size	The cooling rate during solidification
Hatch spacing	A large hatch spacing increases the peak temperature	A large hatch spacing increases the fusion zone size	A large hatch spacing creates a big molten pool that cools slowly during solidification
Substrate thickness	A thick substrate absorbs heat easily and reduces the peak temperature	A thick substrate absorbs heat easily and reduces fusion zone size	A thick substrate absorbs heat easily and increases the cooling rate

Note: The vertical displacement of the powder bed platform controls the layer thickness. This parameter can be adjusted during experiments. In addition, the powder or wire feed rate in the DED process can also be independently controlled. However, there are no clear and universal trends of the effect of these variables on the peak temperature, fusion zone size, and cooling rates.

Worked out example 7.12

Calculate the cooling rate at 800 K on the top surface at the deposit centerline of stainless steel 316 for the following three cases from the data provided below. (a) Process: DED-GMA, arc power: 2130 W and scanning speed: 10 mm/s. (b) Process: DED-L, laser power: 1500 W, and scanning speed: 10.6 mm/s. (c) Process: PBF-L, laser power: 110 W, and scanning speed: 100 mm/s. Ambient temperature: 298 K, and the fraction of energy absorbed: 0.9. Thermal conductivity (k_s): 28 W/m K, density (ρ): 7200 kg/m^3, specific heat (C_p): 611 J/kg K, and thermal diffusivity (α_s): 6.36 × 10^{-6} m^2/s. Explain why the cooling rates for the three cases are so different.

Answer:

The cooling rate on the top surface at the deposit centerline can be calculated using Eq. 7.7 where the temperature (T_c) is 800 K. The computed cooling rates for DED-GMA, DED-L, and PBF-L are 231 K/s, 348 K/s, and 44,759 K/s, respectively. Smaller pools cool at a faster rate. The molten pool in PBF-L is the smallest among the three processes (Figure 7.22) resulting in the fastest rate of cooling. This wide variation in cooling rates significantly affects the resulting microstructure and properties.

Table 7.6 Effects of alloy properties on peak temperature, fusion zone size, and cooling rate.

Properties	Peak temperature	Fusion zone size	The cooling rate during solidification
Density	High-density results in low peak temperature for a given heat input	High-density exhibits a small fusion zone	High-density results in a small molten pool that cools rapidly
Specific heat	Alloys with high specific heat exhibit low peak temperature	Alloys with a high specific heat exhibit a small fusion zone	Alloys with high specific heat create small molten pools that cool rapidly
Thermal conductivity	Alloys with high conductivity dissipate heat rapidly and reduce the temperature	Alloys with high conductivity dissipate heat rapidly and create a small molten pool	Alloys with high conductivity dissipate heat quickly and experience a high cooling rate
Latent heat of fusion	Alloys with a high latent heat exhibit low peak temperature	Alloys with a high latent heat need more heat to melt and create a small fusion zone	Alloys with a high latent heat create a small molten pool that cools rapidly

Worked out example 7.13

Parts of stainless steel 316 and Ti-6Al-4V are manufactured by directed energy deposition using 500 W laser power and 10 mm/s scanning speed. The ambient temperature is 298 K. Calculate the peak temperatures, fusion zone penetration depths, and cooling rates at 800 K and 1000 K for both deposits. Which alloy will result in higher penetration depth and why? Which alloy will exhibit a faster cooling rate and why?

Answer:

Eq. 7.5 can be used to calculate the thermal cycles at the deposit centerline ($y = 0$ m) on the top surface ($z = 0$ m). The maximum temperature of the thermal cycle is the peak temperature. Fusion zone depth is equal to the "z" value in Eq. 7.5 for which the temperature at the deposit centerline ($y = 0$ m) is equal to the solidus temperature of the alloy (1693 K for stainless steel 316 and 1878 K for Ti-6Al-4V). Eq. 7.7 can be used to calculate the cooling rates. Table E7.3 below shows the results. Constant thermophysical properties at room temperature are taken for simplicity.

Table E7.3   Calculated results showing the values of cooling rates for the two alloys.

Process parameters and alloy properties used in Eqs. 7.5 and 7.7		Alloy	Alloy properties		Peak temperature	Fusion zone penetration depth	T_C in Eq. 7.7	Cooling rate on the top surface at the deposit centerline
P	500 W	Stainless steel 316	k_s	28 W/m K	2857 K	1.24 mm	800 K	985 K/s
			ρ	7200 kg/m^3				
V	10 mm/s		C_p	611 J/kg K			1000 K	1926 K/s
			α_s	6.36×10^{-6} m^2/s				
T_0	298 K	Ti-6Al-4V	k_s	30 W/m K	2686 K	1.50 mm	800 K	1055 K/s
			ρ	4000 kg/m^3				
η	0.9		C_p	539 J/kg K			1000 K	2063 K/s
			α_s	13.9×10^{-6} m^2/s				

The lower density of Ti-6Al-4V results in a larger molten pool and higher fusion zone penetration depth. Higher thermal diffusivity of Ti-6Al-4V results in rapid heat conduction and thus faster cooling rates than those for stainless steel 316.

7.8   Emerging numerical simulations of heat transfer in additive manufacturing

The simplified calculations provided in this chapter are helpful to compare approximate values of thermal cycles, maximum velocity, fusion zone dimensions, and cooling rates for different process conditions and alloys. However, they are incapable of correctly estimating the evolution of 3D, transient temperature fields for different AM processes. They ignore convective heat transfer and are useful for understanding trends. Accurate estimation requires the solutions of the equations of conservation of mass, momentum, and energy in 3D transient form. These equations are complex second-order partial differential equations and may only be solved numerically. Several numerical models (see Chapter 12) are being developed worldwide including commercial packages that consider both heat transfer by conduction and convection to numerically simulate heat transfer in AM.

Numerical modeling starts with generating the geometry models for the computational domain consisting of the deposit, feedstock, powder bed for PBF processes, support structures (if any), and substrate. The computational domain is subsequently divided into small cells where a heat source, material properties, and convective and radiative heat losses are applied. The conservation equations are solved iteratively in small time steps. If the temperature field is calculated to estimate residual stresses and distortions for the entire component in subsequent steps, the effect of convective flow inside the liquid pool is often neglected to reduce the calculation time. This is done by solving only the heat conduction equation. However, convective heat transfer calculations are required for the accurate estimation of molten pool geometry, cooling rates during solidification, and peak temperatures inside the molten pool.

Although numerical models are being increasingly used in research and development, they are still not widely used. These calculations are computationally intensive, slow, and unsuitable for real-time applications. Also, the calculations are complex and often require extensive training to use them. Furthermore, the development of numerical models requires knowledge of the AM process and considerable experience in numerical fluid dynamics and heat transfer.

7.9 Summary

Once the energy from a heat source such as a laser or electron beam or an electric arc reaches the feedstock, a molten pool forms quickly and heat is transported to the neighboring regions. The transient temperature field provides important insights into the shape and size of the fusion zone and the heat-affected zone, the solidification rate, the local cooling rates, and the development of the microstructure, properties, and defects. Simple analytical heat conduction equations are often used to estimate the transient temperature fields. Since these equations ignore convective heat transfer, they should be used to estimate trends of important parameters such as the peak temperature but not their absolute values.

Convective heat transfer is often the main mechanism of heat transfer in the fusion zone. Nondimensional numbers are useful for estimating the mechanism of heat transfer in the fusion zone and the relative magnitudes of different driving forces for the flow of liquid metal. With the advancements in computational hardware and software, emerging numerical techniques are being increasingly used to simulate heat transfer in AM. They are useful to understand and control fusion and heat-affected zone geometries, cooling rates, microstructures, and defects. Numerical modeling of heat transfer and fluid flow in AM is a prerequisite for an improved understanding of the AM processes.

Takeaways

Heat sources

- Commonly used heat sources in additive manufacturing include laser beams, electron beams, and electric or plasma arcs.
- The power density of commonly used heat sources follows a Gaussian distribution.
- A heat source is primarily characterized by its radius and power density distribution.

Heat absorption by the feedstock

- In powder bed fusion, heat is absorbed by multiple reflections of the beam among the powder particles.
- In directed energy deposition, powders absorb heat both during their flight from the nozzle to the substrate as well as after they reach the substrate.
- Heat absorption depends on feedstock shape and size, heat source characteristics such as spot radius, power distribution factor, and the shielding gas environment.

(Continued)

Takeaways   (Continued)

Heat conduction through the deposit and substrate

- Heat transfer mechanisms are affected by temperature-dependent thermophysical properties of alloys such as thermal conductivity, specific heat, and density.
- In powder bed fusion, heat conduction is affected by the difference in thermophysical properties of the powder bed, solid deposit, and substrate.
- In both powder bed fusion and directed energy deposition processes, the direction and rate of heat conduction vary while depositing different layers and hatches during the process.
- Back-of-the-envelope calculations are useful to approximately estimate the thermal cycles, fusion zone geometry, and cooling rates in additive manufacturing processes.
- The Fourier number indicates the relative measure of heat dissipation rate to heat storage rate. A high Fourier number indicates rapid heat conduction.

Convective heat transfer within the molten pool

- Fluid flow inside the molten pool is driven by the Marangoni, electromagnetic, and buoyancy forces for arc and e-beam additive manufacturing and droplet or powder impingement. If a laser is used as a heat source, both Marangoni and buoyancy forces affect the convection pattern.
- Force generated due to spatial gradient of surface tension, i.e., Marangoni force is the most dominant driver of convection pattern inside the molten pool in most cases.
- For most alloys, the surface tension decreases with temperature causing a radially outward flow from the middle of the fusion zone to its periphery. This flow pattern results in wide and shallow fusion zone geometry.
- Surface active elements, e.g. sulfur and oxygen in steel change the temperature coefficient of surface tension to positive in certain temperature range and compositions causing a radially inward flow and deep and narrow fusion zone geometry.
- The role of the Marangoni force on convective flow is quantified by the Marangoni number.
- The Peclet number indicates the relative importance of heat transfer by convection and conduction. Convection is often the dominant mechanism of heat transfer within the melt pool as indicated by a Peclet number much higher than 1.

Convective and radiative heat losses from the part

- The convective heat loss is affected by the nature of the shielding gas, its flow rate, and the ambient temperature.
- The radiative heat loss depends on the emissivity of the surface, the temperature distribution on the surface, and the ambient temperature.
- Both the convective and radiative heat losses from the surfaces of the deposit and substrate affect the temperature distribution in the part during the additive manufacturing process.

Temperature and velocity fields, thermal cycles, and heating and cooling rates

- For the same alloy, the fusion zone shape and size can vary significantly depending on the specific additive manufacturing processes used.
- In additive manufacturing, peak temperature and cooling rate can vary significantly depending on the printing process, process parameters, and the alloy used.
- Cooling rates during solidification can vary 10,000-fold depending on the AM process and process parameters. The variation significantly affects microstructure and properties.

Appendix – Meanings of a selection of important technical terms

Conduction mode: A heat transfer mode in fusion welding and additive manufacturing when the maximum temperature inside the molten pool is significantly below the boiling point of the material used.

Electric arc: An arc generated using an electrode in additive manufacturing.

Electron beam: A focused beam of electrons used as a heat source in additive manufacturing.

Fourier number: A dimensionless number representing the ratio of the rate of heat dissipation to the rate of heat storage.

Keyhole: A narrow, deep region inside the molten pool filled with metal vapors that form when the energy density of the heat source is higher than a threshold value.

Laser beam: A focused beam of laser used as a heat source in additive manufacturing. The acronym laser stands for "light amplification by stimulated emission of radiation."

Marangoni number: A dimensionless number that indicates the importance of convective flow driven by the Marangoni force or surface tension force.

Peak temperature: Maximum temperature inside the molten pool.

Peclet number: A dimensionless number to indicate the relative importance of heat conduction and convection inside the molten pool.

Thermal cycles: Variation of the temperature with time at a monitoring location.

Practice problems

1) A heat source of a circular cross-section delivers the maximum power density at its center. How much change in the maximum power density will occur when the radius of the heat source is doubled for a constant heat source power and power distribution factor?

2) In the DED-L of a titanium alloy, how will the maximum temperature on the top surface of the deposit change if the laser power is doubled and the other process parameters are kept constant?

3) The thermal conductivity of alloy "A" is twice that of alloy B. Two identical deposits are made with the two alloys under the same processing conditions. What is the relation between the cooling rates of the two alloys on the top surface at the deposit centerline at 800 K?

4) An alloy with thermal conductivity of 30 W/mK is printed using 500 W laser power and 10 mm/s scanning speed. Absorptivity is 0.9 and the initial temperature of the substrate is 298 K. Show how the cooling rate on the top surface at the deposit centerline varies with temperature.

5) What is the relation between the cooling rate on the top surface at the deposit centerline and the Fourier number during AM for different heat source powers?

6) Compare the cooling rates of SS316 and Ti-6Al-4V alloys at 2000K for the following processing conditions: laser power 300 W, scanning speed 10 mm/s, and make reasonable assumptions about other parameters. The thermophysical properties are available in Table 7.1 as a function of temperature. However, you may assume reasonable constant values of thermophysical properties for this calculation.

7) The lack of fusion defect formation is often calculated using the dimensions of the fusion zone, the layer thickness, and hatch spacing based on geometric consideration. Using constant values of thermophysical properties guided by the values in Table 7.1, compare the fusion zone dimensions of AlSi10Mg and SS316 for 300 W laser power, 10 mm/s scanning speed. Make reasonable assumptions for other processing parameters.

8) How do the Peclet number and Marangoni number vary with heat source power when the other process parameters are kept constant?

9) PBF generally involves rapid scanning of a laser or electron beam to melt thin layers of powders. In contrast, the DED process generally uses a low scanning speed of the heat source. If parts are made from the same alloy, which process provides a faster cooling rate during solidification?

10) In wire-arc additive manufacturing, metallic wires are used as feedstocks that are melted using a high-energy electric arc. What are the forces that drive the convective flow of liquid metal inside the molten pool?

11) The density and thermophysical properties of aluminum alloy and stainless steel are presented in Table 7.1. Under the same processing conditions, which alloys would result in a larger fusion zone? Explain your answer.

12) How does the cooling rate during the solidification vary along the depth of the fusion zone? Justify your answer mathematically.

13) If the temperature of a surface is doubled, how much change in the radiative heat loss would you expect and why?

14) Metal powders often absorb oxygen from the atmosphere during PBF-L. Unmelted powders are often recycled to save money. How would the recycled powders with absorbed oxygen affect the convective flow inside the molten pool?

15) In additive manufacturing, a part is made by depositing multiple layers. How does the molten pool size vary as the deposition progresses? Explain your answer.

References

1 DebRoy, T., Wei, H.L., Zuback, J.S., Mukherjee, T., Elmer, J.W., Milewski, J.O., Beese, A.M., Wilson-Heid, A.D., De, A. and Zhang, W., 2018. Additive manufacturing of metallic components–process, structure and properties. *Progress in Materials Science*, 92, pp.112–224.

2 Mukherjee, T. and DebRoy, T., 2019. Printability of 316 stainless steel. *Science and Technology of Welding and Joining*, 24(5), pp.412–419.

3 Palmer, T.A. and Elmer, J.W., 2007. Characterisation of electron beams at different focus settings and work distances in multiple welders using the enhanced modified Faraday cup. *Science and Technology of Welding and Joining*, 12(2), pp.161–174.

4 Tsai, N.S. and Eagar, T.W., 1985. Distribution of the heat and current fluxes in gas tungsten arcs. *Metallurgical Transactions B*, 16(4), pp.841–846.

5 Wischeropp, T.M., Tarhini, H. and Emmelmann, C., 2020. Influence of laser beam profile on the selective laser melting process of AlSi10Mg. *Journal of Laser Applications*, 32(2), article no.022059.

6 Xie, J., Kar, A., Rothenflue, J.A. and Latham, W.P., 1997. Temperature-dependent absorptivity and cutting capability of CO2, Nd: YAG and chemical oxygen–iodine lasers. *Journal of Laser Applications*, 9(2), pp.77–85.

7 Bergström, D., 2008. The absorption of laser light by rough metal surfaces (Doctoral dissertation, Luleå tekniska universitet).

8 Wang, X.C., Laoui, T., Bonse, J., Kruth, J.P., Lauwers, B. and Froyen, L., 2002. Direct selective laser sintering of hard metal powders: experimental study and simulation. *The International Journal of Advanced Manufacturing Technology*, 19(5), pp.351–357.

9 Wei, H.L., Mukherjee, T., Zhang, W., Zuback, J.S., Knapp, G.L., De, A. and DebRoy, T., 2021. Mechanistic models for additive manufacturing of metallic components. *Progress in Materials Science*, 116, article no.100703.

10 King, W.E., Barth, H.D., Castillo, V.M., Gallegos, G.F., Gibbs, J.W., Hahn, D.E., Kamath, C. and Rubenchik, A.M., 2014. Observation of keyhole-mode laser melting in laser powder-bed fusion additive manufacturing. *Journal of Materials Processing Technology*, 214(12), pp.2915–2925.

11 Mukherjee, T. and DebRoy, T., 2020. Control of asymmetric track geometry in printed parts of stainless steels, nickel, titanium and aluminum alloys. *Computational Materials Science*, 182, article no.109791.

12 Grong, O., 1997. *Metallurgical Modelling of Welding*. Institute of Materials, London, UK.

13 Messler, R.W., Jr, 2008. *Principles of Welding: Processes, Physics, Chemistry, and Metallurgy*. John Wiley & Sons.

14 Kou, S. and Sun, D.K., 1985. Fluid flow and weld penetration in stationary arc welds. *Metallurgical Transactions A*, 16(1), pp.203–213.

15 Oreper, G.M. and Szekely, J., 1987. A comprehensive representation of transient. *Metallurgical Transactions A*, 18(7), pp.1325–1332.

16 Kanouff, M. and Greif, R., 1992. The unsteady development of a GTA weld pool. *International Journal of Heat and Mass Transfer*, 35(4), pp.967–979.

17 Pitscheneder, W., Debroy, T., Mundra, K. and Ebner, R., 1996. Role of sulfur and processing variables on the temporal evolution of weld pool geometry during multikilowatt laser beam welding of steels. *Welding Journal*, 75(3), pp.71s–80s.

18 Kumar, A. and DebRoy, T., 2007. Heat transfer and fluid flow during gas-metal-arc fillet welding for various joint configurations and welding positions. *Metallurgical and Materials Transactions A*, 38(3), pp.506–519.

19 Mukherjee, T. and DebRoy, T., 2018. Mitigation of lack of fusion defects in powder bed fusion additive manufacturing. *Journal of Manufacturing Processes*, 36, pp.442–449.

20 Knapp, G.L., Mukherjee, T., Zuback, J.S., Wei, H.L., Palmer, T.A., De, A. and DebRoy, T.J.A.M., 2017. Building blocks for a digital twin of additive manufacturing. *Acta Materialia*, 135, pp.390–399.

21 Ou, W., Mukherjee, T., Knapp, G.L., Wei, Y., and DebRoy, T., 2018. Fusion zone geometries, cooling rates and solidification parameters during wire arc additive manufacturing. *International Journal of Heat and Mass Transfer*, 127, pp.1084–1094.

22 Manvatkar, V., De, A., and DebRoy, T., 2014. Heat transfer and material flow during laser assisted multi-layer additive manufacturing. *Journal of Applied Physics*, 116(12), article no.124905.

23 Mukherjee, T., Manvatkar, V., De, A., and DebRoy, T., 2017. Dimensionless numbers in additive manufacturing. *Journal of Applied Physics*, 121(6), article no.064904.

24 Kistler, N.A., Nassar, A.R., Reutzel, E.W., Corbin, D.J., and Beese, A.M., 2017. Effect of directed energy deposition processing parameters on laser deposited Inconel® 718: microstructure, fusion zone morphology, and hardness. *Journal of Laser Applications*, 29(2), article no.022005.

25 Antonysamy, A.A., 2012. *Microstructure, texture and mechanical property evolution during additive manufacturing of Ti6Al4V alloy for aerospace applications*. The University of Manchester (United Kingdom).

26 Sames, W.J., Unocic, K.A., Dehoff, R.R., Lolla, T. and Babu, S.S., 2014. Thermal effects on microstructural heterogeneity of Inconel 718 materials fabricated by electron beam melting. *Journal of Materials Research*, 29(17), pp.1920–1930.

27 Shen, N. and Chou, K., 2012. Numerical thermal analysis in electron beam additive manufacturing with preheating effects. In *2012 International Solid Freeform Fabrication Symposium*. University of Texas at Austin. pp.774–784.

28 Roberts, I.A., Wang, C.J., Esterlein, R., Stanford, M., and Mynors, D.J., 2009. A three-dimensional finite element analysis of the temperature field during laser melting of metal powders in additive layer manufacturing. *International Journal of Machine Tools and Manufacture*, 49(12–13), pp.916–923.

29 Li, Y. and Gu, D., 2014. Parametric analysis of thermal behavior during selective laser melting additive manufacturing of aluminum alloy powder. *Materials & Design*, 63, pp.856–867.

30 Elmer, J.W., Allen, S.M. and Eagar, T.W., 1989. Microstructural development during solidification of stainless steel alloys. *Metallurgical Transactions A*, 20(10), pp.2117–2131.

8

Microstructure and Its Control

Learning objectives

After reading this chapter the reader should be able to do the following:

1) Understand different mechanisms of nucleation and factors affecting them.
2) Know solidification morphologies of additively manufactured parts for different processing conditions.
3) Understand grain orientation and grain size for different alloys and processing conditions.
4) Estimate secondary dendritic arm spacing for different processing conditions.
5) Understand the phases present in the additively manufactured components of common alloys.
6) Select and apply microstructure control techniques.
7) Appreciate various microstructure characterization technique for additively manufactured metallic parts.

CONTENTS

Theory and Practice of Additive Manufacturing, First Edition. Tuhin Mukherjee and Tarasankar DebRoy.
© 2024 John Wiley & Sons, Inc. Published 2024 by John Wiley & Sons, Inc.

8.1 Introduction

Common engineering alloys such as steels, titanium, nickel, aluminum, and copper alloys form different grain structures and phases during additive manufacturing. Most parts have polycrystalline structures although single-crystal components of some alloys have also been printed in laboratories. In polycrystalline components, the orientation of the crystals changes at the boundaries of the two adjacent grains. The grains have different shapes, sizes, and orientations. The primary and secondary phases, grains, and their shapes, sizes, and crystallographic orientations and defects define the microstructure of a component that affects the properties and serviceability of the part.

In additive manufacturing (AM), a rapidly moving, intense, high-energy heat source melts the feedstock to form a molten pool (Figure 8.1). The evolution of microstructure starts with the nucleation of solids from the liquid. From the nucleating particles, grains of different crystallographic orientations begin to grow until the liquid transforms into one or more solids. These grains are of different shapes, sizes, and orientations which define the grain structure. Temperature variation in the solid region, often due to deposition of multiple layers, results in a change in shape and size of grains, precipitation of secondary phases such as metal carbides and nitrides, and solid-state phase transformation, such as austenite to ferrite transformation in steels. All of these features of the microstructure affect the properties of the parts. Characterization of these features requires sophisticated equipment to reveal them. The well-known characterization techniques of the microstructure of AM parts include optical microscopy, scanning electron microscopy (SEM), transmission electron microscopy (TEM), X-Ray

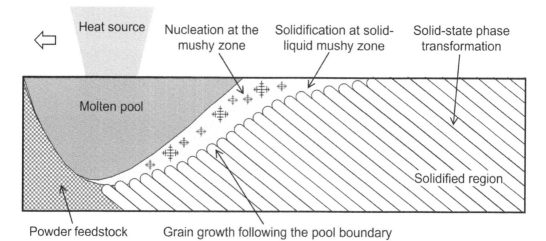

Figure 8.1 Schematic representation of the evolution of microstructure on a longitudinal sectional plane showing nucleation, solidification, grain growth, and solid-state phase transformation. *Source:* T. Mukherjee and T. DebRoy.

and neutron diffraction, and synchrotron-based methods. These techniques are often time-consuming and expensive and do not provide the temporal evolution of the microstructural features in most cases. Therefore, several numerical models are being developed, which when appropriately validated can calculate the evolution of microstructural features in AM parts.

In this chapter, theories and mechanisms of nucleation, solidification, grain growth, and solid-state phase transformations are explained. Grain structures and their wide variety in AM parts depending on the process, process variables, and alloys are discussed. Finally, several techniques for controlling the microstructure of AM parts are highlighted.

8.2 Grain structure

8.2.1 Nucleation

The development of microstructure during additive manufacturing starts with the nucleation of a new phase which involves the formation of small clusters of atoms called nuclei where the atomic arrangement is typical of a crystalline solid. The nuclei then grow when additional clusters of atoms are deposited on it and eventually form a solid structure (Figure 8.1). The types of nucleation are classified as homogeneous or heterogeneous. Homogeneous nucleation is the process of forming a new phase without the need for a preexisting surface and occurs throughout the liquid. It does not require any solid particles within the liquid and this type of nucleation is rather uncommon. The alloy composition and temperature affect the rate of homogeneous nucleation. In contrast, heterogeneous nucleation occurs on the solid-liquid interface such as the surface of a solid particle inside the liquid or at the solid-liquid interface. This type of nucleation is far more common than homogeneous nucleation.

Figure 8.2 shows three possible mechanisms of heterogeneous nucleation, nucleation on fragmented grains, heterogeneous nucleation on foreign particles, and nucleation on impinging powders on the liquid surface. The flow of liquid metal inside the molten pool (Chapter 7) may cause fragmentation of the newly formed grains in the mushy zone [1] [2]. These grain fragments are carried into the molten pool and act as nuclei for the new crystal to nucleate if they do not dissolve in the high temperature regions inside the pool (Figure 8.2 (a)). Foreign particles, such as impurities or deliberately added inoculants or grain refiners can serve as a site for heterogeneous nucleation on which metal atoms can be arranged in a crystalline form (Figure 8.2 (a)). The surface of the molten pool can be cooled to induce nucleation on the surface by applying a stream of cooling gas (Figure 8.2 (b)) and nucleation can occur at the pool surface. In powder-based DED processes, powders supplied from the nozzle on the top of the molten pool can also act as sites for heterogeneous nucleation. For example, Figure 8.2 (c) shows the presence of small grains near the surface owing to the supply of metal powder on the liquid surface during DED-L of a nickel-based superalloy. In the manufacture of most parts, the alloy composition in one layer is the same as that in the previous layer, and nucleation of a new phase is not required. In these situations, the growth occurs readily when the temperature drops below the liquidus. However, during the fabrication of functionally graded composites, nucleation at the melt pool boundary becomes necessary before the growth can occur. Similarly, when a base plate has a different chemical composition from the alloy used to make the part, the nucleation of a new phase is required which requires overcoming an activation energy barrier. Both the nucleation rate and the growth rate of the nuclei depend on temperature. The progress of the solidification process is affected by the difference between the local temperature of the liquid and the liquidus temperature of the alloy, known as undercooling, as described below.

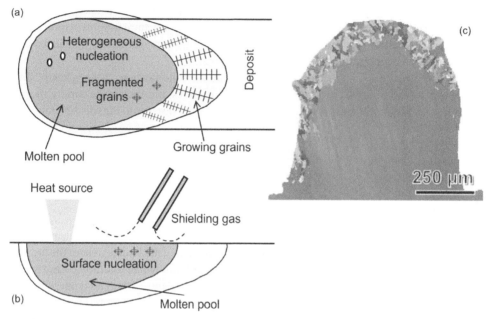

Figure 8.2 Schematic representation of different types of heterogeneous nucleation methods (a) nucleation from the fragmented grains, heterogeneous nucleation from the added foreign particles, and (b) nucleation resulting from surface cooling. Figures (a) and (b) are made by T. Mukherjee and T. DebRoy. (c) Small grains near the top of the deposit are generated due to the addition of powder particles during the DED-L of a nickel-based superalloy [1]. The figure is taken from an open-access article [1] under the terms and conditions of the Creative Commons Attribution (CC BY) license.

8.2.2 Undercooling

Nucleation of crystals does not occur at the liquidus temperature. Liquid alloys have to be cooled below the liquidus temperature for the solidification to occur. Undercooling is the cooling of a liquid alloy below its freezing point without solidification due to compositional and other effects. The difference between the liquidus temperature and the dendrite tip temperature is a measure of the extent of undercooling and serves as the driving force for dendrite growth. The undercooling required for the growth of a solid is the sum of contributions due to curvature undercooling (ΔT_R), thermal undercooling (ΔT_T), kinetic undercooling (ΔT_K), and solute undercooling (ΔT_C):

$$\Delta T = \Delta T_R + \Delta T_T + \Delta T_K + \Delta T_C \tag{8.1}$$

The curvature undercooling, ΔT_R, is the product of Gibbs–Thomson coefficient and the curvature at the dendrite tip. Thermal undercooling, ΔT_T, is the difference in temperature between the dendrite tip and that of the undercooled liquid far away from the tip. Thermal undercooling is most significant when the nucleation barrier affects the transformation from liquid to solid. The kinetic undercooling, ΔT_K, represents a driving force for the atoms from the liquid to be attached to the growing solid. This term is usually small and can be ignored. For most alloys under normal solidification conditions, the solutes rejected by the liquid accumulate near the solidifying interface and cause undercooling which is known as solute undercooling or constitutional undercooling, or constitutional supercooling, ΔT_C. This term is usually much larger than the other undercooling terms on the right-hand side of Eq. 8.1.

The extent of undercooling affects microstructure development because it influences the phase selection and the morphology of the solidification front. Depending on the solidification conditions and the alloy composition the solid-liquid interface may be planer, cellular, columnar dendritic, or equiaxed dendritic as explained below.

8.2.3 Solidification mode

Efforts to correlate the processing parameters with the key solidification parameters can promote an understanding of the relationship between process and microstructure. For the solidification of an alloy, the solidification structure is affected by the temperature gradient and solidification rate. Consider the solidification of an alloy with a planar solid-liquid interface which moves a velocity of "R," as shown in Figure Figure 8.3 (a). The velocity of the interface is known as the growth rate of the solidification front. The liquidus temperature for a given composition can be obtained from the phase diagram. The planar interface is thermodynamically stable if its temperature at the interface (actual liquidus line) is above the liquidus temperature. If its temperature is below the liquidus temperature, solid and liquid should coexist. This means that the planar solid-liquid interface should break down. The shaded area between the equilibrium liquidus line and the actual liquidus line in Figure 8.3 (a) indicates the region where the actual liquid temperature is below the liquidus temperature, that is, the region of constitutional supercooling (also called undercooling). The variation of solidification modes from planar to cellular, columnar dendritic, or equiaxed dendritic is

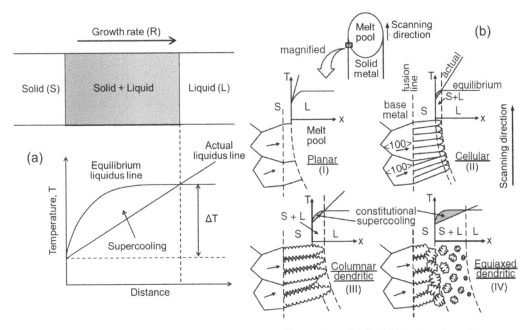

Figure 8.3 (a) Explanation of the constitutional supercooling at the solid-liquid interface. Here, ΔT represents the difference between the solidus and liquidus temperatures. *Source:* T. Mukherjee and T. DebRoy. (b) Different modes of solidification depend on constitutional supercooling [2]. This figure is reproduced with permission from Wiley.

essentially due to the increase in constitutional supercooling at the solidification interface. As constitutional supercooling increases, the solidification mode changes from planar to cellular and from cellular to dendritic (Figure 8.3 (b)). The solidification mode changes from planar to cellular, columnar dendritic, and equiaxed dendritic as the degree of constitutional supercooling at the pool boundary increases. Heterogeneous nucleation aided by constitutional supercooling promotes the formation of equiaxed grains.

During the solidification of alloys, the interface can be planar, cellular, or dendritic depending on the temperature gradient, the solidification growth rate, and the material system involved. To experimentally observe the solid-liquid interface during solidification, transparent organic materials that solidify like alloys have been used. Figure 8.4 shows the four basic types of morphology observed during the solidification of such transparent materials: planar, cellular, columnar dendritic, and equiaxed dendritic. Solidification fronts of different morphologies grow with the solid-liquid interface by segregating the solutes from the liquid to the solid.

8.2.4 Grain growth

From the nucleating particles near the molten pool boundary at the solid-liquid interface, grains begin to grow until the entire liquid transforms into a solid. In AM, grain growth occurs from the substrate or previously deposited layers. The resultant grain structure determines the crystallographic nature of the AM part. The type of grain growth may be epitaxial, non-epitaxial, or competitive growth as explained below.

Epitaxial grain growth
New grains often grow from the liquid metal with the same crystallographic orientations as the grains in the substrate [3]. Such a grain growth process is called epitaxial growth. Figure 8.5 (a) shows an example of epitaxial grain growth in fusion welding where all grains grow following the

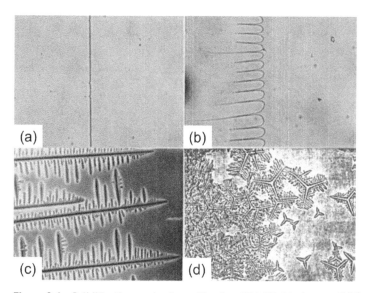

Figure 8.4 Solidification modes (magnification 67×) [2]: (a) planar solidification of carbon tetrabromide (b) cellular solidification of carbon tetrabromide with a small amount of impurity (c) columnar dendritic solidification of carbon tetrabromide with several percent of impurities, and (d) equiaxed dendritic solidification of cyclohexanol with impurity. This figure is reproduced with permission from Wiley.

same crystallographic orientation. The grains tend to grow along preferred directions called "easy growth" directions. As indicated in Table 8.1, the easy growth directions are the <100> directions for cubic materials such as stainless steels, aluminum, and nickel alloys, and the <$10\bar{1}0$> directions for hexagonal close-packed materials such as magnesium and titanium alloys.

Non-epitaxial grain growth

In some cases, the deposit grains do not follow any special orientation relationship with the substrate grains. The crystallographic orientation of the deposit grains can be parallel to any crystallographic direction. Therefore, in non-epitaxial grain growth, grains grow following several different crystallographic orientations as shown in Figure 8.5(b) for fusion welding.

Competitive grain growth:

The easy growth directions indicated in Table 8.1 are the preferred growth directions in different crystal structures. However, grains tend to grow rapidly along the direction of the heat flow. This direction is perpendicular to the local curvature of the fusion boundary plane. In a polycrystalline

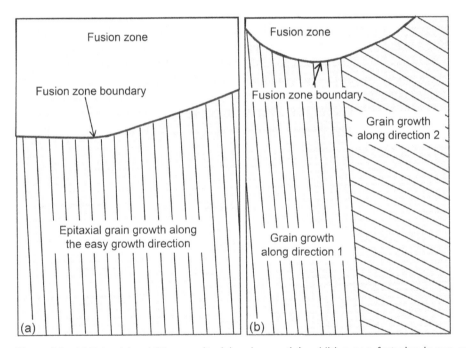

Figure 8.5 (a) Epitaxial and (b) non-epitaxial grain growth in additive manufacturing. In non-epitaxial grain growth, grains grow in two different directions. *Source:* T. Mukherjee and T. DebRoy.

Table 8.1 Easy growth directions for materials with various crystal structures [2].

Crystal structure	Easy-growth direction	Examples
Face-centered-cubic (fcc)	<100>	Aluminum alloys, austenitic stainless steels
Body-centered-cubic (bcc)	<100>	Carbon steels, ferritic stainless steels
Hexagonal-close-packed (hcp)	<$10\bar{1}0$>	Titanium, magnesium
Body-centered-tetragonal (bct)	<110>	Tin

solid, grains are oriented in different directions. The crystallographic easy growth direction and the heat flow direction both affect the solidification structure.

Grain growth near the substrate occurs either by epitaxial growth when the substrate and the deposit have the same crystal orientation or by nucleation of new grains when they have different orientations [4] [5]. However, grain growth is sometimes dominated by a different mechanism known as competitive growth (Figure 8.6 (a)) where grains tend to grow along the maximum heat flow direction. In AM, heat flows mainly from the molten pool to the substrate, and the direction of the heat flow is perpendicular to the boundary of the molten pool. Figure 8.6 (b) shows how grains curve to follow the maximum heat flow direction perpendicular to the melt pool boundary as it incrementally moves along the scanning direction. As the grains curve inward near the center of the molten pool, the angle between the scanning direction and the direction of grain growth decreases. At the same time, the grains also tend to grow in the easy-growth direction. Therefore, during solidification, grains with their easy-growth direction nearly perpendicular to the pool boundary grow more easily and dominate over those less favorably oriented. This mechanism of competitive growth results in the final grain structure of the AM parts.

Grain growth in AM is influenced by the maximum heat flow direction [4]. Figure 8.7 shows the progress of grain growth with time during the solidification of an aluminum alloy where curved columnar grains grow following the direction perpendicular to the molten pool boundary. The maximum heat flow direction at the solid/liquid interface is dependent on the local curvature of the molten pool boundary. Therefore, long grains grow aligned with the maximum temperature perpendicular to the boundary of the moving molten pool. In AM, the solidification and the grain growth process result in a grain structure that affects the final microstructure and properties of the part. The following section describes the evolution, variety, and uniqueness of grain structure in AM parts.

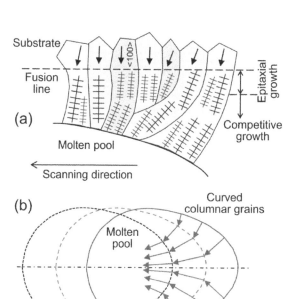

Figure 8.6 (a) Schematic representation of competitive grain growth (b) growth of curved columnar grains following the maximum heat transfer direction along perpendicular to the molten pool boundary. *Source:* T. Mukherjee and T. DebRoy.

Figure 8.7 Progress of grain growth with time during solidification of an aluminum alloy where curved columnar grains grow following the direction perpendicular to the molten pool boundary. Results are shown after 0.5 s, 2.0 s, and 4.0 s after the molten pool starts moving from its initial location. The molten pool moves along the negative direction of X (length direction). Y and Z directions are the width and depth directions, respectively. The results are generated using a numerical model of the grain growth process. *Source:* T. Mukherjee and T. DebRoy.

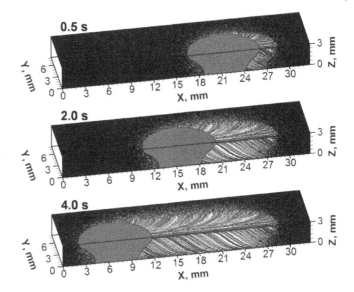

In additive manufacturing, grains grow from the boundary toward the middle of the melt pool. Generally, the grains exhibit a competitive growth between the easy growth direction of the grains in the substrate or previously deposited layers and the maximum heat flow direction (Section 8.4). For polycrystalline materials, the grain growth direction is parallel to the maximum heat flow direction which is normal to the solidifying surface. Therefore, the orientation of the grain structure is affected by the shape and size of the melt pool. The differences in the shapes of the melt pools produced by PBF and DED result in significantly different grain orientations as demonstrated in Figures 8.8 and 8.9, respectively.

Rapid scanning in PBF results in a long and shallow molten pool, which indicates that the solidification involves downward heat flow at the melt pool boundary to the substrate or the previously deposited layer, opposite to the build direction. Therefore, the direction of grain growth may closely align with the build direction [6] [7] [8]. Nearly vertically oriented columnar grains in Inconel 718 parts produced by PBF-EB are shown in Figure 8.8. In contrast, slow scanning speed in DED results in a comparatively short and deep melt pool which is characterized by an obvious curvature at its trailing edge. The heat flow directions in such a case are perpendicular to the local positions at the melt pool boundary. Therefore, unlike the PBF process, the direction of grain growth in DED may significantly deviate from the build direction [9] [10]. Figure 8.9 shows an example of inclined columnar grains in stainless steel 316 part processed by DED-L. Curved columnar grains with a variety of orientations can be observed. In addition, these columnar grains do not align uniformly and grow through multiple layers like those in Figure 8.8.

In both PBF and DED, the shape of the molten pool, which affects the grain orientation, is depicted by the liquidus isotherm. As explained in the previous section, constitutional supercooling plays an important role as a driving force during solidification. The growing grains may not always follow a direction normal to the curved surface determined by the liquidus isotherm when the actual solidifying surface significantly deviates from the liquidus isotherm due to the influence of constitutional supercooling. Figure 8.10 shows an example of the effect of constitutional supercooling on the grain orientation where noticeable discrepancies exist between the actual solidifying surface (black dashed curve) and the liquidus isotherm (black solid curve) where significant

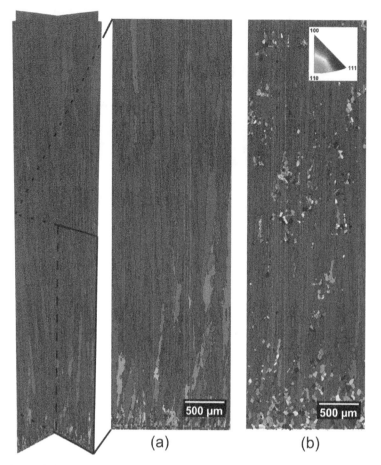

(a)   (b)

Figure 8.8   Grain orientations closely aligned with the (001) build direction [8] in Inconel 718 samples made by PBF-EB. (a) and (b) show simulated and EBSD results, respectively. The figure is taken from an open-access article [8] under the terms and conditions of the Creative Commons Attribution (CC BY) license.

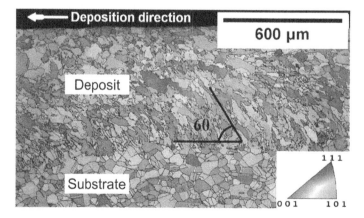

Figure 8.9   EBSD grain orientation map of a stainless steel 316 part made by DED-L [10]. The grains tend to bend towards the deposition direction. The figure is taken from an open-access article [10] under the terms and conditions of the Creative Commons Attribution (CC BY) license.

Figure 8.10 Effect of constitutional supercooling on the grain structure [11] during solidification of stainless steel in fusion welding for scanning speed (a) 1 mm/s, (b) 2 mm/s, and (c) 5 mm/s. The results are generated using a numerical model. The constitutional supercooling is evident from the difference between the solidifying surface (black dashed curve) and the liquidus isotherm (black solid curve). The figure is reprinted with permission from Elsevier.

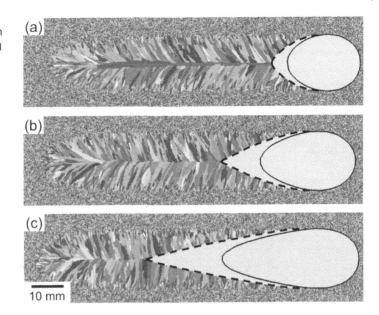

supercooling is observed. The deviation of the actual solidifying surface from the liquidus isotherm increases with scanning speed. In AM, the grain orientation, however, is largely controlled by the competitive grain growth mechanism. However, grains vary significantly in terms of their morphology, shape, and size as discussed in the following section.

8.2.5 Morphology

As discussed in Section 8.2.3, the morphology of the solidification structure, planar, cellular, columnar, or equiaxed dendrites depends on the modes of solidification controlled by constitutional supercooling. Constitutional supercooling is affected by both the growth rate of the solidification front as well as the slope of the liquidus line (Figure 8.11 (a)), also known as the temperature gradient perpendicular to the molten pool boundary. It is well-established in both casting and welding literature that the ratio of the temperature gradient (G) to the growth rate (R) is the indicator of the morphology of the solidifying crystal. A high ratio indicates the presence of planar morphology. The morphology changes from planar to cellular and dendrites as the ratio decreases. Both the temperature gradient and the growth rate vary significantly depending on the shape and size of the pool which may result in a diverse morphology in the grain structure for the same component. For example, growth rates vary significantly along the fusion line (pool boundary) depending on the solid-liquid interface. At any location, the growth rate is perpendicular to the pool boundary.

As shown in Figure 8.11 (a), the growth rate is maximum at the centerline of the molten pool which is equal to the scanning speed. On the fusion line, the normal to the pool boundary is perpendicular to the scanning speed resulting in a growth rate value of zero. Along the molten pool boundary, there exists an angle between the growth rate and the scanning speed, which varies between $0°$ (at the centerline) and $90°$ (at the fusion line). Therefore, the growth rate can be expressed as (scanning speed $\times \cos\alpha$) where "α" is the angle between the growth and the scanning directions. Such variations in growth rates suggest that the solidification mode may change from

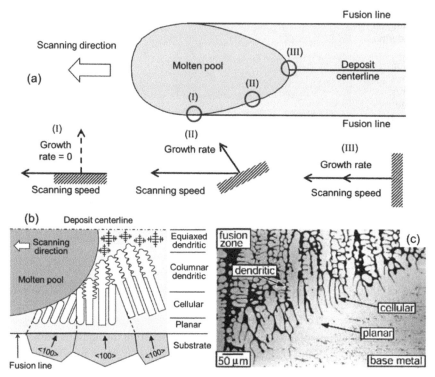

Figure 8.11 (a) A schematic representation of the variation of solidification growth rate along the molten pool boundary and its relationship with the scanning speed. (b) Schematic representation and (c) microstructural evidence [2] to show the variation in solidification modes near the fusion zone. The microstructure shown in figure (c) is for fusion welding of aluminum alloy AA1100 [2]. Figures (a) and (b) are made by T. Mukherjee and T. DebRoy. Figure (c) is reprinted with permission from Wiley.

planar to cellular, columnar dendritic, and equiaxed dendritic across the fusion zone, as depicted in Figure 8.11 (b). At the center of the molten pool, the growth rate is the maximum (Figure 8.11 (a)) indicating a low ratio of the temperature gradient to the growth rate which results in an equiaxed morphology. In contrast, near the fusion line, the growth rate is very small resulting in a high ratio of the temperature gradient to the growth rate and thus a planar morphology. Therefore, along the molten pool boundary the morphology changes from planar to cellular and columnar and equiaxed dendrites (Figure 8.11 (b)). Figure 8.11 (c) shows the planar-to-cellular transition and cellular-to-dendritic transition in 1100 aluminum (essentially pure Al) welded with a 4047 (Al–12Si) filler metal. The local variation of solidification growth rate along the fusion zone boundary is computed in worked out example 8.1.

From the aforementioned discussion, it is clear that the grain morphology depends on the ratio of the temperature gradient (G) to the growth rate (R) and the size of the grain and sub-grain features is controlled by the cooling rate (GR). Essentially, the ratio G/R is the slope of temperature gradient vs. growth rate plot, known as the solidification map in the casting, welding, and AM literature (Figure 8.12). The hyperbolic curves in the plot indicate a particular cooling rate. The figure illustrates the effect of G/R and GR on the solidification microstructure. The solidification morphology can be planar, cellular, columnar dendritic, or equiaxed dendritic with decreasing G/R values. The dimensions of all four solidification morphologies decrease with the increased cooling rate (GR). Solidification maps for particular alloys are widely used to judge what morphologies can form during solidification.

Figure 8.12 A schematic representation of a solidification map [2]. This figure is reprinted with permission from Wiley.

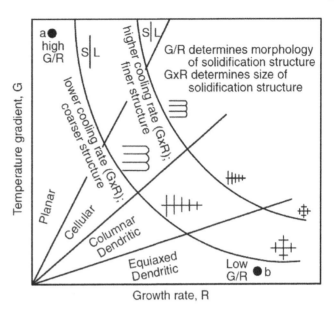

Worked out example 8.1

How does the solidification growth rate vary along the solidifying molten pool boundary during a laser powder bed fusion with a scanning speed of 500 mm/s?

Answer:

Figure E8.1 below shows the top view of the variation in the solidification growth rate for the molten pool. The molten pool boundary at the trailing side (opposite to the scanning direction) is represented by the curve ABC. At locations A and C, the solidification growth rate is zero and at location B, the solidification growth rate is the maximum and is equal to the scanning speed (Figure 8.11). The solidification growth rate is given by scanning speed multiplied by $\cos\alpha$ where "α" is the angle between the growth rate and the scanning speed. The values of "α" at location A, B, and C are $-90°$, $0°$, and $90°$, respectively.

Figure E8.1 Top view of the variation in the solidification growth rate at various locations on the surface of the molten pool. *Source:* T. Mukherjee and T. DebRoy.

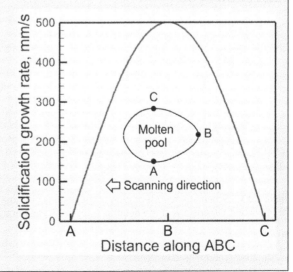

Dendrites grow in branches as shown in Figure 8.13. The main stem of the dendrite which is called the primary dendritic arm grows along the direction of the maximum temperature gradient. The secondary dendritic arms grow perpendicular to the primary dendritic arms. Solidification occurs at the tip (Figure 8.13) of the growing dendrites. The temperature at the dendrite tip is between the liquidus and the solidus temperature of the alloy. The distance between the tip and root of a dendrite increases as the solidification progresses. The distances between two consecutive primary and secondary arms are called primary and secondary dendritic arm spacing, respectively (Figure 8.13). In the solidification literature [12], the tip radius of the primary dendritic arm (r_p) is approximately represented as:

$$r_p = \frac{\lambda_p^2}{3L} \tag{8.2}$$

where λ_p is the primary dendritic arm spacing and L is the distance between the tip and root of a primary dendrite, i.e., the length of the dendrite. An example of the calculation of the tip radius of the dendrite arm is shown in worked out example 8.2. Both the primary and secondary dendritic arms are important characteristics of the solidification morphology, and their dimensions such as dendrite tip radius and dendrite arm spacing, have significant effects on the mechanical properties of parts.

The size of the grains in the additively manufactured components is largely affected by the cooling rates. A finer grain structure is obtained from higher cooling rates because shorter times are available for the growth of the grains, and vice versa. Cooling rates can significantly vary depending on the scanning speed. It is well-known in both fusion welding as well as additive manufacturing literature that rapid scanning results in a faster cooling rate. Therefore, the components fabricated at a faster scanning speed contain smaller grains as shown in Figure 8.14. The figure shows the grain size variations for four scanning speeds of 4 mm/s, 8 mm/s, 10 mm/s, and 16 mm/s, respectively. Grains of different sizes can be observed for all four scanning speeds. The grain size

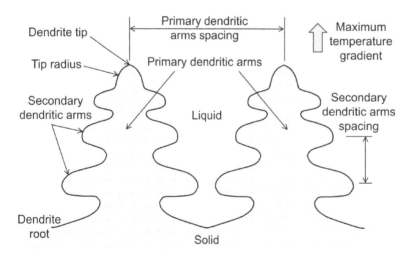

Figure 8.13 Schematic representation of the growth of primary and secondary dendritic arms indicating the important dimensions of the dendrites. *Source:* T. Mukherjee and T. DebRoy.

Figure 8.14 Variation in grain size with scanning speed in stainless steel 304L tubelike structures fabricated by powder-based DED process [13]. The grain size is calculated using a numerical model of grain growth. The figure is taken from an open-access article [13] under the terms and conditions of the Creative Commons Attribution (CC BY) license.

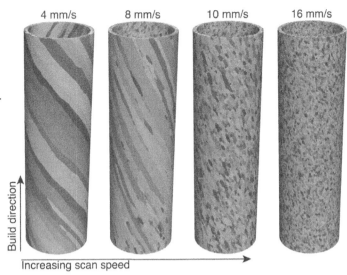

can vary considerably from the long grains along the build direction at low scanning speeds to smaller and more randomly oriented grains at high scanning speeds.

The size of grains can be approximately determined if the cooling rates are known. Cooling rates can be measured from the thermal cycles monitored experimentally or can be calculated analytically as described in Chapter 7 or, more accurately, using a numerical model. Cooling rates can be represented as the product of temperature gradient (G) and the growth rate (R) of the solidification front from the computed values of G and R. There are analytical expressions available in casting, welding, and AM literature to calculate the size of grains or sub-grain features from the cooling rates. For example, the secondary arm spacing (Figure 8.13) of a columnar or equiaxed dendritic structure can be calculated from the cooling rates during solidification as:

$$\lambda_s = A(GR)^{-n} \tag{8.3}$$

where λ_s is the secondary dendritic arm spacing (SDAS) in μm, GR is the cooling rate (in K/s) where G and R are the temperature gradient and the growth rate, respectively, and A and n are material-specific constants whose values for several common alloys are provided in Table 8.2. The correlation between secondary dendrite arm spacing and the cooling rate is explained in worked out examples 8.3 and 8.4.

Table 8.2 Alloy-specific parameters "A" and "n" in Eq. 8.3 for different alloys [14, 15]. The constant "n" is unitless; however, the unit of "A" depends on "n" and units of λ_s (μm) and GR (K/s). For example, for stainless steel 316, the unit of "A" is μm × (K/s)$^{0.28}$.

Alloy	Stainless steel 316	AlSi10Mg	Inconel 718	AISI 1045
A	25	43	34	49
n	0.28	0.32	0.25	0.35

Worked out example 8.2

Columnar dendrites grow on the top surface at the deposit centerline at the trailing edge of the molten pool (Figure 8.7). Calculate the radius of the tip of the primary dendritic arm from the following data. Temperature at the tip of the primary dendrite: 900 K, solidus temperature of aluminum alloy: 850 K, average temperature gradient inside the mushy zone on the top surface at the deposit centerline at the trailing edge of the molten pool: 100 K/mm, and primary dendritic arm spacing: 30 micrometers.

Answer:

The root of the dendrite is the region from where the dendrite starts growing and the temperature at that location is approximated as the solidus temperature (T_s) of the alloy. Therefore, the distance between the tip and root of a primary dendrite (L) can be expressed as:

$$L = \frac{T_t - T_s}{G} \tag{E8.1}$$

where T_t is the dendrite tip temperature and G is the average temperature gradient inside the mushy zone. By rearranging the above equation and Eq. 8.2,

$$r_p = \frac{\lambda_p^2 G}{3(T_t - T_s)} \tag{E8.2}$$

Using the values provided, λ_p = 30 micrometers, G = 100 K/mm, T_t = 900 K, and T_s = 850 K, the tip radius of the primary dendritic arm (r_p) is computed as 0.6 micrometers. As the solidification progresses, the temperature at the tip of the dendritic varies which results in a change in the radius of the tip of the primary dendrite.

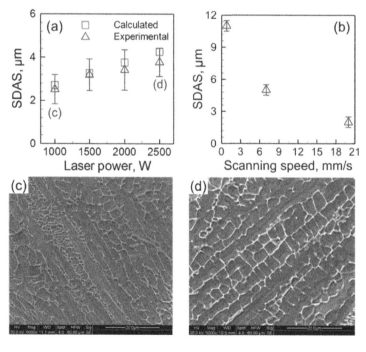

Figure 8.15 Variation in secondary dendritic arm spacing (SDAS) with (a) laser power and (b) scanning speed during DED-L of stainless steel. Microstructure obtained using scanning electron microscopy, indicating the difference in SDAS for two powers, 1000 W and 2500 W in Figure (a) are shown in Figures (c) and (d), respectively. *Source:* T. Mukherjee and T. DebRoy.

High power results in a large molten pool that solidifies slowly exhibiting a low cooling rate which results in large SDAS [17] [18], as observed from Eq. 8.3. An increase in SDAS with heat source power can be observed in Figure 8.15 (a) during DED-L of stainless steel 316. In this case, SDAS is both measured experimentally as well as calculated using Eq. 8.3 where the cooling rates are estimated using a numerical model. Microstructure obtained using scanning electron microscopy, indicating the difference in SDAS for two powers in Figure 8.15 (a) are shown in Figures 8.15 (c) and (d). Unlike heat source power, a faster scanning speed increases the cooling rates and thus reduces the SDAS (Figure 8.15 (b)).

Worked out example 8.3

The microstructure below (Figure E8.2) shows the dendritic morphology with primary and secondary dendrites during solidification of aluminum 4047 alloy part made by DED-GMA. What is the local average cooling rate during solidification?

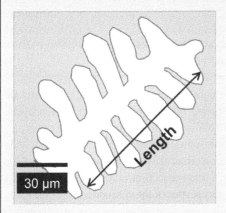

Figure E8.2 Dendritic morphology of an aluminum 4047 alloy part made by DED-GMA. *Source:* T. Mukherjee and T. DebRoy.

Answer:

Form the microstructure, it is evident that there are seven secondary dendrite arms within a length of about 100 μm marked in the figure. Therefore, the average secondary dendritic arm spacing (λ_s) is 100/7 = 14.3 μm. The secondary dendritic arm spacing (λ_s) and the cooling rate (GR) is related by the Eq. 8.3 as:

$$\lambda_s = A(GR)^{-n} \tag{E8.3}$$

The values of A and n for aluminum alloy can be taken as 43 and 0.32, respectively from Table 8.2. Using Eq. E8.3, the local average cooling rate during solidification is calculated as 31.2 K/s. This cooling rate in DED-GMA is significantly lower compared to those observed in laser and electron beam based DED and PBF processes. DED GMA is commonly practiced using a high heat input and low scanning speed which results in low cooling rates [16].

Worked out example 8.4

A stainless steel 316 part is made using laser assisted directed energy deposition process. In this process, 600 W laser power and 10 mm/s scanning speed are used. What would be the average secondary dendritic arm spacing observed in the part? Useful data: laser absorptivity: 0.5, thermal conductivity of stainless steel 316: 28 W/m K, liquidus temperature: 1733 K, and ambient temperature: 298 K. The values of A and n (Eq. 8.3) for stainless steel 316 are 25 and 0.28, respectively.

Answer:

The cooling rate (*GR*) on the top surface at the trailing edge of the molten pool can be calculated as described in Chapter 7 as:

$$GR = \frac{2\pi k_s}{\eta P / V}(T_c - T_0)^2 \tag{E8.4}$$

(Continued)

Worked out example 8.4 (Continued)

where P is laser power, V is scanning speed, η is the laser absorptivity, k_s is thermal conductivity of the alloy, T_c is the temperature at which cooling rate is calculated which is the liquidus temperature in this case, and T_0 is the ambient temperature. The calculated cooling rate (GR) is 12,069 K/s.

The secondary dendritic arm spacing (λ_s) can be estimated from the cooling rate (GR) using Eq. 8.3 as:

$$\lambda_s = A(GR)^{-n} \tag{E8.5}$$

Using the calculated values of the cooling rate (GR) and the values of A and n given in the problem, the secondary dendritic arm spacing is calculated as 1.79 μm.

During AM, the cooling rate decreases as the layer height increases. Thus, the grain structure becomes coarser in the upper layers due to lower local cooling rates than those in the lower layers. The role of local solidification parameters on the grain morphology is illustrated in worked out examples 8.5 and 8.6. Essentially, both G and R exhibit significant spatial variations which result in a wide variety in both morphology and the scale of the solidification structure in the same component. This spatial inhomogeneity in microstructure, discussed in the following section, is very unique to the AM components and can cause anisotropy in the mechanical properties of the component.

8.2.6 Spatial inhomogeneity in grain structure

Spatial inhomogeneity in grain structure in AM parts results in nonuniform microstructure and anisotropy in mechanical properties such as tensile strength. There are three unique attributes of the AM process contributing to the heterogeneous grain growth in AM. First, the evolution of the grain structure and solidification morphology is affected by the shape and size of the moving molten pool which may significantly vary depending on the processing conditions, alloy properties, and physical processes such as the convective flow of liquid inside the pool. Second, already deposited layers experience repeated heating and cooling during the deposition of successive layers which affects the grain structure due to both partial melting and solid-state grain growth. Finally, and most importantly, the local temperature gradient and growth rate of the solidification front determine the grain morphology and size. Both the local temperature gradient and the growth rate vary significantly within a part and their spatial variation can affect the morphology, size, and orientation of grains.

Worked out example 8.5

A stainless steel 316 part is made using laser-assisted directed energy deposition process. In this process, 1000 W laser power and 10 mm/s scanning speed are used. Determine the probable solidification morphology on the top surface of the part at the deposit centerline. Useful data: laser beam absorptivity: 0.5, thermal conductivity of stainless steel 316: 28 W/m K, liquidus temperature: 1733 K, and the ambient temperature: 298 K. Use the solidification map for stainless steel 316 available in reference [19].

Answer:

The cooling rate (GR) on the top surface of the part at the deposit centerline can be calculated as described in Chapter 7 as:

$$GR = \frac{2\pi k_s}{\eta P/V}(T_c - T_0)^2 \tag{E8.6}$$

Worked out example 8.5 (Continued)

where p is laser power, V is scanning speed, η is fraction of the energy absorbed, k_s is thermal conductivity of the alloy, T_c is the temperature at which cooling rate is calculated, which is the liquidus temperature in this case, and T_0 is the ambient temperature. On the top surface of the part at the deposit centerline, the growth rate (R) is equal to the scanning speed (V). Therefore, the temperature gradient G can be expressed as:

$$G = \frac{2\pi k_s}{\eta P}(T_c - T_0)^2 \tag{E8.7}$$

Using the data provided, the calculated temperature gradient is 724,188 K/m. The solidification growth rate is equal to the scanning speed (0.01 m/s). The solidification morphology can be found by plotting the data in the solidification map of stainless steel 316. Figure E8.3 below shows the computed datapoint by a dot in the solidification map [19]. It is evident that the solidification structure is columnar dendrite.

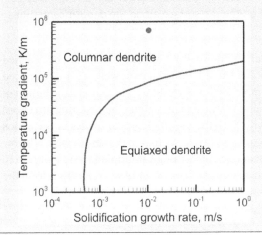

Figure E8.3 The computed datapoint shown by a dot in the solidification map. *Source:* T. Mukherjee and T. DebRoy.

Worked out example 8.6

Two deposits of stainless steel 316 are made using DED-L. In both cases, the linear heat input (laser power/scanning speed) is kept constant to achieve the same peak temperature. However, the scanning speed of one case is twice of that for the other. Which deposit is expected to experience more constitutional supercooling on the top surface along the deposit centerline and why? Which deposit is expected to have a higher probability of forming equiaxed dendrites on the top surface along the deposit centerline?

Answer:

The cooling rate (GR) on the top surface of the part at the deposit centerline can be written as described in Chapter 7 as:

$$GR = \frac{2\pi k_s}{\eta P/V}(T_c - T_0)^2 \tag{E8.8}$$

where P is laser power, V is scanning speed, η is fraction of the energy absorbed, k_s is thermal conductivity of the alloy, T_c is the temperature at which cooling rate is calculated, and T_0 is the

(Continued)

Worked out example 8.6 (Continued)

preheat temperature. For both cases, the linear heat input $\frac{P}{V} = Q$ is kept constant. Therefore, the cooling rate can be rewritten as,

$$GR = \frac{2\pi k_s}{\eta Q}(T_c - T_0)^2 \tag{E8.9}$$

The extent of the constitutional supercooling is indicated by the ratio G/R. On the top surface of the part at the deposit centerline, the growth rate (R) is equal to the scanning speed (V). Therefore, the ratio G/R can be expressed as:

$$G/R = \frac{2\pi k_s}{\eta Q V^2}(T_c - T_0)^2 \tag{E8.10}$$

The deposit with slower scanning speed exhibits higher G/R and more constitutional supercooling. In contrast, the deposit made with faster scanning exhibits lower G/R and is expected to have more equiaxed dendrites (Figure 8.15) on the top surface along the deposit centerline.

The spatial inhomogeneity in the grain structure in multilayer AM components can be revealed by both experimentally characterizing the component at different planes as well as using advanced numerical models of grain growth [20]. For example, Figure 8.16 shows the spatial distribution of the grain structure in a series of longitudinal planes (vertical planes along the scanning direction) obtained from a numerical model of grain growth. Both the morphologies and sizes of the grains appear significantly different in these planes. Figure 8.16 (a) shows that in the central plane (below the heat source axis), elongated columnar grains propagate through multiple layers. Figure 8.16 (b-d) show the longitudinal sectional planes with distances of 60 μm, 240 μm, and 840 μm from the

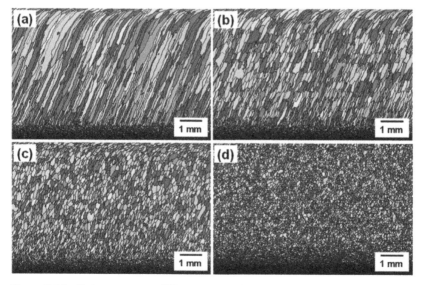

Figure 8.16 Grain structure at different longitudinal planes (vertical planes along the scanning direction) during DED-L of Inconel 718 showing the spatial inhomogeneity in the grain structure. Grain structures are shown in (a) the longitudinal central plan (below the heat source axis) and longitudinal sectional planes with distances of (b) 60 μm, (c) 240 μm, and (d) 840 μm from the longitudinal central plane. The grain structure is calculated using a numerical model of grain growth. *Source:* T. Mukherjee and T. DebRoy.

longitudinal central plane, respectively. It can be observed that the grains appear less elongated as they move away from the central longitudinal plane. Moreover, the average cross-sectional area of the grains decreases with distance from the longitudinal central plane. However, the grains tend to grow perpendicular to the longitudinal plane away from the midsection longitudinal plane because of the curvature of the molten pool. Because of the complex geometry of grain growth, columnar grains may appear to be equiaxed in certain cross-sections and it is important to review grains structures in orthogonal planes to reveal the correct morphology of grains.

8.2.7 Texture

The texture of an AM part depends on both the grain orientation as well as the dimensions of the grains and sub-grain features such as cells or dendrites. Parts having the same shape and size made by different AM processes, processing parameters, and alloys may have different textures. Texture affects the mechanical and chemical properties, reliability, and serviceability of the AM components, and the texture of various alloys processed by DED and PBF and the mechanisms for their formation are of interest.

Figure 8.17 shows the textures formed due to β-columnar grains during PBF-EB of Ti-6Al-4V. The figure schematically illustrates the different surface β-grain structures generated by the different scanning patterns, contour pass, and in-fill hatching. Contour pass scanning is used to make the outer boundary of the part and the inner part of the component is made by in-fill hatching. The grain structure consists of coarse columnar grains parallel to the build direction which is the maximum heat transfer direction in PBF. However, for the contour pass, the heat flows from the molten pool to the powder bed at the side. Therefore, the columnar grains bend sidewise toward

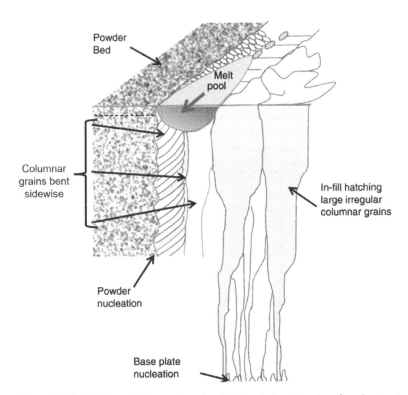

Figure 8.17 A schematic illustration showing the different surface β-grain structures which result in a complex texture in a Ti-6Al-4V part fabricated by PBF-EB [21]. The figure is taken from an open-access article [21] under the terms and conditions of the Creative Commons Attribution (CC BY) license.

the powder bed to follow the maximum heat transfer direction. Therefore, texture in PBF may vary significantly depending on the scanning patterns used.

The texture observed during DED-L of Inconel 718 is shown in Figure 8.18 for a bi-directional scanning strategy where the laser beam changes its scanning direction in each layer. Figure 8.18 (a) schematically shows the different zig-zag grain growth patterns due to the bi-directional scanning. In the first pattern, the directions of the maximum heat flow are perpendicular to the molten pool boundary near the trailing edge in all layers. By comparing this pattern with the experimental results (Figure 8.18 (c)) it is observed that the grain growth does not follow this pattern. In the second pattern, the grains of neighboring layers are perpendicular to each other. However, in the second layer, the maximum heat flow direction deviates 30° from the primary grain growth direction. The 30° misalignment between the dominant heat flow direction and the local grain growth is too large to be practical. The third solidification pattern on the right shows the grain orientation in the first layer to have an angle of 45° with the scanning direction. The

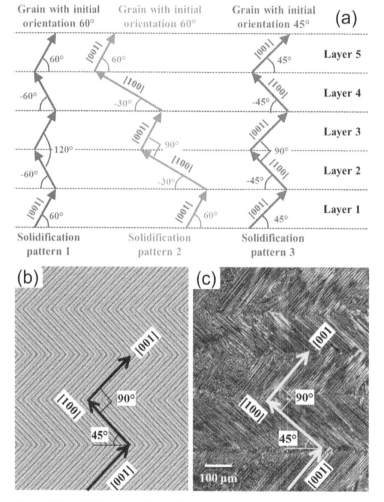

Figure 8.18 (a) Three possible grain growth patterns during DED-L of Inconel 718 [22]. The resultant texture following the third pattern (b) calculated using a numerical model of grain growth and (c) obtained experimentally [22]. The figure is taken from an open-access article [22] under the terms and conditions of the Creative Commons Attribution (CC BY) license.

growth direction makes about a 15° angle with the dominant heat flow direction. The grains in the second layer grow epitaxially from the secondary dendrites in the first layer and makes an angle of 15° with the dominant heat flow direction. This pattern agrees well with the microstructure shown in Figures 8.18 (b) and (c). Therefore, the evolution of texture is affected by both the easy growth directions of the grains and the maximum heat flow directions at the melt pool boundaries.

8.2.8 Effects of process parameters

The temperature gradient and solidification growth rates vary significantly with the important AM process parameters such as heat source power, scanning speed, substrate thickness, substrate preheat temperature, and layer thickness. Therefore, the morphology, orientation, and dimensions of the grain structure are largely affected by the process parameters as summarized in Table 8.3. Worked out examples 8.7 and 8.8 show how the process variables affect the microstructural features.

Table 8.3 Effects of important AM process parameters on solidification morphology (G/R ratio, where G and R are the temperature gradient and solidification growth rate, respectively), grain orientation, and grain size.

Parameters	Solidification morphology	Grain orientation	Grain size
Heat source power	An increase in heat source power results in a higher temperature that enhances the temperature gradient and the G/R ratio.	Grain orientation is affected by the maximum heat flow direction and the shape of the pool. However, the shape of the pool cannot be directly correlated with power.	High heat source power results in a larger molten pool that solidifies with a slow cooling rate and larger grains.
Scanning speed	An increase in scanning speed results in a higher R and thus a lower G/R ratio.	Scanning speed affects the pool size, the heat flow pattern, and grain orientation.	Rapid scanning results in faster cooling rates and consequently small grains.
Substrate thickness	Thicker substrates are better heat sinks. They reduce the temperature gradient and the G/R.	Substrate thickness does not directly affect grain orientation.	Thick substrates increase the rate of heat transfer from the molten pool which results in a high cooling rate and fine grains.
Substrate preheat temperature	A higher preheat temperature reduces the temperature gradient and thus the G/R.	Preheat affects the heat transfer pattern but does not directly affect grain orientation.	Higher preheat temperatures reduce the heat transfer rate which results in a slow cooling rate and coarse grains.
Layer thickness	Thin layers transfer more heat and reduce the temperature gradient and the G/R.	Thin layers tend to have epitaxy with the grains on which they grow.	Thin layers increase the rate of heat transfer from the molten pool which results in a high cooling rate and fine grains.

Worked out example 8.7

Derive an equation to quantitively show how the preheating of the substrate affects the secondary dendritic arm spacing in AM parts.

Answer:

The secondary dendritic arm spacing (λ_s) can be estimated from the cooling rate (GR) using Eq. 8.3 as:

$$\lambda_s = A(GR)^{-n}$$

where A and n are two alloy-specific constants. The cooling rate (GR) on the top surface of the part at the deposit centerline can be calculated as described in Chapter 7 as:

$$GR = \frac{2\pi k_s}{\eta P / V} (T_c - T_0)^2 \tag{E8.11}$$

where P is laser power, V is scanning speed, η is fraction of the energy absorbed, k_s is thermal conductivity of the alloy, and T_c is the temperature at which cooling rate is calculated. The preheat temperature is T_0. Combining the above two equations,

$$\lambda_s = A \left[\frac{2\pi k_s}{\eta P/V} (T_c - T_0)^2 \right]^{-n} \tag{E8.12}$$

For a specific alloy, at a given condition of power and speed, secondary dendritic arm spacing (λ_s) is proportional to $(T_c - T_0)^{-2n}$, where n is a positive number. Therefore, a high preheat temperature will results in a larger secondary dendrite arm spacing.

Worked out example 8.8

During DED-L of an aluminum alloy, the mushy zone length on the top surface at the deposit centerline must not exceed 0.5 mm. In addition, the average secondary dendritic arm spacing on the top surface at the deposit centerline should not be less than 5 µm. What should be the maximum allowable scanning speed to satisfy these criteria? Data: liquidus and solidus temperatures: 900 K and 850 K, respectively, the constants A and n in Eq. 8.3 for aluminum alloy: 43 µm $(K/s)^{0.32}$ and 0.32, respectively.

Answer:

The mushy zone length (Δx) on the top surface at the deposit centerline is equal to the distance between the liquidus (T_L) and solidus (T_s) isotherms along the scanning direction at the trailing edge of the molten pool. The cooling rate (GR) on the top surface at the deposit centerline is equal to

$$GR = \frac{T_L - T_s}{\Delta x} V \tag{E8.13}$$

where the temperature gradient (G) is equal to $\frac{T_L - T_s}{\Delta x}$ and the solidification growth rate (R) is equal to the scanning speed (V). The secondary dendritic arm spacing (λ_s) can be estimated from the cooling rate (GR) using Eq. 8.3 as:

$$\lambda_s = A(GR)^{-n} \tag{E8.14}$$

Worked out example 8.8 (Continued)

where A and n are two alloy specific constants. By combining the above two equations,

$$V = \frac{\Delta x}{T_L - T_s} \left(\frac{A}{\lambda_s}\right)^{1/n} \tag{E8.15}$$

By using the data provided, $\Delta x = 0.5$ mm, $T_L = 900$ K, $T_s = 850$ K, $A = 43$ μm $(K/s)^{0.32}$, $\lambda_s = 5$ micrometers, and $n = 0.32$, the computed maximum allowable scanning speed is 8.3 mm/s.

8.3 Microstructures of common alloys

The previous sections have discussed the evolution of grain structures that form during solidification. However, once the metal has solidified, it will undergo solid-state phase transformations during cooling to room temperature [23–25]. In addition, reheating of previously deposited AM layers due to repetitive thermal cycles can further result in solid-state phase transformations and affect the microstructure. In this section, solid-state phase transformations that occur during AM processing are discussed and the as printed microstructures of several common alloys are examined. A detailed review of the microstructures of common alloys and their control is available in [35].

8.3.1 Steels

Austenitic stainless steels are characterized by their excellent corrosion and oxidation resistance, and good mechanical properties. They contain high concentrations of chromium (>17 wt%) and nickel (typically 8%) and smaller amounts of other alloying elements. Additively manufactured parts of 316 L and 304 contain mainly austenite with or without a small amount of retained delta ferrite and/or martensite. Components made by the directed energy deposition of 316L stainless steel often show 8 to 10% ferrite due to microsegregation of chromium and molybdenum that stabilize ferrite. Parts made by powder bed fusion do not commonly have ferrite in the microstructure because of the high cooling rates. Parts of 316L made by powder bed fusion show a very fine microstructure with several important features in different length scales. Submillimeter scale grains, and subgrain features with many very fine cells, typically smaller than a micrometer, and cell walls of the order of 100 nanometers with dislocation networks are observed. The hierarchical microstructure of austenitic steels is thought to be the source of both high strength and ductility in the same part not commonly observed in other alloys.

Duplex stainless steels such as 2205 (22 wt% Cr, 5% Ni, 3% Mo, 2% Mn) and 2507 (25 wt% Cr, 7% Ni, 4% Mo) contain more than 19% chromium and 3 to 8.5% nickel, have excellent resistance to corrosion and widely used in oil and gas, shipbuilding, petrochemical, and other industries. They contain approximately equal amounts of δ-ferrite and austenite. The alloy composition influences the proportion of ferrite and austenite in the microstructure. The microstructure also contains M_7C_3 and $M_{23}C_6$ carbide precipitates where M is Fe or Cr. The alloying elements also contribute to solid solution strengthening and nitrogen, when present stabilizes austenite and may form nitrite precipitates. Rapid cooling during additive manufacturing, particularly during powder bed fusion, does not allow the formation of optimum amounts of austenite and post-process heat treatment is needed to achieve the desired amount of austenite.

Tool steels contain Cr, V, Nb, W, and often significant amounts of carbon that show carbides dispersed in martensite matrix. Ferritic-martensitic steels such as Grade 91 and HT-9 containing 9 to 12% Cr are often used in power plants. The microstructure contains tempered martensite and ferrite. Maraging steels contain nanoscale intermetallic precipitates in a soft Fe-Ni martensite matrix, have excellent toughness and ductility, and are used in landing gears of aircrafts and other structural applications. Precipitation hardened (PH) stainless steels such as 17–4 PH, 15–5 PH, and 13–8 PH have precipitates of fine intermetallic phases that form during aging in martensitic structure.

8.3.2   Nickel alloys

Solid-solution strengthened nickel alloys contain substitutional elements such as Cr, Fe, Mo, Nb, and W. Precipitation strengthened alloys containing Ti, Al, and/or Nb form precipitates such as γ' ($Ni_3(Ti, Al)$) and γ'' (Ni_3Nb). Microstructures of the as-printed nickel alloys show an FCC (γ) matrix phase and various precipitates such as carbides, Laves, γ' and γ''. Micro-segregation promotes the formation of carbides and Laves phases. For example, in AM of IN718, a Laves phase (Figure 8.19 (a)) having a nominal composition of Ni_2Nb, is commonly present in the as-printed microstructure due to the Nb segregation during solidification. The carbides formed range from NbC in IN718, (Ti, Mo)C in Haynes 282, Mo-rich M_6C and Cr-rich $M_{23}C_6$ in Hastelloy X, to HfC in Rene 142. The matrix phase, γ, tends to solidify in the form of fine, columnar dendrites because of the local solidification conditions. During PBF-L, γ' or γ'' is not formed because of rapid cooling. They form during DED-L and PBF-EB with high preheat where the cooling rates are lower than those in powder bed fusion.

8.3.3   Aluminum alloys

Printed aluminum alloy parts are used in automotive, aerospace, and other industries because of their low weight and good mechanical properties. Precipitates are the most common phases formed during the additive manufacturing of aluminum alloys (Figure 8.19 (b)). For example, in additively manufactured AlSi10Mg alloy, uniformly distributed, nanoscale silicon-rich precipitates are observed within the primary α-grains. Similarly, silicon-rich globular precipitates are also

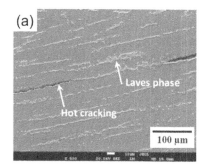 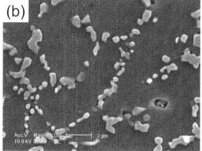

Figure 8.19   (a) Laves phases are observed in Inconel 718 parts made by DED-L which can cause hot cracking [23]. The figure is reprinted with permission from Elsevier. (b) Si-rich precipitates form due to solid-state transformation during DED-L of aluminum alloy 4047 [24]. The figure is reprinted with permission from Springer Nature.

observed in AA 4043 parts made by DED-GMA. Iron-rich precipitates have also been observed in AA 2319 parts made by DED-GMA. These precipitates affect the mechanical properties of parts. Post-process heat treatments are commonly used to control precipitates in additively manufactured aluminum alloy parts.

8.3.4 Titanium alloys

Titanium alloys, such as Ti-6Al-4V, are primarily composed of a hexagonal close-packed α phase and a body-centered cubic β phase at room temperature. In the AM process, after solidification, large columnar grains of the β phase are formed. After that, when the component cools down to room temperature, small grains of α phase with lamellar structure start forming at the grain boundary of the β phase. The α phase starts forming when the part cools down below the β-transus temperature (995°C for Ti-6Al-4V). The growth rate of the α phase and its amount depend on the local thermal cycles. For example, the lamellar spacing of the α phase decreases with the increase in cooling rate. The shape, size, amount, and distribution of the α-grains significantly affect the mechanical properties of the component.

8.3.5 Copper alloys

Copper and its alloys containing Sn, Zn, Ni, Al, Si, and other elements have high strength, wear-resistant, and corrosion resistance. Alloys such as bronze, brass, and cupronickel are difficult to process additively because of their high reflectivity of infrared laser radiation and high thermal conductivity. In addition, molten copper has a high solubility of oxygen, about 1 wt% which makes additively manufactured copper alloys susceptible to porosity. A part of a Cu-Ni-Sn alloy made by PBF-L exhibited a refined microstructure with grain boundary precipitates containing Ni that strengthened the alloy. Post-build heat treating of PBF-L Cu-Sn bronze and DED-GMA bronze parts results in solid solution strengthening. The good strength and ductility of Cu-Sn bronze made by PBF-L and DED-GMA are attributed to solid solution strengthening. Nickel aluminum bronze (NAB), a Cu-Al-Ni alloy that forms a martensitic microstructure during PBF-L is used for naval applications because of its high strength and corrosion resistance.

8.4 Process dependence of microstructure

The microstructure of AM parts is significantly affected by the AM process, process conditions, and the alloy used. For the same alloy, the microstructure can vary widely depending on the process used. This is primarily due to the wide variety of cooling rates, temperature gradients, and solidification parameters observed in different AM processes. For example, the cooling rate during solidification can vary 10,000 times depending on AM process used. Here we discuss the differences in microstructures of stainless steels and titanium alloys processed using DED and PBF processes as examples.

The average grain size of stainless steel 316 parts made by DED-L (Figure 8.20 (a)) is more than 10 times higher than that for components made by PBF-L (Figure 8.20 (b)). The linear heat input (laser power/scanning speed) of DED-L is around 23 times more than that in PBF-L. Low heat input in PBF-L results in rapid cooling which results in smaller grains. Grain orientation maps are compared for

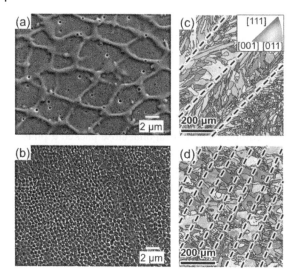

Figure 8.20 Scanning electron microscopy images of the etched surface of stainless steel 316 parts prepared using (a) DED-L and (b) PBF-L. The figures show the difference in grain size for the two AM processes [26]. A laser power of 530 W and 75 mm/s scanning speed are used in DED-L. A laser power of 180 W and 600 mm/s scanning speed are used in PBF-L. Electron beam scattered diffraction orientation maps [27] in stainless steel 316 parts made using (c) DED-L and (d) PBF-L. The track boundaries are indicated in black dashed lines. A laser power of 275 W and 8.47 mm/s scanning speed are used in DED-L. A laser power of 195 W and 1083 mm/s scanning speed are used in PBF-L. The figures are reprinted with permission from Elsevier.

stainless steel 316 parts made by DED-L (Figure 8.20 (c)) and PBF-L (Figure 8.20 (d)). The orientation maps are extracted on the horizontal plane which is along the scanning direction and perpendicular to the build direction. The boundaries of the tracks are shown by black dashed lines which indicate that thinner tracks are used in PBF-L. For both PBF-L and DED-L, columnar grains grow vertically along the build direction. Therefore, the orientation map in the horizontal plane represents the transverse sections of those columnar grains. This grain structure is often called mosaic structure and is commonly observed in AM of stainless steel. The average grain size in the transverse section is 80 µm for DED-L (Figure 8.20 (c)) and 20 µm for PBF-L (Figure 8.20 (d)). The difference in the average grain size is also attributed to the difference in the cooling rates in the two processes.

Large columnar beta grains form during the solidification of titanium alloys [28, 29]. The size (average diameter) of the columnar grains varies significantly for different AM processes depending on the cooling rates (Figure 8.21). DED processes exhibit slower cooling than the PBF processes due to their higher heat input. Slow cooling rates in DED result in large columnar grains. After solidification, alpha (α) grains grow from the grain boundary of the beta grains. However, at a very high cooling rate, α′ martensites may form along with the alpha grains. High cooling rates in PBF-L can form fine acicular and lath-type martensites as observed in commercially pure titanium. However, α′ martensites do not form in DED-L of commercially pure titanium because of slower cooling rates in DED-L than in PBF-L. Platelike alpha grains are observed in DED-L parts.

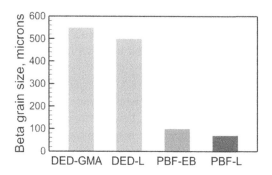

Figure 8.21 (a) Variation in columnar beta grain size in Ti-6Al-4V parts fabricated using different AM processes. The plot is made by T. Mukherjee and T. DebRoy using the data provided in [28].

Microstructures of AM parts can vary significantly depending on the heat input, cooling rates, temperature gradient, and solidification parameters. Therefore, these parameters along with other process variables can be adjusted based on scientific principles to influence the microstructure of AM parts as discussed below.

8.5 Control of microstructure

From the discussions in previous sections, it is evident that the microstructure in AM parts varies widely depending on the AM process, process parameters, and alloys used. Therefore, obtaining the desired microstructure is a challenging task. This section discusses the commonly employed control strategies of microstructure in AM parts. Table 8.4 summarizes the methods and their applications in AM.

8.5.1 Columnar to equiaxed transition

Both columnar and equiaxed grains are widely observed in the solidified region of components made by AM. The columnar grains are generally coarse and often result in anisotropic mechanical

Table 8.4 Commonly used microstructure control techniques in AM.

Microstructure control	Control methods	Explanation
Columnar to equiaxed transition (CET)	Controlling the temperature gradient and solidification growth rate	The ratio of the temperature gradient to the solidification growth rate determines the formation of columnar or equiaxed grains. Therefore, CET can be achieved by adjusting process variables such as power, speed, layer thickness, and preheat temperature.
	Adding inoculants and grain refiners	Inoculants and grain refiners can create heterogeneous nucleation which breaks long columnar grains into small equiaxed grains.
	Ultrasonic vibration	External ultrasonic vibration can break long columnar grains into small equiaxed grains during solidification.
Texture control	Controlling grain orientation	The maximum heat flow direction influences the orientation of grains. Maximum heat flow direction can be controlled by changing the scanning pattern or by adjusting pool shape and size by varying process conditions.
	Controlling dimensions of grains and sub-grain features	Dimensions of grains and sub-grain features such as cells or dendrites depend on the cooling rate which can be controlled by adjusting process variables such as power, speed, layer thickness, and preheat temperature.
Removing unwanted secondary phases	Post-process heat treatment	Unwanted secondary phases such as Laves phases in nickel alloys, carbides, and nitrides of metals affect the mechanical properties of parts and can be removed by post-process heat treatment.
Tempering of martensite		Post-process heat treatment is also used to reduce the amount of martensite that is known to affect the toughness of components.

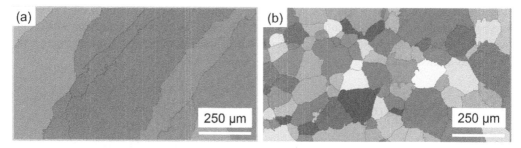

Figure 8.22 CET is achieved by using ultrasonic vibration in PBF-L of Inconel 625 [31]. (a) Long columnar and (b) small equiaxed grains after applying the ultrasound. The figure is taken from an open-access article [31] under the terms and conditions of the Creative Commons Attribution (CC BY) license.

properties. In contrast, equiaxed grains are usually small and contribute to more uniform mechanical properties. Anisotropy of mechanical properties originates from the columnar grains oriented along a certain direction. The anisotropy is detrimental for applications involving multi-directional stresses. The anisotropy can be reduced with an equiaxed grain structure [30, 31] by achieving columnar to equiaxed transition (CET).

There are three techniques to achieve CET in AM. First, CET could be predicted using the ratio of the temperature gradient to the solidification growth rate, G/R, from the solidification maps (Figure 8.12). The G/R value can be controlled using the processing conditions (Table 8.4). Therefore, CET can be achieved by adjusting the AM process conditions in many cases. Second, inoculants or grain refiners are added deliberately to promote heterogeneous nucleation during solidification. Small equiaxed grains nucleate from the surface of the added particles which prevents the formation of long columnar grains. For example, long columnar grains during PBF-L of an aluminum alloy can be avoided by adding zirconia nanoparticles which act as inoculants and form small equiaxed grains [30]. Third, an external ultrasonic vibration has been applied to break the growing columnar grains into small equiaxed grains. Figure 8.22 provides an example of achieving CET in PBF-L of Inconel 625 using ultrasonic vibration. The figure shows that the long columnar grains break and form smaller equiaxed grains when the ultrasound is used. All three methods have been demonstrated in achieving CET in AM parts. However, further research is needed to improve their effectiveness and practical application. For example, a better understanding of the shape, size, and amount of inoculants on CET is needed. Similarly, a methodology for the selection of the power level, amplitude, and frequency of ultrasound to achieve CET would be useful. Numerical models of AM processes can also be used to predict the required set of processing conditions to achieve the desired G/R ratio for CET. How the solidification parameters and CET can be affected by the preheating temperature is illustrated in worked out example 8.9.

Worked out example 8.9

Derive an equation to quantitively show how the preheating of the substrate can help to achieve columnar to equiaxed transition in AM parts.

Answer:

The columnar to equiaxed transition is affected by the ratio of temperature gradient (G) to solidification growth rate (R). A low value of the ratio favors the columnar to equiaxed

Worked out example 8.9 (Continued)

transition. The cooling rate (*GR*) on the top surface of the part at the deposit centerline can be calculated as described in Chapter 7 as:

$$GR = \frac{2\pi k_s}{\eta P / V} (T_c - T_0)^2 \tag{E8.16}$$

where *P* is laser power, *V* is scanning speed, η is fraction of the energy absorbed, k_s is thermal conductivity of the alloy, T_c is the temperature at which cooling rate is calculated, and T_0 is the preheat temperature. On the top surface of the part at the deposit centerline, the growth rate (*R*) is equal to the scanning speed (*V*). Therefore, the ratio *G/R* can be expressed as:

$$G / R = \frac{2\pi k_s}{\eta PV} (T_c - T_0)^2 \tag{E8.17}$$

For a specific alloy and a given power and speed, the ratio *G/R* is proportional to $(T_c - T_0)^2$. Therefore, high preheat temperature results in a low *G/R* ratio which tends to favor the columnar to equiaxed transition.

8.5.2 Texture control

The texture depends on both the dimensions of the grains and sub-grain features such as cells or dendrites as well as the grain orientation. Dimensions of grains are largely affected by the cooling rate which can be controlled by adjusting the process variables such as power, speed, layer thickness, and preheat temperature. Grain orientation is largely affected by the local heat flow directions during the competitive growth of the grains. Local heat flow direction is primarily controlled by the shape and size of the molten pool which can be adjusted by process variables such as power, speed, layer thickness, and preheat temperature. Especially, the scanning speed and layer thickness directly affect the extent of remelting of the previously deposited tracks which also controls the competitive grain growth. Transverse cross-sections (perpendicular to the scanning direction) show a gradual increase in the number of elongated grains with increasing scanning speed [32]. This increase is mostly a consequence of a gradual change in the average angle between the long axis of elongated grains and the scanning direction. The gradual change in the orientation of the elongated grains with the increase in scanning speed is also evident in the longitudinal plane.

The texture can also be controlled by the scanning strategy, printing process, processing conditions, and the alloy used. Figure 8.18 illustrates texture control by changing the scanning strategy in a DED-L process. Different customized textures can also be produced in the PBF process using various types of scanning strategies. In short, the texture is dependent on the evolution of grain structure and their dimensions which are affected by the shape and size of the melt pool. A very important distinction between welding and AM processes is the ability to control texture in AM parts by adjusting scanning strategies. The scanning path can be adjusted in AM. However, in fusion welding, the scanning path depends on the interface where fusion is needed to join parts.

8.5.3 Post-process heat treatment

In AM, post-process heat treatment such as annealing, tempering, and normalizing often help in the recrystallization and growth of grains. In addition, the heat treatment processes can reduce the amount of brittle secondary phases that are detrimental to the mechanical properties.

Post-process heat treatments are regularly used to minimize the precipitates, such as Laves phases in nickel alloys and precipitates in aluminum alloys. High cooling rates in AM often form martensite during AM of steels. Post-process heat treatment can temper martensite to improve the toughness of components. However, like the heat treatment of welded or cast products, the effectiveness of heat treatment of AM parts depends on the selection of appropriate temperature and time.

8.6 Single crystals

A single crystal (SX) has the same crystallographic orientation throughout the part and does not have any grain boundaries. SX components are traditionally made using directional solidification and used in applications where superior high-temperature creep resistance is required such as aero-engine turbine components. Recently, AM is being used effectively in the fabrication and repair of SX components. Both PBF and DED processes have been employed to make SX nickel-base superalloy parts using both laser and electron beam heat sources. For example, Figure 8.23 (a) shows a longitudinal section of a 75 mm long and 12 mm diameter CMSX-4 single crystal cylinder fabricated using PBF-EB. The horizontal section of the SX cylinder and the EBSD mapping shown in Figure 8.23 (b) indicate the single crystallinity of the component.

Directional solidification to obtain single crystallinity is achieved by maintaining an appropriate combination of temperature gradient and solidification growth rate. Values of these parameters necessary to grow SXs are often achieved by preheating the powder bed and using complex combinations of scanning patterns and speed. However, challenges remain in the selection of preheat temperature, heat input, and scanning strategy to achieve directional solidification. Several well-tested mechanistic models are developing to address this issue and facilitate the printing of SX.

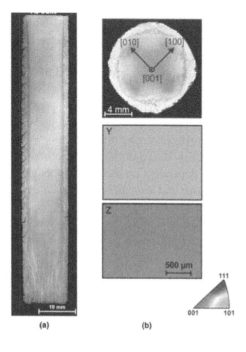

Figure 8.23 A single crystal part of a nickel alloy, CMSX-4 made by PBF-EB. (a) Vertical cross-section, and (b) horizontal cross-section and EBSD mapping [33]. Y and Z directions are along the radial and height directions, respectively. The figure is taken from an open-access article [33] under the terms and conditions of the Creative Commons Attribution (CC BY) license.

(a)　　　　　(b)

8.7 Microstructure characterization techniques

Microstructural features such as grains, dendrites, cells, and secondary phases need high-magnification characterization techniques to characterize their formation. Optical microscopy is often a beneficial and less expensive technique to reveal the microstructure of AM parts. However, very fine features such as fine precipitates need high-resolution techniques such as scanning electron microscopy (SEM) and transmission electron microscopy (TEM). The crystallographic orientation of grains can be revealed by electron beam scattered diffraction (EBSD). The composition and crystal structure of secondary phases are often estimated using the X-ray diffraction technique [34]. In situ X-ray diffraction can provide the phase fraction information during the AM process. This method is well-known in the fusion welding process. For example, Figure 8.24 shows that a synchrotron beamline is used to measure the evolution of ferrite volume fraction in 304 stainless steel during welding. The measurement can provide the temporal and spatial variations of ferrite volume fraction corresponding to local thermal cycles. However, this is an expensive technique and can be done in a few facilities around the world. Microstructural characterization of AM parts is a challenging task because of its complexities. For example, the uniqueness of the texture and grain structure due to multilayer deposition makes the characterization process complex compared to conventional processes such as welding and casting.

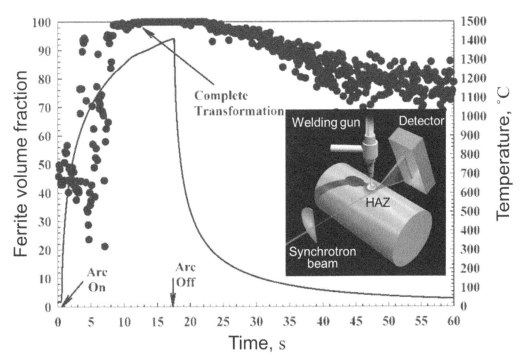

Figure 8.24 In-situ synchrotron measurement of ferrite volume fraction during fusion welding. *Source:* T. DebRoy and J.W. Elmer.

8.8 Summary

The solidification morphologies, orientation, shape, and size of grains, texture, formation of primary and secondary phases, and many sub-grain features define the microstructure of AM parts. The evolution of microstructure relies on nucleation, solidification, grain growth, and solid-state phase transformation and is affected by the process, process variables, and alloy composition. Since AM is a very complex process involving rapid scanning and melting and deposition of multiple tracks, AM parts exhibit unique microstructure which often requires special characterization techniques. In addition, advanced numerical modeling tools are emerging to simulate grain structure and morphologies. A detailed understanding of microstructural features is needed to establish the process-microstructure correlation and can provide a better insight into the properties and serviceability of AM parts.

Takeaways

Nucleation theory and mechanism

- Heterogeneous nucleation is common in additive manufacturing although both heterogeneous and homogeneous nucleations are ways of formation of small clusters of atoms called nuclei where the atomic arrangement is typical of a crystalline solid.
- Three main mechanisms of heterogeneous nucleation are: (1) nucleation on fragmented grains, (2) heterogeneous nucleation on foreign particles, and (3) nucleation on impinging powders and solid-liquid interfaces.

Constitutional supercooling and modes of solidification

- The constitutional supercooling is the cooling of a liquid below its freezing point without solidifying because of the compositional difference near the solid liquid interface.
- Constitutional supercooling controls the solidification modes.
- Four main solidification modes, planar, cellular, columnar dendritic, and equiaxed dendritic occur with increasing supercooling.

Mechanisms of grain growth

- Epitaxial, non-epitaxial, and competitive growth of grains occur depending on the substrate and processing conditions.
- In additive manufacturing, competitive grain growth is the most common.
- In additive manufacturing, grain grow direction is influenced by the dominant direction of heat flow which is perpendicular to the molten pool boundary.

Morphology

- Grain morphology depends on the ratio of temperature gradient to the solidification growth rate. Decreasing ratio changes the morphology from planar to cellular to columnar to equiaxed.
- Dimensions of grains and sub-grain features are controlled by cooling rate. Rapid cooling rates result in smaller grains.

Takeaways   (Continued)

- Solidification map, a plot of temperature gradient vs. solidification growth rate is used to predict the morphology and grain size.

Spatial inhomogeneity in grain structure

- Grain orientation, shape, and size vary spatially in additive manufacturing components due to the variations in the temperature gradient and solidification growth rate.
- Spatial inhomogeneity in microstructure results in anisotropy of mechanical properties.

Texture

- Texture in parts vary significantly depending on the additive manufacturing processes, process variables, scanning strategies, and alloys used.

Solid-state phase transformation

- Solid-state phase transformation during the cooling of additive manufacturing parts affects microstructure and mechanical and chemical properties.
- Solid-state phase transformation involving changes in crystal structure, for example, austenite to ferrite transformation in steels affects evolution of residual stresses.

Phases in common alloys

- Additively manufactured parts of austenitic stainless steels 316 L and 304 mainly contain austenite, occasionally with delta ferrite and/or martensite. Duplex stainless steels are designed to have an equal mix of ferrite and austenite. Rapid cooling hinders the formation of austenite, requiring heat treatment to achieve the desired amount of austenite.
- As-printed nickel alloy microstructures consist of an FCC (γ) matrix phase with various precipitates like carbides, Laves, γ', and γ''. Micro-segregation promotes carbide and Laves phase formation.
- During additive manufacturing of aluminum alloys, precipitates such as silicon-rich ones form in AlSi10Mg and AA 4043.
- During AM of Ti-6Al-4V, large columnar β grains form upon solidification. As the component cools to room temperature, small α grains with a lamellar structure develop at the β grain boundary.

Control of microstructure

- Columnar grains which is detrimental for the mechanical properties can be avoided by forming equiaxed grains by achieving columnar to equiaxed transition (CET).
- CET have been achieved by (1) controlling temperature gradient and growth rate, (2) adding inoculants or grain refiners, and (3) applying external ultrasonic vibration.
- Texture can be controlled by adjusting process variables and scanning strategies.

Appendix – Meanings of a selection of technical terms

<u>Electron beam</u>: A focused beam of electrons used as a heat source in additive manufacturing.

<u>Fusion welding</u>: A welding process where the material is melted by a high-energy source such as an electric arc, laser, or electron beam, and the joint is made after solidification.

Heat source: A high-energy source that is used to supply the heat needed in additive manufacturing.

Laser beam: A laser beam is used as a heat source in additive manufacturing. The acronym laser stands for "light amplification by stimulated emission of radiation."

Mushy zone: A two-phase region containing both solid and liquid.

Nucleation: Random formation of new solid particles inside the liquid during solidification.

Peak temperature: Maximum temperature inside the molten pool.

Phase diagram: A type of plot of temperature versus composition used to show conditions at which thermodynamically distinct phases occur at equilibrium.

Scanning pattern: Pattern of the movement of the heat source to melt the material in additive manufacturing. Often referred to as the deposition pattern for the directed energy deposition process.

Single crystal: A material that has the same crystallographic orientation throughout the part. These parts do not have any grain boundaries.

Texture: Distributions of crystallographic orientation in polycrystalline materials.

Practice problems

1) What is the difference between homogeneous and heterogeneous nucleation?
2) A stainless steel 316 part is made using DED-L using 300 W laser power and 10 mm/s scanning speed. What should be the solidification morphology?
3) Two identical parts are made using DED-L of Inconel 718 using laser powers of 300 W and 600 W. All other process parameters are the same. Which part would have finer grains?
4) How can grain morphology be controlled during additive manufacturing?
5) How does the grain structure vary between different additive manufacturing processes?
6) It has been observed that stainless steel 316 parts made by PBF-EB show much finer secondary dendritic arm spacing than those in parts fabricated using DED-GMA. Explain why.
7) Two Inconel 718 parts "A" and "B" of identical geometry are made using the DED-L process. Both parts are made using the same process parameters except part "A" is built on a substrate of 5 mm thickness while part "B" uses a 20 mm thick substrate. Which part will exhibit smaller secondary dendritic arm spacing?
8) Between laser powder bed fusion and electron beam powder bed fusion, which one is more susceptible to nucleation at the surface? Explain why.
9) 3D printing is not typically used to grow single crystals. However, it has been used to repair single crystal parts and single crystals of some alloys have been grown in laboratories. What are the important factors in the growth of single crystals by additive manufacturing?
10) How can the columnar to equiaxed transition be facilitated during additive manufacturing of metallic components?
11) What are the challenges in achieving the desired microstructure in additively manufactured duplex steel parts? Recommend a strategy for addressing this issue.

References

1 Zhou, Z., Huang, L., Shang, Y., Li, Y., Jiang, L. and Lei, Q., 2018. Causes analysis on cracks in nickel-based single crystal superalloy fabricated by laser powder deposition additive manufacturing. *Materials & Design*, 160, pp.1238–1249.

2 Kou, S., 2003. *Welding Metallurgy*, New Jersey, USA.

3 Savage, W.F. and Hrubec, R.J., 1972. Synthesis of weld solidification using crystalline organic materials. *Welding Journal*, 51(5), pp.S260–S271.

4 Wei, H.L., Elmer, J.W. and DebRoy, T., 2017. Three-dimensional modeling of grain structure evolution during welding of an aluminum alloy. *Acta Materialia*, 126, pp.413–425.

5 Dinda, G.P., Dasgupta, A.K. and Mazumder, J., 2012. Evolution of microstructure in laser deposited Al–11.28% Si alloy. *Surface and Coatings Technology*, 206(8–9), pp.2152–2160.

6 Garibaldi, M., Ashcroft, I., Simonelli, M. and Hague, R., 2016. Metallurgy of high-silicon steel parts produced using selective laser melting. *Acta Materialia*, 110, pp.207–216.

7 Helmer, H., Bauereiß, A., Singer, R.F. and Körner, C., 2016. Grain structure evolution in Inconel 718 during selective electron beam melting. *Materials Science and Engineering: A*, 668, pp.180–187.

8 Koepf, J.A., Gotterbarm, M.R., Markl, M. and Körner, C., 2018. 3D multi-layer grain structure simulation of powder bed fusion additive manufacturing. *Acta Materialia*, 152, pp.119-126. .

9 Yadollahi, A., Shamsaei, N., Thompson, S.M. and Seely, D.W., 2015. Effects of process time interval and heat treatment on the mechanical and microstructural properties of direct laser deposited 316L stainless steel. *Materials Science and Engineering: A*, 644, pp.171–183.

10 Kiran, A., Koukolíková, M., Vavřík, J., Urbánek, M. and Džugan, J., 2021. Base plate preheating effect on microstructure of 316L stainless steel single track deposition by directed energy deposition. *Materials*, 14(18), article no.5129.

11 Chen, S., Guillemot, G. and Gandin, C.A., 2016. Three-dimensional cellular automaton-finite element modeling of solidification grain structures for arc-welding processes. *Acta Materialia*, 115, pp.448–467.

12 Grong, O., 1997. *Metallurgical Modelling of Welding*. Institute of Materials, London, UK.

13 Rodgers, T.M., Bishop, J.E. and Madison, J.D., 2018. Direct numerical simulation of mechanical response in synthetic additively manufactured microstructures. *Modelling and Simulation in Materials Science and Engineering*, 26(5), article no.055010.

14 Mukherjee, T., Wei, H.L., De, A. and DebRoy, T., 2018. Heat and fluid flow in additive manufacturing–Part II: Powder bed fusion of stainless steel, and titanium, nickel and aluminum base alloys. *Computational Materials Science*, 150, pp.369–380.

15 Wolff, S.J., Gan, Z., Lin, S., Bennett, J.L., Yan, W., Hyatt, G., Ehmann, K.F., Wagner, G.J., Liu, W.K. and Cao, J., 2019. Experimentally validated predictions of thermal history and microhardness in laser-deposited Inconel 718 on carbon steel. *Additive Manufacturing*, 27, pp.540–551.

16 Haselhuhn, A.S., Buhr, M.W., Wijnen, B., Sanders, P.G. and Pearce, J.M., 2016. Structure-property relationships of common aluminum weld alloys utilized as feedstock for GMAW-based 3-D metal printing. *Materials Science and Engineering: A*, 673, pp.511–523.

17 Knapp, G.L., Mukherjee, T., Zuback, J.S., Wei, H.L., Palmer, T.A., De, A. and DebRoy, T.J.A.M. 2017. Building blocks for a digital twin of additive manufacturing. *Acta Materialia*, 135, pp.390–399.

18 Elmer, J.W., Allen, S.M. and Eagar, T.W., 1989. Microstructural development during solidification of stainless steel alloys. *Metallurgical Transactions A*, 20, pp.2117–2131.

19 Zhang, K., Wang, S., Liu, W. and Shang, X., 2014. Characterization of stainless steel parts by laser metal deposition shaping. *Materials & Design*, 55, pp.104–119.

20 Wei, H.L., Knapp, G.L., Mukherjee, T. and DebRoy, T., 2019. Three-dimensional grain growth during multi-layer printing of a nickel-based alloy Inconel 718. *Additive Manufacturing*, 25, pp.448–459.

21 Antonysamy, A.A., Meyer, J. and Prangnell, P.B., 2013. Effect of build geometry on the β-grain structure and texture in additive manufacture of Ti6Al4V by selective electron beam melting. *Materials Characterization*, 84, pp.153–168.

22 Wei, H.L., Mazumder, J. and DebRoy, T., 2015. Evolution of solidification texture during additive manufacturing. *Scientific Reports*, 5(1), pp.1–7.

23 Nie, P., Ojo, O.A. and Li, Z., 2014. Numerical modeling of microstructure evolution during laser additive manufacturing of a nickel-based superalloy. *Acta Materialia*, 77, pp.85–95.

24 Dinda, G.P., Dasgupta, A.K., Bhattacharya, S., Natu, H., Dutta, B. and Mazumder, J., 2013. Microstructural characterization of laser-deposited Al 4047 alloy. *Metallurgical and Materials Transactions A*, 44, pp.2233-2242.

25 Baufeld, B., Brandl, E. and Van der Biest, O., 2011. Wire based additive layer manufacturing: comparison of microstructure and mechanical properties of Ti–6Al–4V components fabricated by laser-beam deposition and shaped metal deposition. *Journal of Materials Processing Technology*, 211(6), pp.1146–1158.

26 Revilla, R.I., Van Calster, M., Raes, M., Arroud, G., Andreatta, F., Pyl, L., Guillaume, P. and De Graeve, I., 2020. Microstructure and corrosion behavior of 316L stainless steel prepared using different additive manufacturing methods: a comparative study bringing insights into the impact of microstructure on their passivity. *Corrosion Science*, 176, article no.108914.

27 Rankouhi, B., Bertsch, K.M., de Bellefon, G.M., Thevamaran, M., Thoma, D.J. and Suresh, K., 2020. Experimental validation and microstructure characterization of topology optimized, additively manufactured SS316L components. *Materials Science and Engineering: A*, 776, article no.139050.

28 Neikter, M., Åkerfeldt, P., Pederson, R. and Antti, M.L., 2017, October. Microstructure characterisation of Ti-6Al-4V from different additive manufacturing processes. In *IOP Conference Series: Materials Science and Engineering, IOP Publishing*, 258(1), article no.012007.

29 Attar, H., Ehtemam-Haghighi, S., Kent, D., Wu, X. and Dargusch, M.S., 2017. Comparative study of commercially pure titanium produced by laser engineered net shaping, selective laser melting and casting processes. *Materials Science and Engineering: A*, 705, pp.385–393.

30 Martin, J.H., Yahata, B.D., Hundley, J.M., Mayer, J.A., Schaedler, T.A. and Pollock, T.M., 2017. 3D printing of high-strength aluminium alloys. *Nature*, 549(7672), pp.365–369.

31 Todaro, C.J., Easton, M.A., Qiu, D., Zhang, D., Bermingham, M.J., Lui, E.W., Brandt, M., StJohn, D.H. and Qian, M., 2020. Grain structure control during metal 3D printing by high-intensity ultrasound. *Nature Communications*, 11(1), article no.142.

32 Ocelík, V., Furár, I. and De Hosson, J.T.M., 2010. Microstructure and properties of laser clad coatings studied by orientation imaging microscopy. *Acta Materialia*, 58(20), pp.6763–6772.

33 Körner, C., Ramsperger, M., Meid, C., Bürger, D., Wollgramm, P., Bartsch, M. and Eggeler, G., 2018. Microstructure and mechanical properties of CMSX-4 single crystals prepared by additive manufacturing. *Metallurgical and Materials Transactions A*, 49, pp.3781–3792.

34 Elmer, J.W., Palmer, T.A., Zhang, W. and DebRoy, T., 2008. Time resolved X-ray diffraction observations of phase transformations in transient arc welds. *Science and Technology of Welding and Joining*, 13(3), pp.265–277.

35 Mukherjee, T., Elmer, J.W., Wei, H.L., Lienert, T.J., Zhang, W., Kou, S. and DebRoy, T., 2023. Control of grain structure, phases, and defects in additive manufacturing of high-performance metallic components. *Progress in Materials Science*, 138, article no. 101153.

9

Properties

Learning objectives

After reading this chapter the reader should be able to do the following:

1) Understand the tensile properties, strength, ductility, and toughness of additively manufactured parts and their anisotropic nature.
2) Calculate the yield strength of different additively manufactured parts.
3) Compare different parts based on their tensile properties.
4) Understand hardness and its dependence on process conditions, alloy composition, and microstructure, and calculate its value for different processes and process conditions.
5) Obtain a basic understanding of fracture toughness, fatigue, and creep properties of parts and the important factors that affect them.
6) Compare the mechanical properties of additively manufactured parts with those of cast and wrought products.
7) Learn to select post-processing techniques to improve the mechanical properties of parts.
8) Understand the corrosion resistance of parts and the important factors influencing them.

CONTENTS

Theory and Practice of Additive Manufacturing, First Edition. Tuhin Mukherjee and Tarasankar DebRoy.
© 2024 John Wiley & Sons, Inc. Published 2024 by John Wiley & Sons, Inc.

9.1   Introduction

External loads, high temperatures, and harsh environments affect the serviceability of additively manufactured parts. For example, an additively manufactured hip implant experiences repeated loadings when the patient walks. Prolonged exposure of a printed turbine blade to high temperatures may weaken the blade. Chemicals in fuels can corrode printed aeroengine fuel nozzles. Knowledge of the properties of the printed parts can help engineers to design sound parts that can safely withstand service conditions and avoid premature failure. Mechanical properties such as strength, ductility, toughness, creep, fracture, and fatigue resistance are indicators of the ability of a part to withstand service conditions. The capability of a part to withstand a harsh environment is often indicated by its corrosion resistance. These properties are determined by performing various standardized laboratory tests. In those tests, the nature of the applied load and its duration, as well as the environmental conditions, are carefully considered.

In this chapter, several important mechanical properties of AM parts, tensile strength, ductility, toughness, hardness, fracture toughness, fatigue, and creep are discussed. Important factors controlling these properties are summarized. Several post-processing methods commonly employed to improve the mechanical properties of AM parts are described. Finally, the corrosion resistance of the parts and their dependence on several important process parameters are discussed.

9.2   Mechanical properties

Important mechanical properties of AM components include tensile properties such as strength, ductility, and toughness, as well as hardness, creep and fatigue properties, and fracture toughness. In this section, these mechanical properties, and several factors that influence them for different processing conditions are discussed.

9.2.1   Tensile properties

Tensile loading is very common in engineering applications. During the tensile test, the specimen is mounted by its two ends into the holding grips of the testing apparatus (Figure 9.1). The tensile testing machine applies force or load to elongate the specimen at a constant rate. The applied load and the resulting elongation of the specimen are continuously measured by an extensometer attached to the specimen. Two load cells attached to the two ends of the specimen provide the values of the load applied. The shape and size of the specimen are standardized. For example, a commonly used specimen according to the ASTM standards is a "dog bone"-shaped specimen with a circular cross-section. However, rectangular specimens are also used. The "dog bone"-shaped specimen is selected to restrict the deformation to the center region and to reduce the likelihood of fracture at the ends of the specimen. The standard diameter and gauge lengths are 12.8 mm and 50 mm, respectively. A tensile test typically takes several minutes to perform, and it provides the applied load versus elongation data from which stress versus strain plots are made. From these plots, the important tensile properties, tensile yield strength, ultimate tensile strength, and ductility are estimated as discussed below.

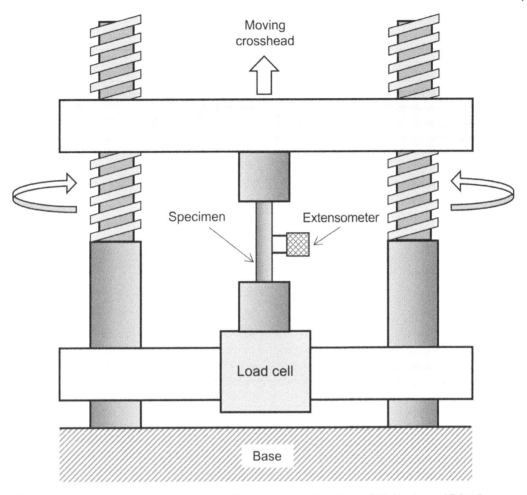

Figure 9.1 Schematic representation of a tensile testing apparatus. *Source:* T. Mukherjee and T. DebRoy.

9.2.1.1 Strength, ductility, and toughness

The degree to which a metallic material deforms or accumulates strains depends on the magnitude of the applied stress. For most metals and alloys at relatively low levels of stress, stress and strain are proportional to each other through a linear relationship. Deformation in which stress and strain are directly proportional is called elastic deformation. Therefore, a plot of stress versus strain results in a linear relationship, as shown in Figure 9.2 (a). In this figure, "A" to "B" represents elastic deformation. The slope of this linear segment is called the modulus of elasticity. This modulus represents the stiffness or resistance to elastic deformation. The higher the modulus, the stiffer the material, or the smaller the elastic strain that results from applied stress. The elastic modulus increases with decreasing temperature. Table 9.1 provides values of the elastic modulus of commonly used engineering alloys at different temperatures. Within the elastic region, the ability of a material to absorb energy is called resilience. The energy is absorbed by the material when it is deformed elastically and is released upon unloading. This is represented as the maximum energy that can be absorbed up to the elastic limit, without creating a permanent distortion. Mathematically, this energy is equal to the area under the stress vs. strain curve (Figure 9.2 (a)) within the elastic limit (point "B").

Table 9.1 Temperature-dependent modulus of elasticity of alloys commonly used in additive manufacturing [1, 2].

Stainless steel 316		Inconel 718		Ti-6Al-4V	
Temperature, K	Modulus of elasticity (GPa)	Temperature, K	Modulus of elasticity (GPa)	Temperature, K	Modulus of elasticity (GPa)
300	191.2	300	156.3	300	125
400	183.6	366.5	151.8	533	110
500	175.9	477.6	144.9	589	100
600	168.1	588.7	138	700	93
700	160.1	699.8	131.4	755	80
800	152.0	810.9	124.7	811	74
900	143.8	922	124	923	55
1000	135.5	1033.2	123.4	1073	27
1100	127.0	1144.3	107.7	1098	22
1200	118.1	1255.4	92.05	1123	18
1300	108.4	1366.5	68.95	1573	12
1400	99.7	1672	23.79	1873	9
1500	89.6	-	-	-	-
1640	59.5	-	-	-	-

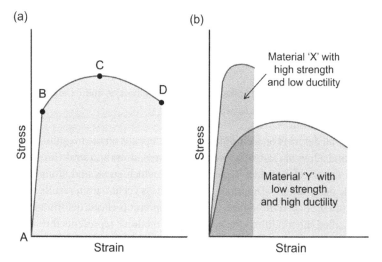

Figure 9.2 (a) Stress versus strain curve of a material. The linear variation, AB, represents the elastic deformation region. The stresses corresponding to "B" and "C" represent the yield stress and ultimate tensile stress (strength) of the material. "D" represents the point of failure. The plastic strain at point "D" represents the ductility. The area under the curve (shaded region) is related to the toughness of the component for a static low strain rate situation [3, 4]. (b) Comparison of stress versus strain curves of two materials "X" and "Y" with different toughness. *Source:* T. Mukherjee and T. DebRoy.

When a material is deformed beyond point "B" in Figure 9.2 (a), the stress is no longer linearly proportional to strain, and irreversible, permanent, plastic deformation occurs. A calculation of maximum elongation that can be sustained by a load-bearing part without any permanent deformation is presented in worked out example 9.1. The tensile stress-strain behavior in the plastic region for a typical material can also be seen in Figure 9.2 (a). Most components are designed to ensure that only elastic deformation will result when stress is applied to avoid permanent deformation. A component that has plastically deformed, or experienced a permanent change in shape, may not be capable of functioning as intended. It is therefore desirable to know the stress level at which plastic deformation begins, or where the phenomenon of yielding occurs (beyond point "B" in Figure 9.2 (a)). The stress corresponding to the yield point (B) is called yield stress. After yielding, the stress necessary to continue the plastic deformation increases to a maximum, point "C" in Figure 9.2 (a). The stress that corresponds to this point is called the ultimate tensile strength or the strength of the material. Beyond this point, the stress decreases to the eventual fracture point "D." The stress versus strain curve depends on the crystal structure, microstructure, and temperature. For example, the modulus of elasticity depends on both the temperature and crystal structure of the alloy. The plastic deformation is controlled by both the temperature and the rate of deformation or strain rate. The plastic deformation as well as the yield and ultimate tensile stresses vary significantly depending on the microstructure. For example, yield strength (σ_y) is related to the average grain size (d) through the Hall-Petch relation [5, 6]:

$$\sigma_y = \sigma_0 + k(d)^{-1/2} \tag{9.1}$$

where σ_0 and k are alloy-specific constants. This relation has been validated experimentally for many alloys over a range of grain sizes.

Worked out example 9.1

A 5 cm long machine part, made of stainless steel 316, is designed to withstand tensile loading. Calculate the maximum elongation it can endure without suffering permanent (irreversible) deformation. The yield strength of 316 stainless steel is 290 MPa and its modulus of elasticity is 193 GPa. If a part of similar size is made of Ti-6Al-4V or aluminum alloy 7075 what would be the maximum elastic deformations possible? The elastic modulus values for Ti-6Al-4V and aluminum alloy 7075 are 113 and 71.7 GPa, respectively and their yield strengths are 880 MPa and 503 MPa, respectively.

Solution:

The yield strength represents the maximum elastic stress a part can endure.
   E = stress/strain = Y/(ΔL/L$_0$) where E is the modulus of elasticity, Y is yield strength, ΔL is the elongation, and L$_0$ is the initial length.
   Solving ΔL = 75 µm which is the maximum elastic deformation possible for 316 stainless steel.
   Following a similar procedure, the maximum possible elastic deformations for Ti-6Al-4V and aluminum alloy 7075 are 389 and 351 µm, respectively. Titanium and aluminum alloys can sustain more deformations than stainless steel for the conditions specified in this problem.

Ductility is a measure of the degree of plastic deformation that has been sustained by the component at fracture. Ductility can be expressed quantitatively as either percent elongation or percent reduction in area. The percent elongation is the percentage of plastic strain at fracture which is the strain corresponding to point "D" in Figure 9.2 (a). Knowledge of the ductility of metallic materials is important for two main reasons. First, it indicates the degree to which a component will deform plastically before fracture. Second, it specifies the degree of allowable deformation during manufacturing.

The toughness of a material is its ability to absorb energy and plastically deform before fracturing. For the static low strain rate situation, toughness is related to the total area under the stress-strain curve (Figure 9.2 (a)) up to the point of fracture (point "D"). Materials with a larger area under the stress-strain curve are tougher. Figure 9.2 (b) shows the stress-strain curves for high- and low-toughness materials. The material "X" has a higher tensile strength than the other material "Y." However, the material "Y" is more ductile and has a greater total elongation. The total area under the stress-strain curve is greater for "Y," and therefore it is a tougher material under a static low strain rate situation. Worked out examples 9.2 and 9.3 allow comparisons of mechanical properties of different alloys from the tensile testing data.

Figure 9.3 summarizes the ultimate tensile strength of Ti-6Al-4V parts fabricated using additive manufacturing and other conventional manufacturing processes. In many cases, the tensile properties of parts fabricated with different AM processes are comparable to those using casting and forging as well as wrought materials. However, rapid scanning in AM often results in a faster cooling rate than the conventional processes which refines the grains and increases the tensile strength. Therefore, the strength of AM parts may in some cases be somewhat higher than those made using conventional processes. The higher strength often causes a reduction in ductility and toughness. Therefore, AM parts are heat-treated (HT) in many cases to regain the ductility. In addition, AM parts are susceptible to have pores due to gas entrapment or lack of fusion (as discussed in Chapter 10). Under tensile stress, cracks may originate from pores which lead to premature failure of the components. Therefore, pores significantly degrade the tensile strength as well as the ductility (elongation) of AM parts as evident from Figure 9.4. These pores are often minimized by post-process hot isostatic pressing (HIP). HIP can enhance ductility by reducing porosity. However, post-process heat treatment and HIP add cost to an expensive AM process.

Long columnar grains affect the tensile properties of AM parts. Therefore, significant efforts have been made to transform long columnar grains into smaller equiaxed grains. This process is called columnar to equiaxed transition (CET). CET was achieved by adding a small amount of Zr to Al 7075. Smaller equiaxed grains formed because of Zr addition result in improvements in strength and

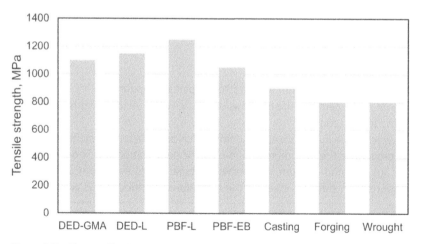

Figure 9.3 The tensile yield strength, ultimate tensile strength, and ductility (% elongation) of Ti-6Al-4V parts fabricated using additive manufacturing and other conventional manufacturing processes. The figure is plotted by T. Mukherjee and T. DebRoy using the data reported in [7].

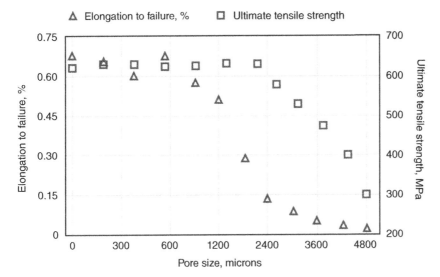

Figure 9.4 Effect of lack of fusion pore size on elongation to failure and ultimate tensile strength of stainless steel 316 parts printed using PBF-L. The figure is plotted by T. Mukherjee and T. DebRoy using the data reported in [8].

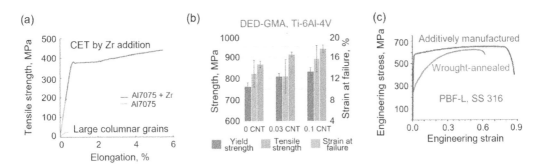

Figure 9.5 (a) A columnar to equiaxed transition (CET) and an improvement of toughness occur on addition of Zr nanoparticles in Al7075 aluminum alloy during PBF-L. The toughness improves because of the formation of equiaxed grains. (b) Ti-6Al-4V with carbon nanotubes fabricated by DED-GMA improves microstructure and mechanical properties. The addition of carbon nanotubes improves the 0.2% proof stress, tensile strength and strain at failure. (c) Both improved strength and ductility can be achieved in a stainless-steel part fabricated by DED-L. The stress-strain plots of additively manufactured components show superior strength and ductility compared to the annealed wrought specimens. The plots are made by T. Mukherjee and T. DebRoy using the data reported in [9].

ductility as observed in Figure 9.5 (a). However, knowledge of the composition, size, and amount of the additives is required to successfully implement this method. The addition of up to 0.1 wt% carbon in Ti-6Al-4V improved tensile strength and ductility (Figure 9.5 (b)) because of a decrease in prior beta grain size and alpha lath length. The addition of carbon formed titanium carbide nanoparticles that acted as sites for heterogeneous nucleation. Such heterogeneous nucleation can promote multiple small equiaxed grains. However, the amount of the additive is determined experimentally and more work is needed to better understand how they can be dispersed uniformly within the molten pool. Both the strength and ductility (Figure 9.5 (c)) of stainless steel components were simultaneously increased by achieving a hierarchal microstructure with the presence of twinning. Twinning is a contributing factor to improved ductility. The high strength has been attributed to fine cellular solidification pinned by solute segregation, low and high-angle grain boundaries, and unique dislocation substructures [9]. The simultaneously increased strength and ductility in a printed part compared with a wrought alloy is also observed in this figure. From the above discussion, it is evident that the tensile properties of AM components depend on the microstructure. The microstructure in AM parts is often inhomogeneous which results in anisotropy in mechanical properties as discussed below.

Worked out example 9.2

Table E9.1 shows the tensile properties of three additively manufactured alloy parts. Which one of them is the most ductile? Which part has the most toughness under a static low strain rate condition?

Table E9.1   The tensile properties of three additively manufactured alloy parts.

Part	Yield strength, MPa	Ultimate strength, MPa	Strain at fracture
A	200	300	0.25
B	200	300	0.40
C	100	200	0.10

Solution:

Part B has the maximum strain at fracture and is the most ductile. Part B is tougher than A since they have the same strength, but B has higher ductility. B is also tougher than C since B has higher strength and ductility than C. An accurate value of the toughness is determined by the impact test.

Worked out example 9.3

An additively manufactured part is to be designed with the most resilient among the following four available alloys. Use the following data in Table E9.2 to select the appropriate alloy.

Table E9.2   The mechanical properties of four alloys.

	Elastic modulus (GPa)	Yield strength (MPa)	Tensile strength (MPa)
316 Stainless steel	193	290	580
Ti-6Al-4V	113	880	950
Al7075	71.7	503	572
IN718	156	1035	1240

Solution:

The reliance is represented as the area under the stress vs. strain curve within the elastic region. Therefore, it is equal to (1/2 × yield strength × strain corresponding to the yield strength) = (yield strength)^2/(2 × elastic modulus). The computed values of resilience for stainless steel, Ti-6Al-4V, Al7075, and IN718 are 2.18×10^5, 34.3×10^5, 17.6×10^5, and 34.3×10^5 J/m^3, respectively. Since IN718 and Ti-6Al-4V have almost the same resilience, either one of these two alloys would be appropriate.

9.2.1.2 Anisotropy in tensile properties

Tensile properties of AM parts are often anisotropic which means the properties vary depending on the direction of the tensile force applied. The tensile properties of a specimen along the build direction may significantly differ from that along the scanning direction. This anisotropy or the directional dependence of tensile properties is primarily attributed to the inhomogeneity in the microstructure resulting from the difference in the local thermal cycles. For example, Figure 9.6 (a) shows a sample with long columnar grains along the build direction. If the sample is pulled in the scanning direction the grain boundaries of the columnar grains may separate from each other and cracks may grow from the grain boundaries. The opening up of the cracks reduces the ductility along the scanning direction. In contrast, long columnar grains along the build direction can elongate if the sample is pulled along the build direction. Therefore, the sample is more ductile along the build direction as shown in Figure 9.6 (b). Since the tensile strength of a specimen depends on many factors, there is often no clear trend of yield strength and ultimate tensile strength depending on the direction. Often the strength in the scanning direction is a little higher than that in the build direction (Figure 9.6 (b)) because of the presence of many grain boundaries along the scanning direction that may hinder the dislocation motion.

To reduce the anisotropy in tensile properties, large columnar grains are often broken into small equiaxed grains. This is called columnar to equiaxed transition (CET). Several common techniques to achieve CET in additive manufacturing are described in Chapter 8. However, anisotropy in tensile properties may also originate from the presence of hard and brittle secondary phases and cracks. Post-process heat treatment and hot isostatic pressing to reduce the amount of cracks and secondary phases can often be beneficial to reduce the anisotropy in tensile properties.

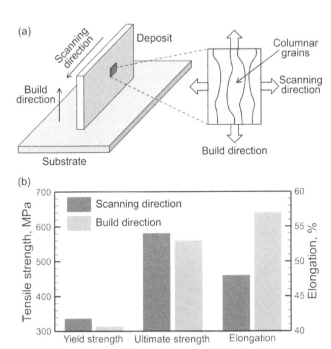

Figure 9.6 (a) A schematic representation of a sample taken from a deposit indicating the presence of long columnar grains along the build direction. *Source:* T. Mukherjee and T. DebRoy. (b) Anisotropy in tensile properties in stainless steel 304 parts printed using DED-L. The plot is made by T. Mukherjee and T. DebRoy using the data from [10].

9.2.2 Hardness

Hardness is an important mechanical property that indicates a material's resistance to localized plastic deformation. It is commonly measured using a small indenter forced into the surface of a material to be tested with a constant rate of force. The depth or size of the resulting indentation is measured from which hardness is calculated. There are different scales to measure hardness that uses different types of indenter, force, and formula to calculate the hardness as summarized in Table 9.2. In these tests, the specimen is neither fractured nor excessively deformed. Only a small indentation is made on the surface.

Table 9.2 Different hardness measurement techniques. Adapted from [4].

Test scale	Indenter	Shape of indentation Side view	Top view	Load	Formula to calculate the hardness
Brinell	10-mm sphere of steel or tungsten carbide			P	$\dfrac{2P}{\pi D\left[D-\sqrt{D^2-d^2}\right]}$
Vickers	Diamond pyramid			P	$1.85\ P/d_1^2$
Knoop	Diamond pyramid	$l/b = 7.11$ $b/t = 4.00$		P	$14.2\ P/l^2$
Rockwell A	Diamond cone			60 kg	
Rockwell B	1/16 inches diameter steel sphere			100 kg	
Rockwell C	Diamond cone			150 kg	

For most of the common alloys used in additive manufacturing, hardness in the Vickers scale is linearly proportional to the average tensile yield strength because both measure the resistance to deformation. Figure 9.7 (a) shows that the hardness for stainless steel 316 parts made with both PBF-L and DED-L has a linear relationship with the average tensile yield strength. Therefore, hardness (H) in the Vickers scale can be calculated from the average yield strength (σ_y) in MPa, as:

$$H = 0.3\ \sigma_y (0.1)^{-m} \tag{9.2}$$

where "m" is an alloy-specific constant. In Eq. 9.2, the calculated hardness is microhardness. Worked out example 9.4 calculates tensile yield strength and Vickers hardness from the average grain size.

Worked out example 9.4

Calculate the hardness in Vickers scale for the stainless steel 316 part made using laser powder bed fusion process. The average grain size of the part is 1.79 µm. Values of the Hall-Petch constants σ_0 and k for stainless steel 316 are 240 MPa and 279 MPa(µm)$^{1/2}$, respectively.

Solution:

Tensile yield strength (σ_y) can be calculated from the grain size (d) in the Hall-Petch relation (Eq. 9.1) as:

$$\sigma_y = \sigma_0 + k(d)^{-1/2}$$

For the given values of σ_0 and k, the average tensile yield strength is obtained from the above equation as 448 MPa. The hardness can be calculated using Eq. 9.2. The value of "m" is taken as 0.25 for stainless steel 316, and the calculated hardness in the Vickers scale is 239.

Several factors influence the hardness of additively manufactured parts as summarized in Table 9.3. For example, alloy compositions significantly govern the hardness of the components as shown in Figure 9.7 (b-d). In the welding literature, there is a weldability parameter (P_{cm}) for steels which is correlated with the Vickers hardness.

Table 9.3 Factors affecting the hardness of additively manufactured components.

Affecting factors	Explanation
Heat source power	High heat source power results in slow cooling and thus large grains. Large grains reduce strength and hardness.
Scanning speed	Rapid scanning causes fast cooling and fine grains which increase both the strength and hardness.
Preheat temperature	High preheat temperature reduces the cooling rate, increases the grain size, and diminishes both strength and hardness.
Chemical composition	For all common alloys, hardness varies significantly with composition. A few examples are provided in Figure 9.7 (b-d).
Microstructure	Small grains result in high strength and hardness. In addition, hard precipitates of secondary phases such as laves phase in nickel alloys may increase hardness.
Residual stresses	The presence of residual stress can affect the deformability of the surface. For example, the presence of high compressive residual stresses requires more force to pierce the indenter to the surface which increases the hardness.

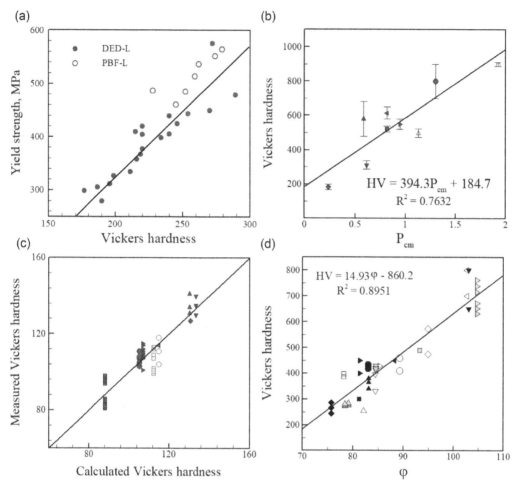

Figure 9.7 (a) Linear correlation between the yield strength and Vickers hardness of additively manufactured stainless steel 316 parts. (b) Linear relation between the Vickers hardness and weldability (P_{cm}) of steels. (c) Comparison of computed and measured hardness in Vickers scale of additively manufactured aluminum alloy parts. (d) A linear relationship between the Vickers hardness and an equivalent term (φ in Eq. 9.5) for nickel alloys. All plots are adapted from an open-access article [11] under the terms and conditions of the Creative Commons Attribution (CC BY) license.

$$P_{cm} = C + Si/30 + (Mn + Cu + Cr)/20 + Ni/60 + Mo/15 + V/10 + 5B \tag{9.3}$$

where C, Si, Mn, etc. represent concentration of these elements in wt.%. The formula is valid for 0.02–0.99 wt% C, 0–13.3 wt% Cr, 0.2–1.62 wt% Mn, 0.06–7.97 wt% Mo, 0.15–18.8 wt% Ni, 0.29–1.02 wt% Si, 0.03–2.01 wt% V, and 0–6.32 wt% B. The same correlation is also applicable to AM parts as shown in Figure 9.7 (b), where hardness in the Vickers scale has been shown to linearly correlate with P_{cm}. Similarly, for aluminum alloys, hardness in the Vickers scale can be represented as a function of chemical composition (in wt.%) as:

$$HV = 37.99 + 19.47Ag + 2.85Cu + 23.36Fe + 24.47Mg$$
$$+ 30.00Mn + 5.43Si + 20.86Ti + 19.06Zn \tag{9.4}$$

where Ag, Cu, Fe, etc. represent concentration of these elements in wt.%. The formula is valid for 0–0.5 wt% Ag, 0–5.3 wt% Cu, 0–0.8 wt% Fe, 0–1.95 wt% Mg, 0–0.55 wt% Mn, 0–12.2 wt% Si, 0–0.064wt % Ti, and 0–0.1 wt% Zn. Hardness calculated using this formula is found to match well with the measured hardness for different aluminum alloys as shown in Figure 9.7 (c). For nickel alloys, an equivalent weldability term (φ) has been defined which is linearly correlated with the hardness in the Vickers scale as shown in Figure 9.7 (d). The term is defined as:

$$\varphi = Ni + 0.65Cr + 0.98Mo + 1.05Mn + 0.35Si + 12.6C - 6.36Al + 3.80B + 0.01Co \\ + 0.26Fe + 7.06Hf + 1.20Nb + 4.95Ta + 5.78Ti + 2.88W \tag{9.5}$$

where Ni, Cr, Mo, etc. represent concentration of these elements in wt.%. Eq. 9.5 is valid for 0–6.5 wt% Al, 0–3.75 wt% B, 0–0.5 wt% C, 0–19.2 wt% Co, 0–21.8 wt% Cr, 0–24.7 wt% Fe, 0–1.5 wt% Hf, 0–0.48 wt% Mn, 0–9.75 wt% Mo, 0–5.1 wt% Nb, 0–4.25 wt% Si, 0–6.35 wt% Ta, 0–4.7 wt% Ti, and 0–4.9 wt% W. Worked out example 9.5 calculates Vickers hardness of aluminum alloys from the chemical composition.

Worked out example 9.5

Parts are made of three aluminum alloys by additive manufacturing. Part A contains 0.5% Cu, 0.01% Fe, and other alloying elements. Part B contains 4.5% Cu and the concentrations of the other alloying elements same as that in A. Part C has the same concentrations of alloying elements as A, except the iron concentration. If the Vickers hardness of B and C are the same, determine the concentration of Fe in alloy C.

Solution:

From Eq. 9.4, $HV_A = HV_0 + 2.85 \times 0.5 + 23.4 \times 0.01$
   $HV_B = HV_0 + 2.85 \times 4.5 + 23.4 \times 0.01$
   $HV_C = HV_0 + 2.85 \times 0.5 + 23.4 \times C_{Fe}$
where HV_A, HV_B, and HV_C are Vickers hardness of parts A, B, and C, respectively, HV_0 is a constant, and C_{Fe} is the concentration of Fe in wt% in part C.
   Since $HV_B = HV_C$, $2.85 \times 4 = 23.4 \times (C_{Fe} - 0.01)$
   $C_{Fe} = 0.497$ wt%
   Note that the concentration of Fe is within the range of validity of the concentration of iron 0 to 0.8 wt%.

In AM parts, the microstructure may change spatially because of the local variations of thermal cycles and the repeated heating and cooling during multilayer deposition. These variations result in spatial and temporal variations in hardness [12]. For example, Figure 9.8 (a) shows the thermal cycle at location "A" for the deposition of four layers of H13 tool steel by DED-L. Location "A" experiences maximum temperature during the deposition of the first layer. The temperature at that location is reduced as the laser beam moves away from this location during the deposition of the subsequent layers. The repeated heating and cooling affect the hardness at "A." For example, after the deposition of the first layer, the hardness at "A" is very high (around 325 VHN in Figure 9.8 (b)) due to the formation of martensite. Repeated heating and cooling during the deposition of subsequent layers transform martensite to tempered martensite (increasing % of tempering in Figure 9.8 (b)) containing carbides rich in vanadium and chromium. The tempering decreases the

(a)

(b)

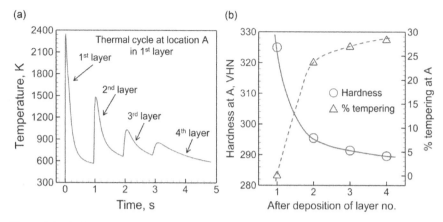

Figure 9.8 (a) Thermal cycle at location "A" in the first layer while depositing the 4 layers. (b) Calculated variations in hardness and percentage tempering of martensite at location "A" during the deposition of 4 layers. The data are for DED-L of H13 tool steel using 250 W laser power and 8.47 mm/s scanning speed. The results are obtained from the model developed by T. Mukherjee and T. DebRoy.

hardness. Therefore, the hardness at location "A" decreases with the deposition of the upper layers as shown in Figure 9.8 (b). AM parts are often heat-treated after the fabrication process to achieve the desired mechanical properties. However, the post-process heat treatment adds an extra cost to an already expensive printing process.

9.2.3 Fracture toughness

Fracture is defined as the separation of a part into two or more pieces in response to applied stress. The fracture process involves two steps, crack initiation and propagation, and the fracture mode largely depends on the mechanism of crack propagation. Fracture toughness is a mechanical property that indicates the ability of a material to resist fracture and can be measured by tests such as the Charpy impact test. Therefore, it depends on the applied stress and the shape and size of the flaw or cracks. For a very thin plate, fracture toughness depends on the thickness of the plate. However, for a metallic AM part, plane strain conditions can be assumed in most cases because the part dimensions are significantly larger than the dimensions of the plasticized region at the crack tip. Under these conditions, negligible plastic deformation occurs at the crack tip, and the fracture toughness (K_{Ic}) can be expressed as:

$$K_{Ic} = Y\sigma\sqrt{\pi a} \tag{9.6}$$

where Y is a geometric constant, σ is the applied stress, and a is related to the size of the crack that would lead to fast fracture for the applied stress and toughness. Two types of cracks are possible, surface cracks and internal cracks. For an internal crack, the length of the crack is 2a and for a crack on the surface, the length of the crack is a. There are nondestructive test methods such as ultrasound, X-ray, and CT scan to detect and measure both internal and surface cracks. In addition, the maximum applied stresses that a component can withstand before fracture is measured. From the measured crack size and applied stress, fracture toughness is predicted using Eq. 9.6. Therefore, it is common practice to plot the applied stress versus the square root of the crack size to compare the fracture toughness of different parts. For example, such a plot is used to compare the fracture properties of AM parts with that of the cast products as shown in Figure 9.9. Log scale is used for both variables for ease of data representation. The plot is made by collecting data from the literature for different AM conditions. The data in the plot shows that the fracture properties of AM parts are comparable to those of cast products and they vary significantly depending on the process conditions used. Several important factors that affect the fracture toughness of AM parts are summarized in Table 9.4. Worked out examples 9.6 and 9.7 relate the size of an internal crack and the maximum allowable stress to prevent fracture.

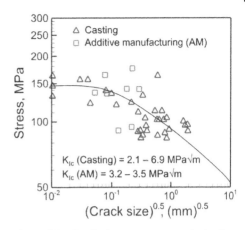

Figure 9.9 Applied stress versus crack size for AlSi10Mg alloy considering various literature data for casting and additive manufacturing. The plot is made by T. Mukherjee and T. DebRoy using the data from [13].

Worked out example 9.6

An aluminum alloy part made by powder bed fusion has a 20 μm long internal crack. If the aluminum has a fracture toughness of 5 MPa√m, what is the maximum stress the part can withstand to prevent fracture?

Solution:

Eq. 9.6 can be used to calculate the maximum stress. Since the crack is an internal crack, the effective crack size "a" should be the crack length/2 = 10 μm. Assuming, $Y = 1$ in Eq. 9.6, the maximum stress can be calculated as 892 MPa.

Worked out example 9.7

If an additively manufactured part is to be designed to withstand a tensile stress of 400 MPa. Assuming that the part contains a 4 mm long internal crack, select an appropriate alloy(s) from Table E9.3. Data: $Y = 1.2$ and the plane strain fracture toughness values are listed in Table E9.3.

Table E9.3 The critical stress and plain strain fracture toughness values.

	Plane strain fracture toughness (MPa(m)$^{1/2}$)	Critical stress (MPa)
316 Stainless steel	195	2050.6
Ti-6Al-4V	95	1000.0
Al7075	27	284.0
IN718	80	841

Solution:

For an internal crack, a = 2 mm. Critical stress is calculated from Eq. 9.6 and listed in the table. The only alloy that would not be able to withstand the stress is Al7075 because it can only withstand a maximum stress of 284 MPa.

Table 9.4 Factors affecting the fracture toughness of additively manufactured components.

Affecting factors	Explanation
Microstructure	The presence of hard and brittle intermetallic compounds or secondary phases may initiate cracks and reduce fracture toughness.
Presence of voids	Voids generated due to gas entrapment or lack of fusion can initiate cracks which can lead to fracture.

9.2.4 Fatigue properties

Fatigue failure may occur when a component is subjected to dynamic and fluctuating loading which is very common in many engineering applications. The failure occurs owing to progressive localized change in the structure of the component under fluctuating stresses or strains. For example, an additively manufactured hip implant experiences fluctuating load while the patient walks. Under such fluctuating loads, fatigue failure occurs by three distinct steps: (1) crack initiation, where a small crack forms inside the part due to the presence of high stress; (2) crack propagation, during which the crack advances incrementally with each cycle of the fluctuating stress; and (3) final catastrophic failure, which occurs very rapidly when the part cannot withstand the propagation of the crack.

A rotating-bending test apparatus, shown in Figure 9.10, is commonly used for fatigue testing. The compression and tensile stresses are imposed on the specimen by vertical loads as it is simultaneously bent and rotated by high-speed motors. The applied stress magnitude is estimated from the applied load. The number of cycles is monitored using a counter attached to the motor. Data are plotted as stress (S) versus the number of cycles (N) to failure for each specimen. The values of N are presented on a logarithmic scale.

The stress vs. the number of cycles (SN) plots are represented schematically in Figure 9.11. The plots indicate that the higher the magnitude of the stress, the smaller the number of cycles the component will withstand before failure. For some alloys, the SN curve (Figure 9.11 (a)) becomes horizontal at higher N values. Therefore, there is a limiting stress level which is called the fatigue limit or the endurance limit below which fatigue failure will not occur. This fatigue limit represents the largest value of stress magnitude that will not cause fatigue failure for an infinite number of cycles. For other alloys, a fatigue limit does not exist (Figure 9.11 (b)) and fatigue failure will ultimately occur regardless of the magnitude of the stress. Fatigue strength is defined as the stress level at which failure occurs for a specified number of cycles (Figure 9.11 (b)). Another important parameter is fatigue life (Figure 9.11 (b)) which is the number of cycles to cause failure at a specified stress level.

The fatigue properties of a component are characterized by the fatigue life, fatigue limit, and fatigue strength obtained from the SN (stress vs. the number of cycles) plots. Several factors influence the fatigue properties of AM parts as summarized in Table 9.5. They are often comparable to those of the wrought and cast products as shown in Figure 9.12. The figure compares the SN plot of AM parts with that for the conventionally processed materials. Post-processing is often used to improve fatigue properties. In some cases, the mechanical properties can even be better than those of conventionally processed materials. For example, hot isostatic pressing (HIP) is often used to reduce the voids in AM parts which significantly improves the fatigue properties. In addition, postprocess stress relieving is used to partially relax the residual stresses in AM components which also enhances fatigue properties.

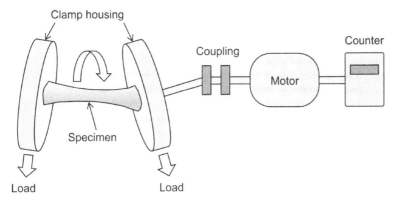

Figure 9.10 A schematic representation of fatigue testing of AM components. *Source:* T. Mukherjee and T. DebRoy.

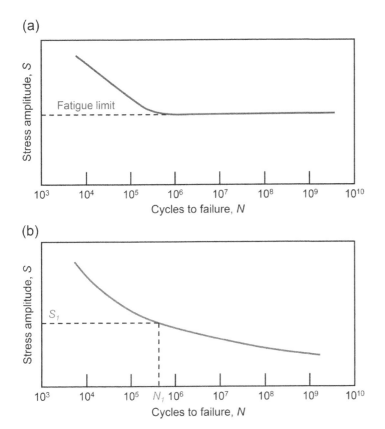

Figure 9.11 Stress amplitude (S) versus the logarithm of the number of cycles to fatigue failure (N) for (a) a material that displays a fatigue limit and (b) a material that does not display a fatigue limit where the component fails after N_1 cycles at a stress amplitude of S_1. *Source:* T. Mukherjee and T. DebRoy.

There is a threshold value of the stress range below which fatigue failure does not occur, when this value is exceeded, fatigue failure is often divided into three stages, crack initiation (step I), crack growth (step II), and final failure (stage III) as shown in Figure 9.13. In the first stage, a small crack forms at some point of high-stress concentration such as a surface scratch or dent. Under

Table 9.5 Factors affecting fatigue properties of additively manufactured components.

Affecting factors	Explanation
Residual stresses	The presence of high tensile residual stresses amplifies the applied stress, helps to propagate the cracks, and accelerates fatigue failure.
Presence of voids	Voids generated due to gas entrapment or lack of fusion can initiate cracks under cyclic loading and lower the fatigue life.
Surface roughness	Rough surfaces may act as sites of stress concentration under cyclic loading and affect fatigue life.
Microstructure	AM parts may contain hard and brittle secondary phases that may serve as crack initiation sites under fluctuating loads and promote fatigue failure.

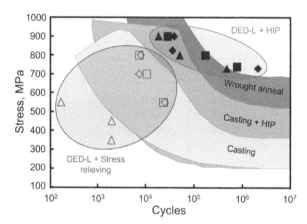

Figure 9.12 Comparison of SN (stress vs. the number of cycles) plot of Ti-6Al-4V parts fabricated using laser directed energy deposition and post-processed with other conventional manufacturing processes. HIP represents hot isostatic pressing. The figure is made by T. Mukherjee and T. DebRoy using the data from [14].

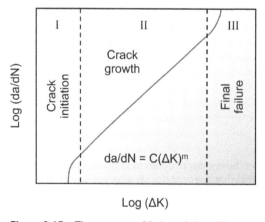

Figure 9.13 Three stages of fatigue failure. The equation in the figure represents the Paris law (Eq. 9.7). *Source:* T. Mukherjee and T. DebRoy.

cyclic loading, the crack grows in each cycle from an initial size a_0 to a critical size a_c at which stage a very rapid final failure occurs. An important task is to characterize the growth of the crack quantitatively from crack size a_0 to a_c with the number of cycles. The driving force of crack growth can be taken as the difference between the stress intensity factors corresponding to the highest and the lowest loads in each cycle. This stress intensity difference $\Delta K = K_{mx} - K_{mn}$ where K_{mx} and K_{mn} are the maximum and minimum stress intensity factors in a cycle of loading. The crack growth rate per cycle, da/dN, is then expressed as a function of the difference in the cyclic stress intensity factor ΔK. The relationship, known as the Paris law, is expressed as:

$$da/dN = C(\Delta K)^m \tag{9.7}$$

Here, a is the crack length, N is the number of cycles, ΔK is the stress intensity factor, and C and m are empirical constants that depend on the composition and geometry of the material, temperature, and other factors. Eq. 9.7 is based on experimental data and is strictly valid for the crack growth under stress which is much lower than the stress corresponding to the material's fracture toughness.

The critical crack size at the end of the second stage is determined by the fracture toughness (K_{Ic}). The number of cycles needed from the start to finish of the second stage of crack growth, N, is given by:

$$N = \int_{a_0}^{a_c} \frac{da}{c(\Delta K)^m} = \int_{a_0}^{a_c} \frac{da}{c[Y\Delta\sigma(\pi a)^{1/2}]^m} = \frac{1}{c[Y\Delta\sigma(\pi)^{1/2}]^m} \int_{a_0}^{a_c} \frac{da}{(a)^{m/2}}$$

$$= B[a_c^{1-m/2} - a_0^{1-m/2}] \text{ where } B = \frac{1}{c\left(1 - \dfrac{m}{2}\right)[Y\Delta\sigma(\pi)^{1/2}]^m} \tag{9.8}$$

If the stress range, $\Delta\sigma$, geometric factor, Y, and the constants c and m are known, the number of cycles of stress necessary for a crack to grow from an initial size a_0 to a critical crack size a_c can be calculated from the integrated form of the Paris equation (Eq. 9.8). Worked out example 9.8 illustrates this calculation method.

Worked out example 9.8

The crack growth rate in an alloy component under cyclic stress can be expressed by the Paris law as da/dN (m/cycle) $= 0.6 \times 10^{-8}(\Delta K)^{2.3}$ where ΔK is expressed in MPa m$^{1/2}$. Assume an initial 1 mm long crack, a maximum tensile stress of 200 MPa, and a minimum stress of 0 MPa. The K_{IC} for this material is 40 MPa m$^{1/2}$. Assume a geometric factor of 1.18. Estimate the final crack size for fatigue failure and the number of cycles for failure.

Solution:

The number of cycles is given by Eq. 9.8:

$$N = B[a_c^{1-m/2} - a_0^{1-m/2}] \text{ where } B = \frac{1}{c\left(1 - \dfrac{m}{2}\right)[Y\Delta\sigma(\pi)^{1/2}]^m}$$

Here $c = 0.6 \times 10^{-8}$ expressed in a fairly complex unit of [m/cycle]/[MPa (m$^{1/2}$)]^m, $m = 2.3$, $Y = 1.18$, $a_0 = 10^{-3}$ m, a_c: to be calculated from fracture toughness and the maximum tensile stress $\Delta\sigma = 200$ MPa.

(Continued)

> **Worked out example 9.8 (Continued)**
>
> The final crack size from $K_{IC} = Y\,\sigma_m(\pi a_c)^{1/2}$ where σ_m is the maximum stress.
>
> $a_c = (40 / (1.18 \times 200))^2 / 3.14 = 0.00915\ m =$ final crack size for fatigue failure
>
> $$B = 1 / [0.6 \times 10^{-8}(1 - 1.15)(1.18 \times 200 \times 3.14^{1/2})^{2.3}] = -1686$$
>
> $$a_c^{1-m/2} - a_0^{1-m/2} = -0.573$$
>
> $$N = (-1686) \times (-0.573) = 966\ \text{cycles}$$

9.2.5 Creep

Additively manufactured components are often exposed to static mechanical stresses at elevated temperatures. Examples of such parts include turbine blades and aeroengine components. They may undergo permanent deformation after prolonged service. The permanent deformation at elevated temperature is called creep which is time-dependent and is affected by the magnitude of the static stress and temperature. Creep is often a limiting factor in the lifetime of a component at elevated temperatures.

In the creep test, a constant load or stress is applied to a specimen at an elevated temperature. The resultant strain is measured and plotted as a function of elapsed time. Figure 9.14 (a) represents a typical creep strain versus time plot at various stress and temperature combinations. The curve consists of three regions, each of which has its distinctive strain–time feature. At first, the slope of the curve diminishes with time indicating a continuously decreasing creep rate. This is called primary creep where the material experiences creep resistance due to strain hardening and deformation becomes more difficult as the material is strained. For secondary creep, the slope is a constant. The constant creep rate is attributed to the balance between strain hardening and recovery, where the recovery is a process to make a material soft and retain its ability to deform. Finally, for the tertiary creep, there is an acceleration in the creep rate indicated by the increasing slope of the curve with time and the component exhibits rupture or creep failure. The time taken for the rupture to occur is called rupture lifetime (Figure 9.14 (a)) for a combination of applied stress and temperature. Rupture lifetime varies significantly with both stress and temperature. Therefore, the results of creep rupture tests are presented as the stress versus rupture lifetime plot at different temperatures (Figure 9.14 (b)). For a constant temperature, high stresses result in a rapid rupture. Therefore, the rupture lifetime decreases with increasing stress. In addition, a high temperature softens the material which also exhibits rapid creep rupture. Worked out example 9.9 calculates the minimum temperature for creep to become an important concern for components made from various alloys.

The creep of a component depends on the applied stress and temperature. Table 9.6 summarizes several factors that affect the creep of additive manufactured parts. The creep of AM parts can differ from those of conventionally manufactured parts. For example, Figure 9.14 (c) shows the creep strain versus time plot for a nickel-base superalloy. Additively manufactured components exhibit fine grains due to rapid cooling which degrades the creep properties as compared to the cast products as shown in the figure. Post-process heat treatment can coarsen the grains and improve creep strength at a higher cost. Similarly, poor creep properties of AM parts compared to those of cast parts are evident from the applied stress versus rupture lifetime plot shown in Figure 9.14 (d).

The effect of temperature and stress on creep-rupture lifetime is often expressed in plots of stress versus a parameter known as the Larson-Miller parameter. This parameter, L, depends on both temperature and creep rupture life:

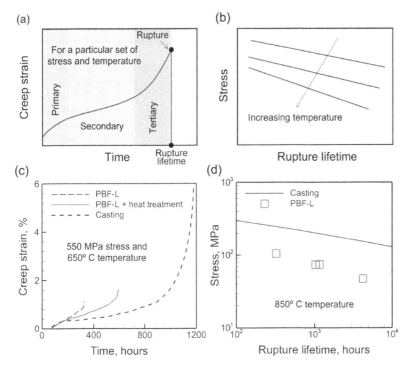

Figure 9.14 (a) A schematic representation of creep strain versus time plot for a particular set of applied stress and temperature. (b) A schematic representation of applied stress versus rupture lifetime plot. Source of figures (a) and (b) is T. Mukherjee and T. DebRoy. (c) Comparison of creep strain versus time plot for a nickel-base superalloy fabricated using powder bed fusion with laser (PBF-L) and casting. The figure is made by T. Mukherjee and T. DebRoy using the data from [15]. (d) Comparison of applied stress versus rupture lifetime plot of a nickel-base superalloy fabricated using PBF-L and casting. The figure is made by T. Mukherjee and T. DebRoy using the data from [16].

Table 9.6 Factors affecting creep of additively manufactured components.

Affecting factors	Explanation
Grain size	Smaller grains permit more grain boundary sliding which accelerates the creep rate and results in a rapid creep rupture.
Presence of voids	Voids generated due to gas entrapment or lack of fusion can weaken the component and accelerate creep rupture.
Solidification morphology	Equiaxed grains are smaller compared to columnar grains. Small equiaxed grains permit more grain boundary sliding which accelerates the creep rate and results in a rapid creep rupture. Therefore, a component with equiaxed grains shows poor creep properties compared to that having columnar grains.

$$L = T\left(C + \log t_r\right) \tag{9.9}$$

where T is the temperature in K, C is a constant, and t_r is the rupture lifetime in hours. A plot of stress in the logarithmic scale versus the Larson-Miller parameter is shown in Figure 9.15. The use of this parameter for solving a problem is presented in Worked out example 9.10 and shows the important effect of temperature on creep rupture life.

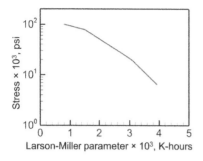

Stress × 10³, psi
Larson-Miller parameter × 10³, K-hours

Figure 9.15 A plot of stress versus Larson-Miller parameter (L). The scale for the x-axis shows the parameter in 1000 hours segments. *Source:* T. Mukherjee and T. DebRoy.

Worked out example 9.9

High temperatures and stresses are important factors in the deformation of materials by creep. Calculate the minimum temperature for creep to become an important concern for components made of the following alloys: 316 stainless steel, 7075 aluminum alloy, Nickel alloy 718, Ti-6Al-4V, and H13 tool steel. Their solidus temperatures are 1693, 750, 1533, 1878, and 1585 K, respectively.

Solution:

Creep is important when the temperature exceeds $0.4T_m$ where T_m is the solidus temperature in K. The values of $0.4T_m$ for these alloys are the following:
316 Stainless steel: 677.2 K
7075 Al: 300 K
IN 718: 613 K
Ti-6Al-4V: 751 K
H13: 634 K

Worked out example 9.10

The Larson-Miller parameter for an alloy is plotted in Figure 9.15. Calculate the creep rupture time if a component is subjected to a pressure of 200 MPa at 973K. Recalculate the rupture time if the temperature is raised to 1023 K. The constant C in Eq. 9.9 is 20.

Solution:

We use Eq. 9.9 to calculate creep rupture life. From Figure 9.15, the value of the Larson-Miller parameter is 22,500 h for 200 MPa.

$22,500 = 973(20 + \log t_r)$ or $20 + \log t_r = 23.12$ or $\log t_r = 3.12$
$t_r = 10^{3.12} = 1318.2$ hours $= 1318.2/24$ hour $= 54.9$ days
If the temperature is 1023 K, $22,500 = 1023(20 + \log t_r)$ or $20 + \log t_r = 21.99$ or $\log t_r = 1.99$
$t_r = 10^{1.99} = 97.7$ hours

A 50 K increase in temperature decreases the creep rupture lifetime from about 55 days to about 98 hours. A small change in temperature has a significant effect on the creep rupture lifetime.

9.3 Post-processing to improve mechanical properties

AM parts often undergo significant post-processing to improve their properties. Table 9.7 summarizes several commonly used techniques. Although these processes can improve properties, they add cost to an already expensive process of AM.

Table 9.7 Commonly used post-processing methods to improve mechanical properties of AM parts.

Property	Post-processing methods	Explanation
Tensile properties	Heat treatment	Post-process heat treatment can coarsen grains and increase ductility and toughness. In addition, heat treatment can dissolve unwanted brittle phases and improve tensile properties.
	Hot isostatic pressing	Hot isostatic pressing can reduce internal pores and improve tensile strength of parts.
Hardness	Surface hardening	Surface hardening methods, such as carburizing and nitriding, are often used to provide a hard coating on the part surface.
	Heat treatment	Heat treatment can dissolve some unwanted brittle phases and increase grain size both of which can decrease hardness.
	Stress relieving	Stress-relieving can alleviate tensile residual stresses on part surfaces and change hardness. Relaxation of high tensile stress on part surface improves hardness. Similarly, relaxation of the compressive residual stresses will reduce hardness.
Fatigue	Shot peening	Shot peening is a process in which small, hard metallic balls are impinged on the part surface to impart compressive residual stresses and resist crack growth.
	Stress relieving	Stress-relieving methods partially alleviate accumulated residual stresses and improve fatigue properties.
	Surface grinding	Grinding makes the surface smooth which reduces probable sites for stress concentration that may initiate fatigue cracks.
	Hot isostatic pressing	Hot isostatic pressing reduces internal flaws and cracks and improves fatigue properties.
Creep	Heat treatment	Heat treatment coarsens grains resulting in less grain boundary sliding which reduces the creep rate.
	Hot isostatic pressing	Hot isostatic pressing reduces internal flaws and cracks and improves creep properties.
Fracture toughness	Heat treatment	Heat treatment reduces the presence of hard and brittle intermetallic compounds or secondary phases that may initiate cracks and improve fracture toughness.
	Hot isostatic pressing	Hot isostatic pressing (HIP) reduces internal flaws and cracks that can initiate fracture. Therefore, HIPing increases fracture toughness.

9.4 Corrosion resistance

Alloys often react with materials in their environment which degrade their properties. In extreme cases, metallic parts fail in service because of corrosion which involves electrochemical reactions. However, some metals are more susceptible to corrosion than others. We know that platinum and gold are more inert than iron and aluminum. The resistance to corrosion of metals and alloys is determined by their relative susceptibility to react with seawater. The test results are presented in the form of rankings known as the galvanic series shown in Figure 9.16.

More cathodic (inert)

Platinum
Gold
Graphite
Titanium
Silver
Stainless steel 316
Nickel (passive)
Copper
Nickel (active)

More anodic (active)

Tin
Lead
Stainless steel 316
Iron/steel
Aluminum alloys
Cadmium
Zinc
Magnesium

Figure 9.16 Galvanic series for several metals and alloys. *Source:* T. Mukherjee and T. DebRoy.

A more formal laboratory test for determining the propensity of metals and alloys to corrode is to evaluate their standard electrode potential or the electromotive potential based on the electrochemical nature of corrosion in aqueous media. There are four components of an electrochemical cell. (a) An anode where the corrosion occurs. Here the corroding metal gives up electrons to form metal ions. (b) The electrons are received at a cathode, and a product of the reaction forms here. (c) An electrolyte remains in contact with both the anode and the cathode for ionic transport to occur. (d) An electrical contact is necessary to transport the electrons from the anode to the cathode. The relative propensity of a metal or alloy to give up electrons is a measure of its corrosion resistance. A reference electrode is placed within one molar solution of its ions. The two electrolytes are separated by a membrane. A hydrogen electrode is taken as a standard electrode which is immersed in an electrolyte that contains one molar solution of hydrogen ions at 25°C. Here the hydrogen gas is bubbled at a pressure of 1 atmosphere around an inert electrode such as platinum. The reaction at the hydrogen electrode is $2H^+ + 2e^- \rightleftharpoons H_2$ where e^- represents an electron. The potential difference between the two electrodes is measured for various metals and alloys. Since the hydrogen electrode is assigned a potential of zero, the measured potential difference is the standard electrode potential of the metal being tested. Table 9.8 shows the standard emf series for several selected metals.

In Table 9.8 the reactions are written as reduction reactions, i.e., the metal ions reacting with electrons. For oxidation reactions, the directions need to be reversed and the sign of the electrode potentials has to be changed. In this convention, the metals become more susceptible to corrosion as their standard electrode potentials become more negative. Also, when the solutions are not at 1 molar concentration, the relation between electrode potential and the concentration is given by the Nernst equation:

$$E = E_0 + \left(RT / nF\right) \ln\left(C_{ion}\right) \tag{9.10}$$

where E is the electrode potential (in Volts) in a solution of concentration C_{ion} (in moles), E_0 is the standard electrode potential (in Volts) in a 1 molar solution, and n is the valence of the metal ion. The coefficient (RT/F) has a value of 0.0592 Volts at 25°C. The standard electromotive series gives a thermodynamic assessment of the driving forces for different metals to corrode under standard conditions when equilibrium is reached. However, in practice, the equilibrium may not be reached. The electromotive force data do not provide any information about the corrosion rate. The rate is evaluated by defining a parameter known as the corrosion penetration rate which is the thickness decrease resulting from the loss of a material per unit time expressed as KW/(ρAt) where W is weight loss after time t, A is the cross-sectional area, and ρ is density. The symbol K is a constant whose value depends on the units chosen for the variables. The value of K is 86.7 mm per year when the weight loss per unit time is expressed in mm/year, when W, ρ, A, and t are expressed in milligrams, grams per centimeter cube, square inches, and hours, respectively.

In practice, corrosion may occur uniformly over a surface or in a localized manner. Depending on the mechanism of corrosion, localized corrosion may be divided into several categories such as crevice, pitting, intergranular, stress, and galvanic corrosion. When a crevice is filled with water,

Table 9.8 Standard emf series. The values of the standard electrode potentials are valid for 25°C and 1 molar solution.

Reaction	Standard electrode potential (V)
$Au^{3+} + 3e^- = Au$	1.42
$Pt^{2+} + 2e^- = Pt$	1.20
$Fe^{3+} + e^- = Fe^{2+}$	0.77
$Cu^{2+} + 2e^- = Cu$	0.34
$2H^+ + 2e^- = H_2$	0.00
$Ni^{2+} + 2e^- = Ni$	−0.25
$Co^{2+} + 2e^- = Co$	−0.28
$Fe^{2+} + 2e^- = Fe$	−0.44
$Cr^{3+} + 3e^- = Cr$	−0.74
$Zn^{2+} + 2e^- = Zn$	−0.76
$Ti^{2+} + 2e^- = Ti$	−1.63
$Al^{3+} + 3e^- = Al$	−1.66
$Mg^{2+} + 2e^- = Mg$	−2.36

the oxygen concentration at the tip of the crevice is lower than that at the other locations within the crevice and the tip of the crevice acts as an anode where the corrosion occurs. Pitting corrosion occurs when there is a scratch or a small pore on the surface and the bottom of the pit acts as an anode and corrodes progressively. Intergranular corrosion occurs in many alloys and environments. For example, in stainless steels, certain heat treatment results in the precipitation of chromium carbide at the grain boundaries, and the resulting depletion of chromium near the grain boundary acts as an anode and corrodes. Applied tensile stresses or residual stresses combined with a corrosive environment may result in stress corrosion cracking. Small cracks form and grow perpendicular to the applied stresses resulting in part failure. In a corrosive environment, cracks may form at stress levels well below the tensile strength. Galvanic corrosion occurs when two alloys are electrically connected in presence of an electrolyte. A galvanic cell forms with the less noble metal (lower standard electrode potential) as the anode which corrodes.

Irrespective of the mechanism of corrosion, corrosion is very important for many additively manufactured components [17, 18]. For example, stents [19] made by additive manufacturing are exposed to body fluids (Figure 9.17). Similarly, hip and knee implants are exposed to body fluids and are susceptible to degradation. Often complex and intricate parts with small internal cavities are made using additive manufacturing. Those components are susceptible to crevice corrosion. Special features of microstructures of additively manufactured parts such as finer microstructures, surface roughness, porosity, residual stresses, grain structure, and inclusions and precipitates affect the corrosion resistance of additively manufactured components as discussed in Table 9.9.

The surface roughness of parts produced by the powder bed fusion process show roughness values on the scale of about 10 to 30 μm. Roughness here is the arithmetic average of a set of absolute values of the profile elevations measured with reference to the mean value. The roughness of AM parts depends on the process variables such as power density and is affected by surface defects such as balling. An understanding of how process parameters affect roughness is still developing. As expected, surface roughness has a significant impact on corrosion as indicated in Table 9.9. Lack of fusion and porosity in parts made by AM depend on the process parameters and the feedstock and they affect the corrosion properties. Large hatch spacing, insufficient heat input, and dissolved gases in the feedstock contribute to porosity and lack of fusion defects. A columnar grain structure

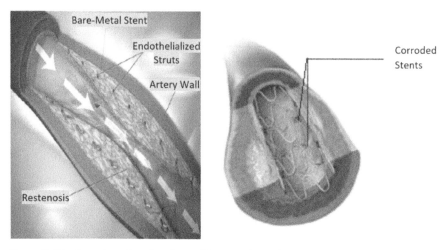

Figure 9.17 Corrosion of a printed metallic stent placed inside a human artery [19]. The figure is reprinted with permission from Elsevier.

Table 9.9 Influence of microstructural features on corrosion resistance of additively manufactured parts.

Factors	Impact on corrosion
Surface roughness of parts	Aluminum alloy AlSi10Mg and 17–4 PH stainless steels processed using powder bed fusion show increased corrosion.
Lack of fusion and porosity	The corrosion resistance of 17–4 PH and 316 stainless steels show detrimental effects of voids in stress corrosion cracking susceptibility.
Grain structure	Columnar grain structures along the build direction produced by both the directed energy deposition and the powder bed fusion processes adversely affect the corrosion resistance.
Residual stress	Corrosion susceptibility increases with residual stresses for stainless steel. However, a similar trend was not observed for Ti-6Al-4V. Distortion of the parts often accompanies residual stress. The areas of compressive and tensile residual stresses make micro galvanic cells and adversely affect corrosion properties.
Alloying elements	Nickel and chromium have beneficial effects on the corrosion performance of iron alloys because they form passive layers. High carbon content may result in carbide precipitates under some conditions and may adversely impact corrosion performance.

not only degrades mechanical properties but also adversely affects corrosion resistance. Also, finer grains are thought to have a beneficial effect on corrosion resistance. The presence of both compressive and tensile residual stresses can form micro-galvanic cells and adversely affect corrosion performance. The chemical composition of the alloy is an important factor in corrosion performance.

The surface roughness of the components produced by AM can be mitigated by electropolishing which is a form of dissolution [20]. During electropolishing, the local elevated peaks on the rough surface experience higher electrical current density and undergo preferential anodic dissolution which is known as anodic leveling. The selective removal of elevated portions of the surface results in a smoother metal surface. Figure 9.18 (a) shows the rough surface morphology of Ti-6Al-4V alloy before polishing and smooth morphology after polishing. There were many pores and unmelted powder particles after printing and the surface was rough and dull. When the chloride electrolyte concentration was 0.1 moles per liter (Figure 9.18 (b)), the polishing rate was low but

Figure 9.18 (a) Surface morphologies of Ti–6Al–4V alloy before electropolishing and (b) after electropolished at 0.5 A cm^{-2} in (b) 0.1 moles per liter, (c) 0.2 moles per liter, (d) 0.3 moles per liter, (e) 0.4 moles per liter, and (f) 0.5 moles per liter chloride ion concentrations [20]. The figure is reprinted with permission from Springer Nature.

the rate improved when the concentration was increased to 0.2 moles per liter (Figure 9.18 (c)) or 0.3 moles per liter (Figure 9.18 (d)). A bright and flat surface was obtained when the specimen was polished in 0.4 moles per liter chloride (Figure 9.18 (e)). The flatness of the surface was optimal with minimum surface roughness. A higher chloride concentration (Figure 9.18 (f)) resulted in aggressive dissolution and did not achieve satisfactory surface smoothness.

The quantity of metal corroded at the anode, or the amount of metal deposited at the cathode, is given by Faraday's equation:

$$w = ItM / (nF) \tag{9.11}$$

where w is the weight of the metal dissolved in the solution (corroded) or deposited at the cathode (plated), I is the current in A, M is the atomic mass in kg/kg-mole, n is the valence of the metal ion, t is time in s, and F is Faraday constant, 96,500 C/mol. The calculation of corrosion rate using Faraday's equation is explained in worked out example 9.11. For the oxidation of a metal, the ratio of the volume of metal oxide to that of the metal from which the oxide forms gives a measure of the ability of the oxide to provide passivation. The ratio, called the Pilling-Bedworth ratio is given by:

$$PB = V_O / V_M \tag{9.12}$$

where PB is the Pilling-Bedworth ratio, V_O, is the volume of the oxide formed from the volume of metal, V_M. The volumes of the oxide and metal can be calculated from their respective mass and density. Consider, for example, the formation of Al_2O_3:

$$2Al + 3O_2 = Al_2O_3$$

Al has an atomic weight of 27 gm/gm-atom and Al_2O_3 has a molecular weight of 102 gm/gm-mole, their densities are 2.7 and 3.95 gm/cm^3 respectively. As a result, the PB for the oxidation of aluminum is $(102/3.95)/(54/2.7) = 1.29$. It was suggested that when the ratio is less than 1, the film is unable to cover the entire surface and is not protective. Also, when the ratio is more than 2, the volume of the oxide is much larger than optimum, and the oxide may crack or peel and may not offer much protection from oxidation. When the PB is between 1 and 2, the oxide may be adherent and provide passivation. The value of PB = 1.27 for the oxidation of aluminum is a good example. Similarly, the PB = 1.58 for ZnO is another example of an adherent oxide film that can protect against oxidation. Worked out example 9.12 shows how the nature of the oxidation products is important for corrosion of metals.

Worked out example 9.11

A commercially pure titanium part made by AM exhibited considerable surface roughness and needed to be electropolished to remove a layer of 0.1 mm thickness. How long would it take to electropolish the part? The current density was 0.5 × 10^4 A/m^2, the density of titanium is 4500 kg/m^3, the valence of titanium is 4, and its atomic weight is 48 × 10^{-3} kg/mole.

Solution:

Eq. 9.11 gives time = wnF/(IM) where the symbols have their usual meanings. If the area electropolished is A,
  w/A = (0.1 × 10^{-3}) × 4500 kg/m^2 = 0.45 kg/m^2
  I/A = 0.5 × 10^4 A/m^2
  M = 48 × 10^{-3} kg/mole
  Time = (0.45 × 4 × 96,500) / (0.5 × 10^4 × 48 × 10^{-3}) = 723.7 s = 12 minutes.

Worked out example 9.12

When iron and aluminum are exposed to the atmosphere, iron corrodes easily, especially in moist air and aluminum seems to resist pronounced degradation. However, both the Galvanic series and standard electrode potentials indicate aluminum to be more susceptible to corrosion than iron. Explain the relative resistance to corrosion of iron and aluminum. Assume that the oxidation product of iron is (a) $Fe(OH)_3$ or (b) Fe_2O_3 depending on the environmental conditions. Densities of $Fe(OH)_3$, Fe_2O_3, and Fe are 3400, 5240, and 7874 kg/m^3.

Solution:

First, we assume the oxidation product of iron to be $Fe(OH)_3$ which has a molecular weight of 107. The ratio of the volume of $Fe(OH)_3$ to the volume of Fe is $(107/3400)/(56/7874) = 4.4$. The increase in the volume of the oxidation product is so large that it would be impossible to accommodate the oxidation product within the same volume that the metal occupied. As a result, the oxidation product is likely to flake from the surface and expose the surface to further oxidation. Next, let us consider a case where the reaction product is Fe_2O_3. Here the PB = $(160/5240)/(112/7874) = 2.15$. Here the volume of Fe_2O_3 is much larger than that of the iron from which it forms, and the oxide is not going to be adherent. Therefore, in both (a) and (b), the oxidation product will not protect the iron from further oxidation. In contrast, for the oxidation of aluminum, the PB is 1.27 which means that Al_2O_3 will protect aluminum from further oxidation. So, the Galvanic series provides a thermodynamic driving force for corrosion, but the nature of the oxidation product is an important consideration.

9.5 Summary

Mechanical properties of AM components such as strength, hardness, ductility, toughness, creep, and fatigue properties, as well as corrosion resistance, vary significantly depending on the alloy, process, and processing conditions. The microstructure of AM parts often plays a major role in determining the properties. Since AM is a very complex process involving rapid scanning and melting and deposition of multiple tracks, AM parts exhibit unique properties. Apart from the experimental measurements, a quantitative understanding of the properties of AM parts is also evolving. A detailed understanding of properties can provide better insight into the reliability and serviceability of AM parts, which needs more research and development in this field.

Takeaways

Strength and ductility

- Tensile yield strength, ultimate tensile strength, ductility, and toughness are important mechanical properties of additively manufactured components.
- Average yield strength can be calculated from the average grain size using the Hall-Petch equation.
- Long columnar grains affect tensile strength. Technologies are emerging to break the columnar grains into smaller equiaxed grains (columnar to equiaxed transition) and to a lesser extent achieve both high strength and toughness in some additively manufactured parts.
- Tensile properties of parts are often anisotropic which means they vary depending on the direction of the measurement.

Hardness

- Hardness is linearly correlated with the average tensile yield strength and can be calculated from the yield strength.
- For steel, nickel, and aluminum alloys, hardness can be roughly estimated from the chemical composition of alloys.
- The hardness of parts is governed by chemical composition, heat source power, scanning speed, preheat temperature, microstructure, and residual stresses.
- Temporal evolution and spatial variation of the hardness of parts are affected by the local changes in temperature with time, also known as the thermal cycle.

Toughness

- Fracture toughness can be measured by Charpy and other tests and calculated from the applied stress, crack size, and geometry.
- If the crack is present inside the part, the effective crack size used in the calculation is taken as half of the crack length.
- Fracture properties are often represented by applied stress versus the square root of the crack size plots.
- Fracture toughness is affected by the presence of cracks and voids as well as microstructure.

(Continued)

Takeaways   (Continued)

Fatigue properties

- Fatigue properties are evaluated from the applied stress versus the number of cycles plots.
- Fatigue properties are largely affected by residual stresses, the presence of voids, surface roughness, and microstructure.
- After appropriate heat treatment, AM parts can exhibit almost similar fatigue properties as cast or wrought products in many cases.

Creep

- Creep resistance is evaluated from two types of plots (1) creep strain versus time plot at constant applied stress and temperature and (2) applied stress versus rupture lifetime plot at different temperatures.
- Creep properties are largely affected by grain size, the presence of voids, and solidification morphology.
- Smaller grains permit more grain boundary sliding which accelerates the creep rate and results in a rapid creep rupture. Therefore, single crystal parts exhibit the best creep resistance. Recently, additive manufacturing is being used to make single crystals.

Post-processing to improve mechanical properties

- Commonly used methods are post-process heat treatment, hot isostatic pressing, stress relieving, surface treatment, and surface finishing.
- Post-processing improves mechanical properties but adds an extra cost to an already expensive process.

Corrosion resistance

- Chemical properties and corrosion resistance are largely degraded at high temperatures and the presence of a reactive environment.
- The chemical composition of the alloy used, microstructure, and presence of precipitates along the grain boundaries affect the corrosion resistance of parts.

Appendix – Meanings of a selection of technical terms

Casting: A manufacturing process where a liquid material is poured into a mold and the material takes the shape of the mold after solidification.

Dislocation: A linear defect within the crystal structure where some atoms are misaligned.

Electron beam: A focused beam of electrons used as a heat source in additive manufacturing.

Grinding: A machining process generally to obtain a good surface finish. Generally, a wheel of abrasive material rotating at a very high speed is used as the machining tool.

Heat treatment: A process of heating a part in a furnace at a high temperature less than the melting of the material followed by a subsequent cooling process. The rate of cooling can vary depending on the requirement.

Hot isostatic pressing: A manufacturing process where uniform pressure is applied on a part to reduce the number of internal cracks and pores.

Laser beam: A focused beam of laser used as a heat source in additive manufacturing. The acronym laser stands for "light amplification by stimulated emission of radiation."

Plain strain: A two-dimensional state of strain in which the shape change or the distortion of the material occurs on a single plane. A plane strain condition can be observed when the region of the strain accumulation is very small compared to the dimension of the entire component.

Single crystal: A material that has the same crystallographic orientation throughout the part. Therefore, ideally, these parts do not have any grain boundaries.

Twinning: Anatomic arrangement within a crystal where atoms on one side of a grain boundary are positioned as a mirror image of the other side.

Voids: Hollow pores formed during additive manufacturing due to gas entrapment or lack of fusional bonding among neighboring tracks.

Weldability: Ability of a material to weld without producing any defect.

Practice problems

1) A stainless steel 316 part is made using the laser-assisted powder bed fusion process. In this process, 500 W laser power and 15 mm/s scanning speed are used. Estimate approximately the tensile yield strength and Vickers hardness of the part.

2) A nickel alloy part made by powder bed fusion has an internal crack, 30 μm in length. If the nickel alloy has a fracture toughness of 5 MPa√m, what is the maximum stress the part can withstand to resist fracture failure?

3) The table below provides the chemical composition (in wt%) of three different alloys. Estimate the approximate hardness in the Vickers scale of AM parts made by these alloys.

Alloys	Ti	Al	V	Fe	Ni	Cr	Mn	Mg	Si	Mo
Stainless steel 316	_	0.005	_	Bal.	8.26	17.2	1.56	_	0.33	_
Inconel 718	1.02	0.50	_	Bal.	53.4	18.8	0.07	_	0.12	2.99
Aluminum alloy 6061	0.15	Bal.	_	0.7	_	_	0.15	1.2	0.8	_

4) Two stainless steel 316 parts are made using the laser-assisted powder bed fusion process with a laser scanning speed of 15 mm/s. However, two parts are made using 300 W and 600 W laser power, respectively. Which part will have better creep resistance?

5) Calculate the creep rupture time of an alloy component subjected to a pressure of 200 MPa at 873K. Recalculate the rupture time if the temperature is raised to 1023 K. The value of the Larson-Miller parameter is 22,500 h for 200 MPa.

6) A commercially pure nickel part made by AM exhibited considerable surface roughness and needed to be electropolished to remove a layer of 0.2 mm thickness. How long would it take to electropolish the part? The current density was 0.6×10^4 A/m^2, the density of nickel is 8900 kg/m^3, the valence of nickel is 2, and its atomic weight is 59×10^{-3} kg/mole.

7) Parts are made of three aluminum alloys by additive manufacturing. Part A contains 0.4% Cu, 0.02% Fe, and other alloying elements. Part B contains 5.0% Cu and the concentrations of the other alloying elements same as that in A. Part C has the same concentrations of alloying elements as A, except the iron concentration. If the Vickers hardness of B and C are the same, determine the concentration of Fe in alloy C.

8) Explain how preheating the powder bed can affect the creep resistance of a part made by powder bed fusion.

9) The solidus temperature of alloys "A" and "B" are 1200 K and 2000 K, respectively. Parts made of these alloys are used at a temperature of 600 K. For which alloy part, creep is an important concern?

10) An additively manufactured part in a power plant is subjected to a pressure of 200 MPa during service. How much the creep rupture lifetime will change if the service temperature is increased by 20 K?

References

1 Mukherjee, T., Zuback, J.S., Zhang, W. and DebRoy, T., 2018. Residual stresses and distortion in additively manufactured compositionally graded and dissimilar joints. *Computational Materials Science*, 143, pp.325–337.

2 Mukherjee, T., Zhang, W. and DebRoy, T., 2017. An improved prediction of residual stresses and distortion in additive manufacturing. *Computational Materials Science*, 126, pp.360–372.

3 Dieter, G.E., 1988. *Mechanical Metallurgy*. Mc Graw-Hill Book Co., New York.

4 Callister, W.D. and David, G.R., 2010. *Materials Science and Engineering*, 8th Edition. John wiley & sons, New York.

5 Santos, G.A., de Moura Neto, C., Osório, W.R. and Garcia, A., 2007. Design of mechanical properties of a Zn27Al alloy based on microstructure dendritic array spacing. *Materials & Design*, 28(9), pp.2425–2430.

6 Manvatkar, V.D., Gokhale, A.A., Jagan Reddy, G., Venkataramana, A. and De, A., 2011. Estimation of melt pool dimensions, thermal cycle, and hardness distribution in the laser-engineered net shaping process of austenitic stainless steel. *Metallurgical and Materials Transactions A*, 42(13), pp.4080–4087.

7 Lewandowski, J.J. and Seifi, M., 2016. Metal additive manufacturing: a review of mechanical properties. *Annual Review of Materials Research*, 46, pp.151–186.

8 Wilson-Heid, A.E., Novak, T.C. and Beese, A.M., 2019. Characterization of the effects of internal pores on tensile properties of additively manufactured austenitic stainless steel 316L. *Experimental Mechanics*, 59(6), pp.793–804.

9 DebRoy, T., Mukherjee, T., Wei, H.L., Elmer, J.W. and Milewski, J.O., 2021. Metallurgy, mechanistic models and machine learning in metal printing. *Nature Reviews Materials*, 6(1), pp.48–68.

10 Wang, Z., Palmer, T.A. and Beese, A.M. 2016. Effect of processing parameters on microstructure and tensile properties of austenitic stainless steel 304L made by directed energy deposition additive manufacturing. *Acta Materialia*, 110, pp.226–235.

11 Zuback, J.S. and DebRoy, T., 2018. The hardness of additively manufactured alloys. *Materials*, 11(11), article no.2070.

12 Mukherjee, T., DebRoy, T., Lienert, T.J., Maloy, S.A. and Hosemann, P., 2021. Spatial and temporal variation of hardness of a printed steel part. *Acta Materialia*, 209, article no.116775.

13 Romano, S., Brückner-Foit, A., Brandão, A., Gumpinger, J., Ghidini, T. and Beretta, S., 2018. Fatigue properties of AlSi10Mg obtained by additive manufacturing: Defect-based modelling and prediction of fatigue strength. *Engineering Fracture Mechanics*, 187, pp.165–189.

14 Kobryn, P.A. and Semiatin, S.L., 2001. Mechanical properties of laser-deposited Ti-6Al-4V. In *2001 International Solid Freeform Fabrication Symposium*, pp.179–186.

15 Kuo, Y.L., Horikawa, S. and Kakehi, K., 2017. Effects of build direction and heat treatment on creep properties of Ni-base superalloy built up by additive manufacturing. *Scripta Materialia*, 129, pp.74–78.

16 Kunze, K., Etter, T., Grässlin, J. and Shklover, V., 2015. Texture, anisotropy in microstructure and mechanical properties of IN738LC alloy processed by selective laser melting (SLM). *Materials Science and Engineering: A*, 620, pp.213–222.

17 Örnek, C., 2018. Additive manufacturing–a general corrosion perspective. *Corrosion Engineering Science and Technology*, 53(7), pp.531–535.

18 Kong, D., Dong, C., Ni, X. and Li, X., 2019. Corrosion of metallic materials fabricated by selective laser melting. *NPJ Materials Degradation*, 3(1), pp.1–14.

19 Ettefagh, A.H., Guo, S. and Raush, J., 2021. Corrosion performance of additively manufactured stainless steel parts: A review. *Additive Manufacturing*, 37, article no.101689.

20 Zhang, Y., Li, J., Che, S. and Tian, Y., 2020. Electrochemical polishing of additively manufactured Ti–6Al–4V alloy. *Metals and Materials International*, 26(6), pp.783–792.

10

Common Defects in Additively Manufactured Parts

Learning objectives

After reading this chapter the reader should be able to do the following:

1) Appreciate the common types of defects in metal additive manufacturing processes.
2) Understand the mechanisms of formation of lack of fusion, keyhole instability porosity, gas porosity, cracking, chemical composition change, surface roughness, and balling.
3) Know the effects of process variables on the different types of defects.
4) Be familiar with the sensing techniques for recognizing the formation of defects during the manufacturing process.
5) Recognize the common techniques for mitigating defects such as machining, surface treatments, and hot isostatic pressing.
6) Comprehend the emerging tools of mechanistic modeling and machine learning for understanding and mitigating defect formation.

CONTENTS

Theory and Practice of Additive Manufacturing, First Edition. Tuhin Mukherjee and Tarasankar DebRoy.
© 2024 John Wiley & Sons, Inc. Published 2024 by John Wiley & Sons, Inc.

10.1   Introduction

Part quality, reliability, and serviceability of additively manufactured parts are often affected by many types of common defects. They include cracking, porosity due to gas bubbles, keyhole instability-induced porosity, lack of fusion, surface roughness and waviness, balling, loss of alloying elements, residual stress, distortion, and delamination. Cracking may occur via multiple mechanisms. For example, solidification cracking can occur toward the end of the solidification process. When the solid fraction in the two-phase region is high, the liquid metal cannot be easily transported to the crack site because of the high solid fraction of the two-phase region and cracks grow because of the solidification stress. Common gases such as nitrogen, oxygen, or hydrogen may dissolve in the liquid metal at high temperatures from the environment. When the metal cools, the solubility of the gases may diminish at lower temperatures and the dissolved gases can desorb to form gas bubbles through heterogeneous nucleation and growth. When a high power density heat source is used, a vapor cavity known as the keyhole forms. If the keyhole is not stable, it can collapse and a vapor cavity may be trapped in the liquid pool forming a cavity, in many cases near the bottom of the keyhole. The lack of fusion defects originates from the inadequate overlap of adjacent tracks and layers leaving some gaps between deposits. A consequence of the lack of adequate fusion is a lower density of the part compared to the theoretical density of the feedstock material. Depending on the extent of the lack of fusion and porosity, some of the mechanical properties such as fatigue property can be degraded, seriously compromising the serviceability of the part.

Many types of surface defects are common in additive manufacturing depending on the variant of additive manufacturing, the alloy feedstock, process parameters, nature of the part, and process variables used. For example, in wire arc additive manufacturing, the part surface often exhibits surface waviness (SW), which is a periodic difference between the maximum to minimum elevation on a surface. The surface roughness depends on several factors including the powder size, process parameters, and the nature of the surface, flat or curved. The balling defects originate from the breakup of a track of liquid into small balls and are often the result of high scanning speed. Molten pools reach a very high temperature in many cases and a significant amount of volatile alloying elements may be lost by vaporization. Because different alloying elements vaporize at different rates, the composition of the parts may differ from that of the feedstock, affecting the microstructure and mechanical properties of the parts. The residual stress, distortion, and delamination originate from the nonuniform heating and cooling and these issues are addressed in detail in Chapter 11.

Defects are a serious problem because they affect mechanical properties, particularly under multiaxial or cyclic loading. Microstructure, properties, and defects in the manufactured parts depend on the variant of additive manufacturing, process parameters, the alloy, and the specific geometry of the build. However, the manufacture of defect-free, high-quality parts in a cost-effective way is a major challenge because of the large number of process variables, their wide range of values, and the inefficient trial-and-error optimization of process variables to avoid defects. In many cases, post-processing is used to reduce defects in additively manufactured materials. For example, shot penning, laser penning, abrasive polishing, and machining are often undertaken to improve surface roughness. However, they add to the cost of manufacturing. Defects are often cited as an important factor why after several decades of research and development, only a handful of over 5500 commercial alloys are now additively manufactured and the market value of all the products now amounts to a negligible portion of the manufacturing economy. However, progress is being made to improve the quality consistency and cost competitiveness of printed parts.

One of the ways to mitigate the defects is to undertake real-time surveillance of the formation of defects so that appropriate intervention can be made as soon as possible. Such sensing may enable remedial action to adjust process parameters where possible or abandonment of the printing where such action may be appropriate. The various types of sensing and control are discussed in Chapter 6.

Unlike welding and casting where the technology matured largely by trial-and-error testing, the high cost of machine and feedstock make this approach an expensive undertaking. Also, the business culture of rapid production of high-quality parts does not allow the time necessary to do a large volume of trial-and-error tests. Finally, the empirical trial approach does not always result in the desired defect-free, high-quality parts. A new path of controlling defects based on scientific principles is now evolving. If the underlying mechanisms are well understood, the physical processes can be quantified based on the laws of physics. As a result, a connection can be established between the process variables and the underlying physical processes that define the formation of defects. However, in many cases, the underlying physical processes are still developing, and the evolution of defects cannot be quantified based on the laws of physics. In those cases, machine learning provides a framework for the mitigation of defects where an adequate volume of data is available. Scientific principles based on the laws of physics and machine learning are now being increasingly used to address the mitigation of defects. Here we review the common defects, the mechanisms of their formation, characterization, and mitigation strategies.

10.2   Cracking

Solidification cracking, liquidation cracking, and ductility dip cracking are the three types of cracking that occur in many additively manufactured alloy parts. Alloys with poor weldability typically exhibit high cracking susceptibilities during additive manufacturing. Understanding the mechanisms of the formation of these cracks is important for their prevention.

10.2.1   Solidification cracking

Parts of aluminum, titanium, and nickel alloys and steels, made by different additively manufactured processes, have shown solidification cracking, also known as hot cracking [1] [2]. The shrinkage of the solidifying metal is significantly higher than the surrounding metal and as a result, tensile stresses develop in the solidifying region. Solidification cracking occurs when the tensile stress due to shrinkage exceeds the yield strength and there is insufficient liquid metal to fill the crack. Figure 10.1 shows solidification cracking in a nickel alloy [1]. Many factors are responsible for the solidification cracking and these are discussed below.

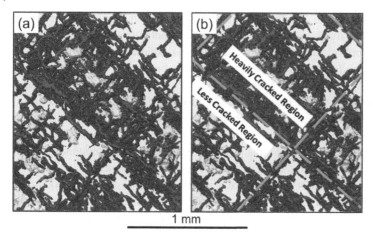

Figure 10.1 Solidification cracking in a non-weldable nickel base superalloy CM247LC processed by PBF-L [1] in two regions of a part in (a) and (b). The figure is taken from an open-access article [1] under the terms and conditions of the Creative Commons Attribution (CC BY) license.

10.2.1.1 Tensile stress due to shrinkage

The solidifying fusion zone behind the heat source dissipates heat in the adjacent volume. As the temperature drops, a two-phase region containing both the solid and the liquid forms. The solid fraction in the two-phase region increases with the drop in temperature. The alloy tries to shrink due to both solidification and the temperature drop. At the same time, it is constrained by the adjacent region which does not shrink as much because the temperature change is less pronounced further away from the fusion zone. Also, the temperatures of the previously deposited layers are lower than those of the depositing layers and the contraction of the depositing layer is higher than the lower layer. As a result, the contraction of the upper layer is constrained by the previously deposited layers. If the strength of the solidifying region does not exceed the yield strength of the alloy, solidification cracks form.

10.2.1.2 Solidification temperature range

Several properties of an alloy are important for evaluating its susceptibility to solidification cracking. For example, the solidification temperature range which is the difference between the liquidus and solidus temperatures affects an alloy's susceptibility to solidification cracking. For a given temperature gradient, the alloy with a larger solidification range is more likely to form long, liquid-filled channels. During the completion of solidification, solid and liquid phases coexist and the areas between the dendrites can be replenished by the liquid. A large solidification range results in long and thin liquid films being present between the gaps between the columnar dendrites. This liquid film containing a large two-phase mushy region is weak and cannot resist stresses generated during solidification, and makes the alloy with a large solidification range susceptible to hot cracking. The equilibrium freezing range assumes a nominal composition of the alloy. However, during additive manufacturing, the solute rejection during solidification enriches the liquid and the composition of the liquid is enriched with alloying elements that segregate. As a result, the equilibrium solidification range is extended because of the segregation of alloying elements during solidification. In many cases, the solidification temperature range is calculated considering no solute diffusion in the solid and rapid diffusion of the solutes in the liquid and equilibrium at the solid/liquid interface known as Scheil assumptions.

10.2.1.3 Temperature–solid fraction relation

The availability of the liquid to heal the cracks depend on how the fraction solid changes with temperature. This variation depends on the specific alloy. A steep variation in the temperature versus solid fraction plot means that a very low liquid fraction near a solid fraction of 1 may prevail over a relatively large temperature range [3]. This depletion of the available liquid to heal the cracks over a large temperature range makes alloys susceptible to cracking. Figure 10.2 shows the variation of temperature with solid fraction for Al7075 and AlSi10Mg alloys. It is observed that the change in temperature with the solid fraction is much steeper for Al7075 than AlSi10Mg. Thus for Al7075, a solid fraction range close to 1 can

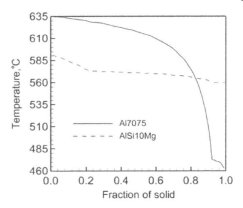

Figure 10.2 Temperature versus fraction solid during solidification of Al7075 and AlSi10Mg alloys. *Source:* T. Mukherjee and T. DebRoy.

prevail over a relatively large temperature range making the entire region susceptible to cracking. Sometimes instead of plotting temperature versus fraction solid, plots are made of temperature versus (fraction solid)$^{1/2}$ and the maximum slope of the plot is a measure of the solidification cracking susceptibility. Worked out example 10.1 illustrates this concept.

10.2.1.4 Solidification morphology

The ability of the solidification microstructure to accommodate strain due to shrinkage and the ease of transporting solidifying liquid between grains are two important criteria for preventing cracking. Both these requirements depend on the grain structure, making grain structure a significant factor in cracking susceptibility. As has been discussed in Chapter 8 on microstructure, high temperature gradients (G) and low solidification growth rates (R) favor columnar grains while the equiaxed grains are preferred with high R and low G. Since the temperature gradient is commonly large in many additive manufacturing processes, columnar grain structures are often observed. During additive manufacturing, long columnar grains grow across many layers unless their growth is interrupted by changes in scanning patterns and other interventions. The growth of solidification cracks along the boundary of columnar grains across multiple layers is quite common. For this reason, significant efforts are made to achieve columnar to equiaxed transition (CET) of the solidification microstructure. Equiaxed grains are smaller in size and their morphology allows easier transport of liquid in the two-phase region. Also, their mechanical properties are isotropic and they can tolerate strains more easily than columnar structures.

The evolution of the solidification structure is affected by the nucleation and growth during the cooling process. Significant efforts have been made to examine various routes to achieve CET. They include additions of nucleating agents that promote the nucleation of equiaxed structures and have been widely used in the foundry industry. The type of additives, their amount, and uniform distribution throughout the fusion zone are important factors and this procedure adds both cost and complexity in making parts. External vibration to break up columnar grains has also been tried. However, the addition of an energy source close to the part inside the deposition chamber is a challenging task. Altering scanning strategies have been successful in controlling the direction of growth of columnar grains in different layers. Figure 10.3(a) shows the evolution of grain structure for the unidirectional scanning from left to right

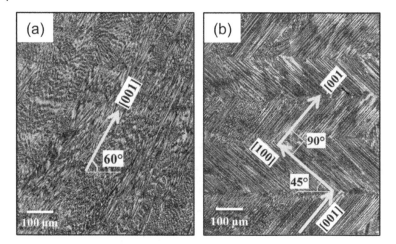

Figure 10.3 Optical micrograph in a longitudinal midsection of as-deposited Inconel 718 sample for (a) unidirectional left to right laser scanning in all layers and (b) bidirectional laser scanning showing dendrite growth directions [4]. The figure is taken from an open-access article [4] under the terms and conditions of the Creative Commons Attribution (CC BY) license.

and Figure 10.3(b) shows the grain structure resulting from alternating/bidirectional scanning. The figures show that the long columnar grains obtained by unidirectional scanning can be altered by changing the scanning direction (bidirectional scanning) for IN718. The control of microstructure for IN 718 provides a low-cost and straightforward solution to achieve CET but the evolution of microstructure is highly alloy dependent and more work is needed on the effectiveness for achieving CET in other alloys.

10.2.2 Liquation cracking

Liquation cracking can occur outside and adjacent to the fusion zone or in previously deposited metal where the temperature remains below the solidus temperature of the alloy. It is caused by localized melting of low melting regions at a grain boundary under the influence of the thermal strains associated with additive manufacturing. During solidification, the segregation of alloying elements often results in the formation of low melting equilibrium phases that may cause liquation cracking. The chemical compositions of the alloys, the temperature range of solidification, solidification shrinkage, thermal contraction during cooling, and the presence of intermetallic compounds and precipitates are all important factors in liquation cracking. Liquation cracking has been observed during the additive manufacturing of nickel-based alloys, aluminum alloys, and austenitic stainless steels.

In nickel alloy IN 740, the segregation of boron (B), niobium (Nb), and silicon (Si) during solidification may result in the formation of a low melting equilibrium phase such as MB_2 where M is a metallic element and intermetallic Laves phase that has a composition AD_2 where A and D are both metals. The segregation can significantly increase the solidus temperature of the alloy and increase the susceptibility to liquation cracking. The susceptibility of liquation cracking also depends on the grain structure, particularly the orientation of the grain boundaries since liquid films are more stable at high-angle grain boundaries. Higher local stress concentration during the final stage of solidification can initiate cracking in the partially melted region. Liquid films at the grain boundaries may fail to endure the thermal or mechanical strain during the cooling process resulting in cracking. Evidence of liquid films is often observed in fractured surfaces of liquation cracking.

10.2.3 Ductility dip cracking

Ductility dip cracking (DDC) is a solid-state intergranular cracking in the temperature range between 60% and 90% of the solidus temperature where the ductility of many alloys shows a marked drop. This type of cracking is influenced by the grain structure, particularly the large size of grains and the nature of the grain boundary, the type of precipitates at the grain boundary, temperature, and restraint. Although this type of cracking has been reported in many alloys, nickel-based alloys and stainless steels are particularly susceptible to ductility dip cracking. Figure 10.4 shows a microstructure of nickel base alloy CM247LC processed by PBF-L [1]. This alloy is known to be susceptible to DDC and the microstructure shows pronounced cracking. Unlike solidification cracking and liquation cracking, ductility dip cracking occurs without the involvement of any liquid metal.

Although the mechanism of ductility dip cracking is not well understood, it is thought that the formation of voids involving grain boundary slides is involved in this type of cracking. Tortuous grain boundaries resist such slides and resist DDC while DDC occurs at high-angle grain boundaries. Similarly, some types of precipitates at the grain boundaries resist the sliding of the grains and DDC. The presence of ferrite in the austenite matrix creates tortuous boundaries and resists DDC. In addition, restraint must be present for DDC to occur.

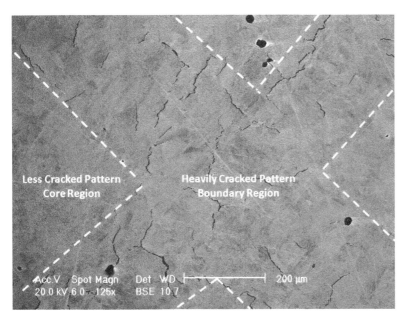

Figure 10.4 A SEM micrograph of a nickel-based alloy CM247LC processed by PBF-L showing the heavy and less cracked regions due to a repeated scanning pattern [1]. The figure is taken from an open-access article [1] under the terms and conditions of the Creative Commons Attribution (CC BY) license.

10.3 Voids and pores

10.3.1 Gas porosity

Hydrogen, oxygen, and nitrogen are often present in small quantities in the deposition environment, and they are soluble in the alloys to different extents. For example, molten aluminum alloys absorb different amounts of all these gases from the environment. The solubility of oxygen in aluminum alloys at temperatures prevailing in the fusion zone is small. However, aluminum oxide is stable at elevated temperatures and oxide inclusions may form during solidification and porosity due to oxygen bubbles do not form in aluminum alloy parts. Nitrogen also reacts with liquid aluminum to form aluminum nitride (AlN). Figure 10.5 shows the equilibrium solubility of hydrogen in (a) aluminum and (b) magnesium in liquid and solid states. Hydrogen has a higher solubility in both liquid aluminum and magnesium than in their respective solid states. This difference in the solubility results in the rejection of hydrogen from the metal during cooling. As a result, they nucleate tiny bubbles of hydrogen at various heterogeneous nucleating sites such as solid-liquid interfaces and inclusions. The nucleated tiny bubbles then tend to grow by diffusion. Hydrogen and nitrogen also dissolve in steels and can be trapped in solid alloys as porosity.

The solubility of hydrogen in aluminum is affected by the presence of alloying elements. The presence of lithium, magnesium, and titanium increases the solubility of hydrogen in liquid aluminum. In contrast, zinc, silicon, copper, and iron reduce it. The strong attractive interactions between hydrogen and lithium, magnesium, and titanium increase the solubility of hydrogen and the strong bonding of aluminum atoms to zinc, silicon, copper, and iron reduces the solubility of hydrogen in aluminum alloys.

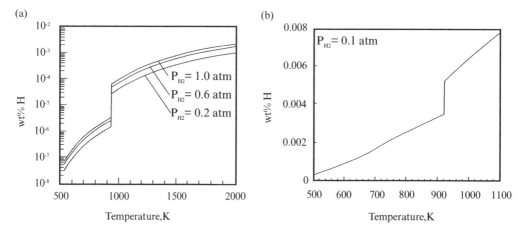

Figure 10.5 Solubility of hydrogen in (a) aluminum and (b) magnesium in liquid and solid states showing significant drops in solubility of hydrogen in both metals when solidification occurs. Figures (a) and (b) are made by T. Mukherjee and T. DebRoy using the data from [6] and [7], respectively.

Worked out example 10.1

The selection of aluminum alloys for additive manufacturing often involves the consideration of solidification cracking that depends on both the type of alloy and processing conditions.

Worked out example 10.1   (Continued)

Under similar processing conditions, the parameter $|dT/d(f_S)^{1/2}|$ where T is temperature and f_S is the solid fraction has been suggested as an indicator of the propensity of solidification cracking of different alloys. Use Figure E10.1 to compare the solidification cracking propensity of the five aluminum alloys shown in the figure.

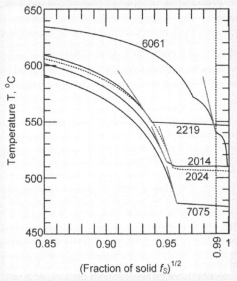

Figure E10.1   Solidification curves of five aluminum alloys [5]. The figure is taken from an open-access article [5] under the terms and conditions of the Creative Commons Attribution (CC BY) license.

Solution:

Maximum slopes of the plots in Figure E10.1 are to be compared for high fraction solids. The slopes are also shown on the plot at this solid fraction by tangential lines. The computed slopes are the following: alloy 6061 : 4923 K, alloy 2219 : 1626 K, alloy 2014 : 2531 K, alloy 2024 : 3389 K, and alloy 7075 : 4255 K. Thus the solidification cracking susceptibilities of the five alloys from highest to lowest are as follows: alloys 6061, 7075, 2024, 2014, and 2219. The slopes calculated from the graph are approximate, but they are within 2% of the values quoted in reference [5].

The growth and escape of the bubbles depend on the size of the fusion zone, thermophysical properties of the alloy, and process parameters of the additive manufacturing that determine the cooling rate. Bubbles can move within the fusion zone with the prevailing velocity of the liquid metal in the fusion zone. In addition, the bubbles can rise due to the buoyancy force. The rising velocity of the spherical bubbles can be approximately estimated by the Stokes law [8] which is valid for low Reynolds number (< 2) and is given by:

$$u = 2r^2\Delta\rho g / (9\mu) \tag{10.1}$$

where r is the radius of the bubble, $\Delta\rho$ is the density difference between the liquid and the gas bubble, g is the acceleration due to gravity, and μ is the viscosity of the liquid. The size of the bubble and the time available for the growth and escape of the bubbles are important factors for the formation of porosity. Eq. 10.1 shows that low liquid viscosity and large bubble radius increase the chance of bubbles escaping the fusion zone by floatation. In contrast, rapid cooling allows less time for bubble escape. Small circular pores typically smaller than about 50 μm are often observed in additively manufactured metallic parts if the gas bubbles are unable to escape the fusion zone. Aluminum alloy parts are particularly prone to hydrogen-induced porosity which often originates from the difference in the hydrogen solubility in aluminum alloys between the liquid and the solid phases during cooling. This principle is illustrated in worked out example 10.2. Time taken for bubbles of different sizes to escape the fusion zone is calculated in worked out example 10.3.

The main sources of hydrogen in the fusion zone are the feedstock and the shielding gas. Gases are trapped in powders and wires during their manufacture. When powders are made during gas or plasma atomization or wires are drawn, gases can be acquired from the environment. In addition, depending on how the feedstocks are stored, gases and moisture on their surface can be introduced in the fusion zone. When the feedstocks containing gases are melted gases are introduced in the fusion zone. Removing hydrogen from these sources is a good practice to control porosity formation due to hydrogen.

Worked out example 10.2

Gas porosity in aluminum is often attributed to hydrogen. The solubility of hydrogen in liquid aluminum can be expressed by: $\log S = -2760/T + 1/2 \log P + 1.356$ and that in solid aluminum by $\log S = -2080/T + 1/2 \log P - 0.652$ where S is the solubility of hydrogen in cm^3 measured at 273 K and 760 Torr per hundred grams of aluminum, T is the temperature in K, and P is the partial pressure of hydrogen in Torr. Determine (a) the relative solubility of hydrogen in liquid and solid aluminum and (b) the volume of hydrogen gas evolved per cm^3 of aluminum at the melting point of aluminum (933K) during solidification if the partial pressure of hydrogen is 0.001 atmosphere.

Solution:

(a) Hydrogen solubility in liquid aluminum is obtained from the following expression: $\log S = -2760/933 + 0.5 \log (0.001 \times 760) + 1.356$ or $S = 2.179 \times 10^{-2} cm^3$ at 273 K per 100 gm of Al at 760 torr. Similarly, the solubility of hydrogen in solid aluminum is $\log S = -2080/933 + 1/2 \log (0.001 \times 760) - 0.652$, or $S = 1.146 \times 10^{-3} cm^3$ at 273 K per 100 gm of Al at 760 torr. Thus, the solubility of hydrogen in liquid is 19 times higher than that in solid aluminum.

(b) The volume of hydrogen gas evolved per cm^3 of solid aluminum at room temperature = $1.146 \times 10^{-3} cm^3 / (100/2.7) = 3.09 \times 10^{-5} cm^3$.

Worked out example 10.3

During additive manufacturing, hydrogen gas bubbles may escape from the fusion zone depending on the upward velocity of the bubble and the time available for the bubble to reach the top surface of the fusion zone. Use the following data and make reasonable assumptions to approximately estimate the possibility of bubbles of different sizes to escape being trapped in the part. Discuss the assumptions you make. Density of aluminum alloy: 2700 kg/m^3, viscosity of the liquid alloy: 1 mPa-s, length of the fusion zone: 2 mm, depth of the fusion zone: 0.5 mm, scanning speed: 0.1 m/s.

Worked out example 10.3 (Continued)

Solution:

We assume the following for simplicity: (a) The bubble is at the bottom of the fusion zone halfway along the length of the fusion zone. (b) The properties of the liquid alloy in the fusion zone are homogeneous and independent of temperature. (c) The motion of the bubble is not affected by the flow of liquid metal in the fusion zone. This assumption is made to avoid tracking the bubble under the influence of both the drag force and the buoyancy force which will require a comprehensive heat transfer and fluid flow model and bubble tracking.

The approximate time that any point on the surface remains liquid is obtained from the length of the fusion zone and the scanning speed. If a bubble is positioned at midlength of the fusion zone, the time the bubble has to rise to the surface to avoid being frozen in the metal is half of the length of the fusion zone divided by the scanning speed, i.e., 1 mm/100 mm/s = 0.01 s.

The time for a bubble to rise vertically along the depth of the fusion zone will depend on the depth of the fusion zone (0.5 mm) and the rising velocity of the bubble. The velocity can be computed from the Stokes equation (Eq. 10.1) for any assumed value of bubble size. The computed velocities and times needed for bubbles of different sizes to rise vertically and escape the fusion zone are indicated in Table E10.1 below:

Table E10.1 The computed velocities and times needed for bubbles of different sizes to rise.

Bubble radius (micrometers)	Terminal velocity (m/s)	Time to reach the surface (s)
10	0.147×10^{-03}	3.40
25	0.920×10^{-03}	0.54
50	0.368×10^{-02}	0.14

The times needed for the bubbles to reach the top surface for the assumed bubble sizes are significantly more than the times available before the fusion zone freezes with the bubbles contained within it.

10.3.2 Porosity due to keyhole collapse

When a high-power density laser or electron beam impinges on a metallic surface, the alloy melts instantaneously and alloying elements vaporize at appreciable rates. The recoil pressure produced by the metal vapors displaces the liquid metal and forms a cavity within the fusion zone. The cavity, filled with metal vapors can be quite deep, many times the width of the cavity and it is called a keyhole. The depth of the keyhole depends on the power density (power/area over which the beam impinges), scanning speed, and the thermophysical properties of the alloy. Higher power density and slower scanning speed tend to increase the depth of the keyhole. In-situ X-ray studies have shown that deep keyholes fluctuate with high frequency and amplitude and are often unstable. The lower portion of the keyhole collapses from time to time and traps a portion of the vapors within the liquid pool forming a void space.

The pocket of metal vapors, separated from the main keyhole, moves within the fusion zone under the influence of both the drag force of the flowing liquid alloy stream surrounding the vapor packet and the gravitational force. The gravitational force helps to push it near the surface of the fusion zone. If the vapor packet fails to escape the fusion zone before it solidifies, porosity resulting from the instability of the keyholes appears in the part. These pores differ in shape and size from those of gas pores. The keyhole instability-induced pores are larger than the typical gas porosity, i.e., larger than 50 micrometers in size and they are also not as spherical as the gas pores.

Pressure on the keyhole wall is balanced by the forces that want to expand and contract the keyhole as follows:

$$P_r + (P_v - P_a) = P_s + P_h \tag{10.2}$$

where P_r, P_v, P_a, P_s, and P_h are recoil, vapor pressure, ambient, surface tension, and hydrostatic pressures, respectively. The terms on the left side of the equality are responsible for keeping the keyhole open while the terms on the right-hand side tend to close the keyhole. Since the keyhole is immersed in the fusion zone containing moving liquid alloy with fluctuating velocities, the local temperatures and pressures can change with time resulting in the collapse of the lower part of the keyhole. The instability is particularly pronounced when the depth of the keyhole exceeds the circumference and this type of instability is called Raleigh instability.

A good starting point in understanding keyhole-induced porosity is the estimation of the depth of the keyhole that depends on the process parameters and the thermophysical properties of the alloys. The keyhole depth to laser beam radius ratio was found to scale with a dimensionless parameter, Keyhole number Ke, expressed as [9–11]:

$$Ke = \eta P / [(T_l - T_0)\pi\rho C_p(\alpha u r^3)^{1/2}] \tag{10.3}$$

where η is absorption coefficient, P is power, T_l is the liquidus temperature of the alloy, T_0 is the substrate temperature, ρ is density, C_p is specific heat, α is thermal diffusivity, u is scanning velocity, and r is beam radius. Figure 10.6 shows a plot of the minimum and maximum keyhole depths in titanium and aluminum alloys and stainless steels over a 2 mm length as vertical error bars [9]. The results show that keyholes form when the Keyhole number Ke exceeds 6. The keyhole number, Ke, is related to the aspect ratio (keyhole depth to laser beam radius ratio) of the keyhole:

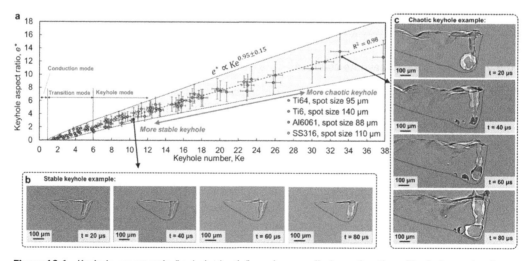

Figure 10.6 Keyhole aspect ratio (keyhole depth/laser beam radius) as a function of keyhole number for four alloys shown in the figure [9]. The figure is taken from an open-access article [9] under the terms and conditions of the Creative Commons Attribution (CC BY) license.

$$e^* = 0.4(Ke - 1.4) \tag{10.4}$$

Eq. 10.4 indicates that e^* is positive when Ke is greater than 1.4. Figure 10.6 shows that the fluctuation of the keyhole increases with the increase in keyhole number consistent with Eq. 10.3. The extent of fluctuation of the keyhole wall and the resulting instability can be examined from the stability of a cylindrical column of fluid. The Rayleigh instability criterion [12] provides the geometric condition for stability as:

$$e^* < \pi \tag{10.5}$$

The propensity to form keyhole-induced porosity increases with Ke and the porosity is most pronounced [9] when the keyhole number exceeds 30. The use of Keyhole Number, Ke to compute the formation of a keyhole is explained in worked out examples 10.4 and 10.5.

Worked out example 10.4

Dimensionless correlations are often used to determine when a keyhole forms during laser materials processing. Use an appropriate dimensionless correlation and the following data to determine (a) if the processing occurs in keyhole or conduction mode and (b) if a keyhole forms, determine its aspect ratio. Alloy: stainless steel 316, power: 400 W, beam radius: 40 micrometer, absorptivity: 0.9, scanning speed: 1 m/s, liquidus temperature: 1723K, substrate temperature: 298 K, specific heat: 834.3 thermal conductivity at the liquidus temperature: 26.9 W/(m K), thermal diffusivity at liquidus temperature: 4.48×10^{-6} m^2/s.

Solution:

(a) The keyhole number (Eq. 10.3) indicates the possibility to form a keyhole when Ke is larger than 6. The keyhole number Ke is obtained as:

$$Ke = \eta P / [(T_l - T_0) \pi \rho C_p (\alpha u r^3)^{1/2}]$$

$$= 0.9 \times 400 / [(1723 - 298) \times 3.14 \times 7200 \times 834.3 \times \{(4.48 \times 10^{-6} \times 1 \times (40 \times 10^{-6})^3\}^{1/2}]$$
$$= 25.02$$

(b) The aspect ratio is determined from the keyhole number using Eq. 10.4:
$$e^* = 0.4(Ke - 1.4) = 9.45$$

Worked out example 10.5

The keyhole-induced porosity is most pronounced when the keyhole number, Ke, is higher than 30. For the following processing conditions and thermophysical properties, determine the maximum laser power that can be used without the risk of pronounced keyhole-induced porosity. Alloy: stainless steel 316, beam radius: 40 micrometers, absorptivity: 0.9, scanning speed: 1 m/s, liquidus temperature: 1723K, substrate temperature: 298 K, specific heat: 834.3, thermal conductivity at the liquidus temperature: 26.9 W/(m K), thermal diffusivity at liquidus temperature: 4.48×10^{-6} m^2/s.

(Continued)

Worked out example 10.5   (Continued)

Solution:

Pronounced keyhole instability induced porosity is observed when Ke is higher than 30.

So $\eta P / [(T_l - T_0)\pi\rho C_p(\alpha u r^3)^{1/2}] = 30$ where η is 0.9, T_l is 1723 K, T_0 is 298 K, ρ is 7200 kg/m^3, α is 4.48 × 10^{-6} m^2/s, u is 1 m/s, and r is 40 micrometer. Solving, one gets P = 479.6 W and the aspect ratio, e*, is obtained as 11.4 which is much larger than π and pronounced keyhole induced porosity is expected.

10.3.3   Lack of fusion voids

Insufficient overlap of deposits between adjacent tracks and successive layers may result in gaps between deposits. The undesirable space between both adjacent tracks and successive layers is commonly attributed to insufficient melting of the feedstock owing to inappropriate choices of process variable–alloy combinations. A weak heat source, inappropriately fast scanning speed, thick layers, or excessive hatch spacing (space between adjacent tracks) can result in such flaws. Often the adjustment of process parameters such as the heat source power and scanning speed can avoid this type of flaw and achieve high part density. However, such defects are common in metallic parts made by additive manufacturing. Assuming that the lack of fusion results only from the insufficient overlap of hemispherical fusion zones in adjacent tracks and layers, the geometric criteria for avoiding lack of fusion has been suggested as [13]:

$$(h/w)^2 + (l/d)^2 \leq 1 \qquad\qquad (10.6)$$

where h is the hatch spacing, w is melt-pool width, l is layer thickness, and d is melt-pool depth. Eq. 10.6 indicates that the lack-of fusion porosity can be avoided by forming fusion zone cross-sections of adequate depth and width in comparison with the hatch spacing and layer thickness. If the width and the depth of the fusion zone are both 41.4% larger than the hatch spacing and layer thickness, respectively, lack of fusion can be avoided. The use of Eq. 10.6 requires the estimation of w and d. Models of varying complexity have been used to estimate fusion zone geometry. Depending on the mechanism of heat transfer within the fusion zone either a heat conduction model or a more rigorous heat transfer and fluid flow model [14] may be used. The calculation of lack of fusion voids based on fusion zone geometry is illustrated in worked out examples 10.6 and 10.7.

A visual representation of geometry-based evaluation of lack of fusion defects is shown in Figure 10.7. The figure shows computed cross-sections of the fusion zone in various layers and hatches. The lack of overlap is observed as void space of almost uniform shape and size. Such uniform void patterns are not observed in parts. The lack of fusion voids may have elongated and irregular shapes and sharp edges that vary in size considerably as shown in Figure 10.8. Excessively large hatch spacing and layer thicknesses do contribute to the void space in parts and the molten pool size does affect the lack of fusion. However, the mismatch between the observations and geometry-based estimation of symmetrical, uniform, repetitious lack of fusion voids point to mechanisms in addition to the just small fusion zone cross-sections unable to provide sufficient overlap for a given set of hatch spacing and layer thickness. Inadequate supply of feedstock, faulty powder packing, and powder ejection due to rapid vaporization of alloying elements are some of the potential difficulties.

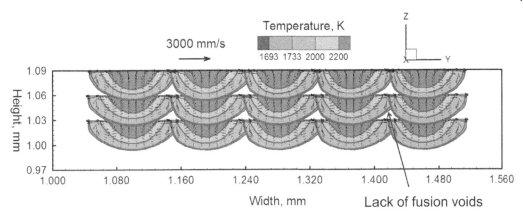

Figure 10.7 Lack of fusion voids inside a transverse section of a five-layer and five-track deposit of stainless steel part built using 1000 mm/s speed and 80 micrometer spacing between parallel tracks. The results are generated using a model developed by T. Mukherjee and T. DebRoy.

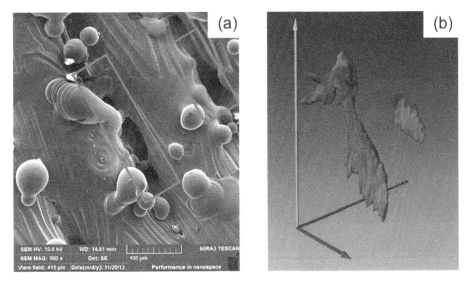

Figure 10.8 (a) Lack of fusion voids observed in an SEM image and (b) a 3D reconstructed synchrotron image. Both images show irregular shapes of the voids [15]. This figure is reprinted with permission from Elsevier.

Eq. 10.6 provides a binary choice of whether or not a lack of fusion void is likely to form. However, it does not provide any quantitative correlation between the extent of the lack of fusion and the important variables. A nondimensional correlation has been proposed considering the important process variables and alloy properties for the laser powder bed fusion process. The laser power (P), scanning speed (u), layer thickness (l), hatch spacing (h), melt pool width (w), and depth (d) are considered in the correlation. High laser powers facilitate the melting of the powder particles and reduce the propensity of lack-of-fusion porosity. In contrast, high scanning speeds reduce the heat input per unit length necessary for the melting of the powder.

Large layer thickness and hatch spacing require the melting of a sufficient volume of powder to achieve sufficient overlap between the adjacent tracks and layers to avoid any lack of fusion void. Similarly, the enthalpy of melting per unit volume of metal (H_m) of the alloy is considered an important variable since different alloys need different amounts of heat for melting. It was suggested [16] that the following correlation could accurately describe the extent of lack of fusion porosity, ϕ.

$$\phi = 0.14 \left[\frac{(h/w)(l/d)}{\sqrt{[(P/H_m)/(uwd)]}} \right]^{0.94} \tag{10.7}$$

The correlation was tested [16] by calculating the porosity fractions in parts made by laser powder bed fusion of five commonly used alloys for a wide range of process conditions and comparing them with the corresponding independent experimentally measured results as shown in Figure 10.9.

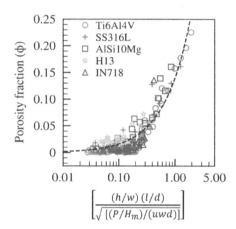

Figure 10.9 Experimentally measured porosity fraction and computed porosity fraction (ϕ) for PBF-L of Ti6Al4V, SS316L, AlSi10Mg, IN718, and H13 tool steel. *Source:* P.R. Zagade, B.P. Gautham, A. De, and T. DebRoy.

Table 10.1 Ranges of process conditions and measured porosity fraction for PBF-L of Ti6Al4V, SS316L, AlSi10Mg, IN718, and H13 and the thermophysical properties of the alloys [16].

Parameter	Ti6Al4V	SS316L	AlSi10Mg	IN718	H13
Laser power, P (W)	40–200	90–300	150–350	90–370	150–300
Scanning speed, u (m/s)	0.12–1.56	0.3–2.8	0.5–2.5	0.4–1.6	0.4–1.0
Layer thickness, l (mm)	0.03–0.05	0.025–0.040	0.03–0.09	0.025–0.060	0.040
Hatch spacing, h (mm)	0.10	0.08–0.12	0.045–0.130	0.08–0.12	0.080–0.120
Porosity fraction, ϕ	0.001–0.220	0.001–0.160	0.002–0.160	0.002–0.130	0.001–0.070
Density, ρ (kg/m^3)	4200	7400	2610	7700	7100

Table 10.1 (Continued)

Parameter	Ti6Al4V	SS316L	AlSi10Mg	IN718	H13
Solidus, liquidus temperatures, (T_S, T_L) (K)	1878, 1923	1658, 1723	823, 850	1533, 1609	1585, 1723
Conductivity, k (W/mK)	$8.7 \times [1 + (1.18 \times 10^{-3} \times T)]$	$11.3 \times [1 + (0.89 \times 10^{-3} \times T)]$	$118 \times [1 + (0.01 \times 10^{-3} \times T)]$	$11.5 \times [1 + (1.06 \times 10^{-3} \times T)]$	$22.1 \times [1 + (0.76 \times 10^{-3} \times T)]$
Specific heat, C (J/kgK)	$260 \times [1 + (1.18 \times 10^{-3} \times T)]$	$280 \times [1 + (0.89 \times 10^{-3} \times T)]$	$980 \times [1 + (0.01 \times 10^{-3} \times T)]$	$280 \times [1 + (1.06 \times 10^{-3} \times T)]$	$310 \times [1 + (0.76 \times 10^{-3} \times T)]$
Enthalpy of melting per unit volume, H_m (J/m^3)	6.86×10^9	9.47×10^9	2.84×10^9	8.47×10^9	7.94×10^9

Worked out example 10.6

Inappropriate combinations of process parameters are often thought to be responsible for the lack of fusion defects in additively manufactured parts. Estimate if the following processing conditions will result in a lack of fusion defect in an SS316 part. If a lack of fusion is anticipated, estimate the void fraction. Alloy: SS316, scanning speed: 1 m/s, power: 200 W, layer thickness: 0.025 mm, hatch spacing: 0.035 mm, fusion zone width and depth: 65 micrometers and 25 micrometers, respectively. Use the necessary data from Table 10.1.

Solution:

Using Eq. 10.6, the estimated value of $(h/w)^2 + (l/d)^2$ is estimated as $(0.035/0.065)^2 + (0.025/0.025)^2 = 1.28$. Since this number is much larger than 1, a lack of fusion defect is expected to form.

The extent of lack of fusion void can be estimated from Eq. 10.7 using the following data: h = 0.035 mm, w = 0.065 mm, l = 0.025 mm, d = 0.025 mm, P = 200 W, u = 1 m/s, H_m is 9.47×10^9 J/m^3. Using these data, the lack of fusion void fraction is estimated to be 0.023.

Worked out example 10.7

For a given set of processing conditions, the extent of the lack of fusion porosity of parts made using different alloys depends on the thermophysical properties of the alloys. Use the data provided in Table 10.1 to estimate the relative propensities of lack of fusion voids in SS316 and Ti6Al4V alloys for the following processing conditions. Scanning speed: 1 m/s, power: 200 W, layer thickness: 0.025 mm, hatch spacing: 0.035 mm Fusion zone width and depth depend on the alloy. Their values are 65 micrometers and 25 micrometers, respectively for SS316, and 0.13 mm and 0.065 mm, respectively for Ti6Al4V.

Solution:

For stainless steel 316, the previous problem shows that a lack of fusion voids will form and the lack of fusion void fraction was estimated to be 0.023. A similar exercise for Ti6Al4V gives $(h/w)^2 + (l/d)^2$ as $(0.035/0.130)^2 + (0.025/0.065)^2 = 0.22$. Since this number is much smaller than 1, a lack of fusion defect is not expected to form in the Ti6Al4V alloy part.

10.4   Surface defects

10.4.1   Surface roughness

The surface roughness of the additively manufactured metallic parts affects their performance and impedes the successful inspection of components. The rough surfaces provide crack initiation sites which reduces fatigue resistance and fracture toughness. Figure 10.10 (a) and (b) show typical surface unevenness [17]. Surface roughness is measured by determining the height of a peak or the depth of a valley at various locations on the surface using a profilometer. For example, if f_n is the deviation of a location from the average elevation of N points on a surface, the average surface roughness (R_a) is calculated as [18]:

$$R_a = \frac{1}{N}\sum\nolimits_{i=1}^{N} |f_n| \tag{10.8}$$

Surface roughness of less than 1 micrometer is considered to be very smooth for metallic surfaces. The roughness of most as-deposited metallic components, however, is much greater than 1 micrometer. For example, the powder bed fusion process provides the smoothest surface, with typical roughness between 5 and 50 micrometers. Directed energy deposition results in large melt pools and rougher surfaces. Depending on the application, parts may require post-processing such as surface machining, grinding, shot penning, or chemical polishing to reduce roughness. However, there are limitations to each of these processes. For example, shot peening is inappropriate for thin and intricate structures where the local surface deformation cannot be tolerated.

The roughness may result from multiple mechanisms. These include spattering, uneven melting of the powder and wire feedstock, surface porosity, steps due to the construction of a curved 3D object by progressive deposition of 2D layers, and unevenness and surface marks due to the removal of the support structure after parts are made. AM process selection, process parameters, and alloy type affect the quality of the surface. Thus, the surface roughness is affected by a large number of variables. Of these variables, the manufacture of parts with curved and inclined surfaces presents a special problem because these surfaces are approximated by small steps [19] as shown in Figure 10.11 (a). The layer thickness and the build angle shown in Figure 10.11 (a) play important roles since larger steps deviate considerably from the continuous surface curvature. Figure 10.11 (b) shows how the thicker layer increases the surface roughness.

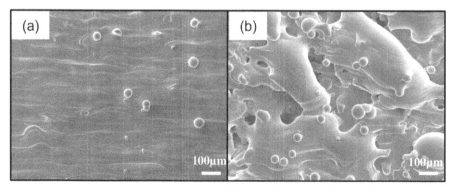

Figure 10.10   Scanning electron micrographs [17] showing the roughness of the top surface of Ti-6Al-4V samples fabricated at (a) 400 W and 2300 mm/s and (b) 400 W and 3500 mm/s, with a powder layer thickness of 20 μm. The figure is taken from an open-access article [17] under the terms and conditions of the Creative Commons Attribution (CC BY) license.

(a)

(b)

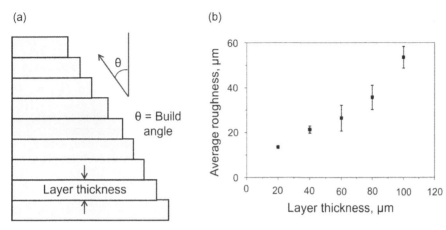

Figure 10.11 (a) In a layer-by-layer deposition, curved surfaces are approximated by multiple small steps causing the staircase effect that contributes to surface roughness. *Source:* T. Mukherjee and T. DebRoy. (b) Thicker layers increase surface roughness [17]. The figure is taken from an open-access article [17] under the terms and conditions of the Creative Commons Attribution (CC BY) license.

The surface roughness (R_a) has been related to the layer thickness (t_l) and the build angle (θ) by the following equation [19]:

$$R_a = t_l sin\left(\frac{90-\theta}{4}\right) tan(90-\theta) \tag{10.9}$$

where R_a is the arithmetic mean of the surface roughness in micrometer, layer thickness (t_l) is expressed in micrometer, and the build angle (Figure 10.11 (a)) is expressed in degrees. Eq. 10.9 shows that the surface roughness increases linearly with the layer thickness and with the steepness of the inclination. Rapid building of parts with higher layer thickness increases surface roughness. Similarly, a smaller build angle (θ) results in rougher surfaces. An application of this methodology is illustrated in worked out example 10.8.

Worked out example 10.8

Surface roughness of a curved surface has been related to the layer thickness and build angle in the literature. The maximum possible surface roughness of a part is specified as 50 micrometers. (a) If the geometry of the part requires a build angle of a curved surface to be 45 degrees, what is the maximum layer thickness that can be used? (b) Determine how the surface roughness will vary with the build angle if the layer thickness is 100 micrometers.

Solution:

Eq. 10.9 shows that both the layer thickness and the build angle affect the surface roughness resulting from the approximation of a curved surface with steps. Rearranging Eq. 10.9, one gets the expression for layer thickness t_l as:

$$t_l = R_a / [sin\left(\frac{90-\theta}{4}\right) tan(90-\theta)] \tag{E10.1}$$

$= 50.0/[sin(45/4)tan(45)] = 102.8$ micrometer

(Continued)

Worked out example 10.8 (Continued)

(b) A plot of surface roughness as a function of build angle can be obtained using equation 10.10 for a layer thickness of 100 micrometers for different values of build angle as shown in Figure E10.2 below.

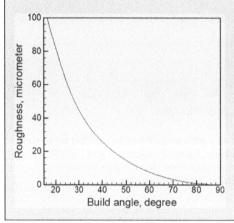

Figure E10.2 A plot of roughness versus build angle for a layer thickness of 100 micrometers obtained from Eq. 10.9. A lower inclination angle means a more slanted side. The build angle is shown schematically in Figure 10.11(a).

Many variables such as part design, process selection, feedstock, and process parameters affect the roughness of all build surfaces, inclined or horizontal. Part design factors include part orientation and the support structure. The powder bed process, when operated with fine powder particles, is capable of producing the best possible surface finish while the directed energy deposition wire-arc process normally requires post-processing because of the considerable roughness or waviness of build surfaces. Powder particle size distribution, wire diameter, and feed rate also affect the surface quality. Process variables and the thermophysical properties of the alloy affect the melting of the feedstock, the distribution of the molten metal, and the ejection of the metal powder particles to form spatters, a portion of which may impinge on the build surface. Because of the large number of variables that affect surface roughness, it is difficult to quantitatively understand surface roughness based on physics-based mechanistic models.

For the powder bed fusion process, a nondimensional scaling analysis considering many important process variables and alloy properties has been proposed. The variables include heat input, fusion zone aspect ratio of pool length to depth, Marangoni force, powder diameter, layer thickness, enthalpy of melting of the alloy, and the contact angle. A surface roughness index (SRI) constructed using the Buckingham pi theorem considering these variables provides a scale for comparing roughness for different alloys and processing conditions because the material properties and process variables are considered in the index. The surface roughness index (SRI) is expressed as [20]:

$$SRI = Et^2 (\varepsilon)^n \sqrt{\frac{\theta}{HF}} \tag{10.10}$$

where E, t, ε, θ, H, and F are enthalpy of melting (J/m^3), layer thickness (m), pool aspect ratio (pool length/depth), contact angle (radian), heat input (laser power/scanning speed, J/m), and Marangoni force (N). The symbol n is a positive fraction. The computed values of SRI for four commonly used alloys, SS316, Ti6Al4V, IN738, and AlSi10Mg, are shown in Figure 10.12. For the same

Figure 10.12 Measured and computed surface roughness for four alloys made by PBF-L. *Source:* Y. Du, T. Mukherjee, N. Finch, A. De, and T. DebRoy.

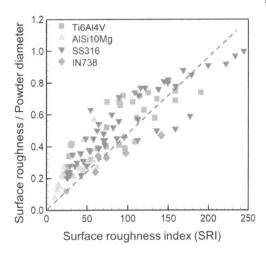

processing variables, the AlSi10Mg alloy has the lowest value of surface SRI, indicating the highest likelihood to produce a smooth surface among the four alloys, while the SS316 is the most vulnerable to surface roughness among the four alloys. The correlation is valid for the range of data used to validate it and specifically for the four alloys. However, it is believed that the variables represented in Eq. 10.10 play important roles in determining surface roughness.

The current solutions for reducing surface roughness include post-processing, trial-and-error optimization of process variables, hybrid manufacturing, and machine learning. However, each of these techniques has its limitations. For example, post-processing adds extra cost and cannot always improve the internal surfaces. Surface roughness can be minimized by adjusting the process variables, such as the power and scanning speed and the layer thickness. However, this process is time-consuming and expensive because it needs numerous tests to explore a large parameter space. Hybrid manufacturing can reduce the surface roughness but it adds a step such as machining after the printing of each layer. Also, it slows down the production process and increases part costs. Printing parts with a smooth surface considering both theory and experiments based on scientific principles provides a solution to this issue. This approach allows consideration of all the important variables that affect surface roughness, is easy to use, and can be verified with experiments. The surface roughness is affected by the powder size and process variables and this is discussed in worked out example 10.9.

Worked out example 10.9

The surface roughness index of four alloys during the powder bed fusion process has been correlated with surface roughness divided by the powder diameter as shown in Figure 10.12. The surface roughness index considers the effects of heat input, layer thickness, fusion zone aspect ratio, Marangoni force, contact angle, and the enthalpy of melting of the alloy. Explain why these variables and the powder diameter affect the surface roughness.

Solution:

Heat input per unit length (H): The heat input per unit length is the ratio of laser power to scanning speed. It affects the amount of molten liquid and the stability of the molten pool. Low

---

(Continued)

> **Worked out example 10.9   (Continued)**
>
> laser power and a high scanning speed may result in an insufficient amount of molten liquid, and sometimes a discontinuous deposit of liquid metal, thus causing surface roughness. Also, too high a heat input may result in the ejection of liquid metal particles in the form of spatters.
>
> <u>Powder diameter (*D*)</u>: Larger powders need more heat to melt than small particles. Therefore, large particles are sometimes partially melted and they are often found at the edge of the tracks.
>
> <u>Layer thickness (*t*)</u>: Thicker layers increase the roughness due to the "staircase effect" when parts have curved surfaces. Thin layers help to produce parts with relatively smooth curved surfaces. However, they also slow down the deposition of parts.
>
> <u>Pool aspect ratio (*ε*)</u>: Long molten pools with relatively small depth form at high scanning speeds. The resulting high length-to-depth aspect ratio often separates into discontinuous small liquid pools, known as balling, and causes surface roughness. A small length-to-depth ratio at high powers and low scanning speeds can avoid balling.
>
> <u>Marangoni force (*F*)</u>: At the surface of the fusion zone, liquid metal flows from locations with low surface tension to locations with high surface tension because of the Marangoni effect. A large Marangoni force results in a strong convective flow, good spreading of the molten metal, and a smooth surface.
>
> <u>Contact angle (*θ*)</u>: The contact angle between the molten liquid droplet and the previously deposited layer or the substrate affects the spreading of the liquid on the surface, wetting, and therefore, the surface roughness. A low contact angle is desirable because of the liquid's ability to spread easily on a solid surface creating smooth surfaces.
>
> <u>Enthalpy of melting (*E*)</u>: Alloys with a low enthalpy of melting are easy to melt and create smooth surfaces.

10.4.2   Balling and surface waviness

10.4.2.1   Balling

The fusion zone becomes elongated at high scanning speeds and depending on the processing conditions and alloy properties it can break up into small islands or balls. Rapidly moving thin long molten pools suffer from instability. Several small disconnected beads or balls appear on the surface known as balling as shown in Figure 10.13. Balling hinders the powder from spreading evenly during the powder bed fusion process. It also degrades fatigue property and requires post-processing. High heat input achieved by high laser power and low scanning speed often prevents balling.

The mechanism of the formation of balls is not well understood although the instability of the liquid pool has been suggested as a possible factor. Several factors are believed to be responsible for the balling. They include the volumetric energy density, Kelvin Helmholtz hydrodynamic instability expressed by the dimensionless Richardson number, the strength of the liquid metal velocity in the fusion zone expressed by the dimensionless Marangoni number, solidification time of the liquid pool, and the surface tension force. These are explained below.

The volumetric energy density: The volumetric energy density, E, is the amount of energy supplied from the heat source per unit volume of material deposited. It is expressed by the following equation.

Figure 10.13 Scanning electron microscopy image [21] of 316 stainless steel showing the balling of single scan tracks for different scan speeds at a constant power of 190 W. The figure is reprinted with permission from Springer Nature.

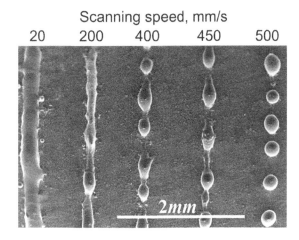

$$E = \frac{P}{v\left(\pi r^2\right)} \tag{10.11}$$

where P is laser power, v is scanning speed, and r is laser beam radius. The lack of energy, E, necessary to maintain a stable molten pool may break it up into several small balls.

Kelvin Helmholtz instability expressed by Richardson number: The difference in the velocities of the gas and the liquid on the surface of the liquid pool causes this instability. This hydrodynamic instability can be represented by a dimensionless number called the Richardson number. The susceptibility to the balling defect increases with the decrease in Richardson number, R:

$$R = \frac{gL}{\left(U_g - U_l\right)^2} \tag{10.12}$$

where g is the acceleration due to gravity, L is the length of the liquid pool, U_g is the velocity of shielding gas, and U_l is the velocity of the liquid metal at the surface of the fusion zone. The layer is considered unstable if the Richardson number is less than 0.25.

Pool aspect ratio: Capillary instability, sometimes called Raleigh instability, can also break up a molten pool and cause balling because breaking up a long fusion zone into drops reduces the total surface energy. Capillary instability of the molten pool depends on the ratio of pool length to pool depth (aspect ratio). When the aspect ratio is high and the length of the molten pool is significantly larger than its depth, the pool becomes unstable and breaks up into small balls.

Surface tension force: Surface tension force opposes the liquid to escape the fusion zone. This force depends on the surface tension of the alloy and the fusion zone geometry. At low surface tension force, the molten pool may be disintegrated to form balls.

Solidification time of molten pool: A molten pool that solidifies rapidly may not allow the molten pool to spread uniformly and may break up into small isolated balls. Solidification time is represented by the ratio of pool length to the scanning speed.

Marangoni number: Convective flow of liquid metal, primarily driven by the surface tension gradient on the molten pool's surface, aids in the spreading of the liquid metal. Marangoni number

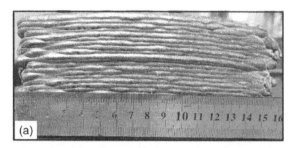

Figure 10.14 (a) Surface waviness of a 304 stainless steel thin walled structure deposited by the wire arc process using 250 A, 21.5 V at a speed of 15 mm/s [22]. The figure is reprinted with permission from Taylor and Francis. (b) A carbon steel tube transition sleeve for an oil and gas facility printed using wire arc additive manufacturing showing a characteristic waviness. Figure courtesy: Dr. Luiz Paes from the Center for Research and Development of Welding Processes and Additive Manufacturing - Uberlândia – Brazil.

is a measure of the strength of this flow known as Marangoni flow. A low Marangoni number indicates a weak flow that is insufficient for the uniform spreading of liquid metal which may cause balling.

10.4.2.2 Surface waviness

Wetting of the substrate and the previously deposited surface by a relatively large liquid metal pool and its uneven spreading causes waviness of the surface [22] [23]. Figure 10.14 shows typical surfaces of wire arc-produced parts made at a high deposition rate using a relatively thick layer thickness showing surface waviness.

10.5 Loss of alloying elements

Many metallic feedstocks contain volatile alloying elements. Familiar examples are magnesium and zinc in aluminum alloys, chromium and manganese in steels, and aluminum in titanium alloys. When the feedstock is heated and melted, volatile alloying elements may vaporize at appreciable rates. The rate of vaporization of an alloying element is influenced by its vapor pressure above the liquid alloy, which depends on the local temperature and the composition of the alloy, particularly the concentrations of the volatile alloying elements. The ambient pressure is also an important factor. When electron beams are used as a heat source, the chamber pressure is maintained very low and the vaporization of alloying elements occurs at appreciable rates. Different alloying elements vaporize at different rates at a given temperature. As a result, the chemical composition of the part made by additive manufacturing may differ from that of the feedstock. Since the chemical composition affects the microstructure, corrosion resistance, and mechanical properties, it is important to understand and mitigate the loss of alloying elements and the composition change.

When a laser or an electron beam is used as a heat source, melt pool temperatures can reach much higher than the liquidus temperatures of alloys. When the power density is high, the temperature directly under the heat source can reach as high as the boiling point of the alloy. The vaporization can be so intense that a vapor cavity known as the keyhole may form directly under the heat source. Metal vapors come out of the vapor cavity and the keyhole moves with the motion of the heat source. The vaporization process is particularly intense when an electron beam is used because they are operated at very low ambient pressures. A small portion of the vaporized materials recondenses on the melt pool surface. At relatively low power densities, keyholes do not form and the interaction between the heat source and the melt pool is characterized as conduction mode. Appreciable vaporization of volatile alloying elements can still occur in the conduction mode. The temperature directly under the heat source on the melt pool surface is much higher than that near the periphery of the melt pool. As a result, most of the vaporized material originates from the middle of the melt pool directly under the heat source. Powder bed additive manufacturing processes have been operated in both keyhole and conduction modes.

Both in-situ and ex-situ measurements have been made to understand the vaporization behavior during laser-materials interaction. Measurements of the chemical compositions of the feedstock and the part are helpful to understand the alloying element vaporization since the difference in the composition is related to the loss of volatile alloying elements. Consider an electron beam additively manufactured part of Ti-6Al-4V. The aluminum concentration showed a considerable decrease of about 0.8 wt% in comparison with that of the feedstock (Figure 10.15). The difference in the composition can be attributed to the selective vaporization of aluminum compared to titanium. Since vanadium does not vaporize at any appreciable rate, the decrease in the aluminum composition was compensated by an increase in the titanium concentration. Titanium was lost by vaporization but the loss was much smaller than that of aluminum.

Loss of volatile alloying elements has been studied extensively in welding [25] since it affects both the composition and properties of welds. For example, pronounced changes in the concentrations of manganese and chromium take place during laser welding of 204 stainless steel welds (Figure 10.16). In the keyhole mode welding of stainless steels, the temperature field, the internal surface area of the keyhole, and the top surface area of the molten pool affect the rates of vaporization of the alloying elements. These variables, in turn, are affected by the laser power, power density, welding speed, and chemical composition and properties of the steel. Because of the involvement of many variables, the role of welding variables on the chemical composition of the alloying elements cannot be predicted intuitively. However, computational models have been used to correctly predict the roles of laser power and welding speed on composition changes.

The observed composition change of alloying element i, Δ%i, can be related to its rate of evaporation through the following mass balance.

$$\Delta\%i = 100 \, r \, A \, / \, (\rho v) \tag{10.13}$$

where r is the vaporization rate of i per unit surface area, A is the surface area, ρ is the density of the weld metal, and v is the volume of the alloy melted per unit time. Of these parameters, the volume of metal melted per unit time can be calculated by multiplying the deposit cross-section by the scanning velocity. The rate of evaporation per unit area, r, depends on local temperature and varies spatially. Thus the product rA needs to be spatially integrated. Composition change depends on the vaporization flux and the surface-to-volume ratio of the fusion zone. Since the surface area to volume ratio is nearly inversely proportional to a length that may be approximately taken as the depth of the fusion zone, this ratio becomes more important for small melt pools, typically obtained at low heat input (power/speed). For the conditions of the

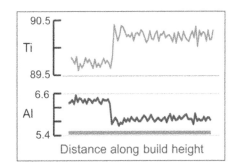

Figure 10.15 Measured chemical compositions of Ti and Al along the build height for a Ti-6Al-4V sample made by PBF-EB showing depletion in Al concentration and an increase in Ti concentration [24]. The concentrations are in weight percentages. Image credit: NASA.

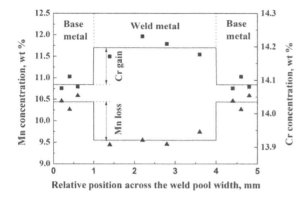

Figure 10.16 Composition change of stainless steel 204 during keyhole mode laser welding. The solid lines are the simulated results. The laser power was 1250 W and the welding speed was 8 mm/s. *Source:* T. Liu, L.J. Yang, H.L. Wei, W.C. Qiu, and T. DebRoy.

experiments presented in Figure 10.16, a small melt pool was formed and a significant composition change was observed. This fact may appear counterintuitive at first. However, it is consistent with the law of mass conservation. The composition change is most pronounced at low powers because of the small size and, consequently, a high surface-to-volume ratio of the melt pool. In conduction mode processing when a keyhole does not form, the alloying elements vaporize from the surface of the molten pool. However, the alloying elements are depleted from the entire volume of the liquid pool since the molten pool is well mixed. Therefore, the surface area-to-volume ratio significantly affects the composition change. Higher temperatures result in more intense vaporization and increase the size of the fusion zone which means that the loss of alloying elements is distributed over a larger volume. Both the rate of vaporization and the volume of the fusion zone are needed to estimate the composition change.

A simple model to calculate the evaporation rate is given by the Langmuir equation [26]:

$$J_i = a\, P_i / (2\pi M_i R T)^{0.5} \qquad (10.14)$$

where J_i is the vaporization flux of element i, a is a positive fraction that is related to the recondensation of the vaporized species that depends on the total pressure, P_i is the equilibrium vapor pressure of i, M_i is the molecular weight of i, R is the gas constant, and T is the temperature. The value of a is close to 1 at very low ambient pressures used typically when an electron beam is used as a heat source. When a laser beam is used and the ambient pressure is 1 atmosphere, a value of much lower than 1 is appropriate. The temperature field affects the vaporization rate significantly since vapor pressures are strongly temperature-dependent. As a result, both the temperature field and the geometry of the molten pool are important for understanding the change in composition during AM. The variables that influence the composition change are those that affect molten pool geometry and temperature distributions such as power, scanning speed, feedstock feed rate, and beam diameter. The estimation of vaporization rate requires vapor pressure versus temperature data at high temperatures, sometimes above the boiling point. Worked out example 10.10 shows how vapor pressures can be estimated above the boiling point of an element.

The measurements of the composition change between the feedstock and the part provide a measure of the alloying element loss [27, 28]. Characterization tools such as energy dispersive spectroscopy (EDS) or electron probe microanalysis (EPMA) are often used for nondestructive analysis. Since elemental segregation is common in additively manufactured parts, care needs to be taken to have a statistically significant volume of data for accurate estimation. Accurate estimation of composition can also be obtained using destructive methods such as inductively coupled plasma mass spectrometry. The peak temperature attained during laser processing has been approximately estimated from the vapor composition emitted from the fusion zone [27, 28]. The principle for the estimation is illustrated in worked out example 10.11.

Worked out example 10.10

During additive manufacturing, the evaporative loss of an alloying element depends on its equilibrium vapor pressure and the local temperature. Understanding evaporative loss requires vapor pressure data as a function of temperature. These data are not always available at high temperatures close to or higher than the boiling point of the elements. However, if the enthalpy of vaporization is known, the properties may be estimated from the Clausius Clapeyron equation. Assume that the boiling point of iron is 3134 K and the enthalpy of vaporization is 354 kJ/mol. Estimate the equilibrium vapor pressure of iron as a function of temperature up to 10 atmosphere pressure.

Solution:

If the vapor pressure (p_1) of iron is known at one temperature (T_1), the Clausius Clapeyron equation allows us to estimate the vapor pressure (p_2) at another temperature (T_2) if the enthalpy of vaporization is known.

$$\ln(p_2/p_1) = (H_v/R)(1/T_1 - 1/T_2) \tag{E10.2}$$

In other words, if the equilibrium vapor pressure, p_1 at a temperature T_1 is known, then p_2 at any temperature can be calculated if the enthalpy of vaporization, H_v and gas constant, R, are known. For iron $T_1 = 3134$ K, $p_1 = 1$ atm, the heat of vaporization is 354 KJ/mol, and R = 8.3145 J/(K mol). The equilibrium vapor pressure of iron is estimated from Eq. E10.2 and plotted in Figure E10.3 below.

Figure E10.3 A plot of the equilibrium vapor pressure of iron estimated from the Clausius Clapeyron equation as a function of temperature.

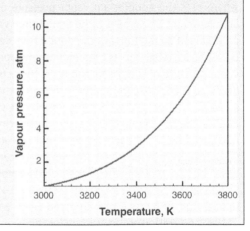

Worked out example 10.11

The composition of the vapors emitted from the fusion zone during additive manufacturing depends on the alloy composition and the temperature profile on the surface of the fusion zone. In an experiment [27, 28] in which a laser beam was irradiated on the surface of a stainless steel plate, a portion of the vapors was condensed on the inner surface of a both-end open quartz tube. The molar ratio of nickel to manganese was found to be about 0.05, indicating

(Continued)

Worked out example 10.11   (Continued)

that the flux of manganese was about 20 times that of nickel. Use the following data to interpret this observation. Composition of the stainless steel by weight %: Fe: 69.3, Mn:7.5, Cr: 17.8, Ni: 4.7, other elements: 0.7. Atomic weights (kg/kg mole): Fe: 56, Mn: 55, Cr:17.8, Ni: 58.8. Equilibrium vapor pressures of pure elements are given by: p^0_i (atmosphere) = $-A_i/T + B_i + C_i$ log(T) + D_i T/1000, where T is in K where the coefficients A_i, B_i, C_i, and D_i for Mn and Ni are given in Table E10.2 below.

Table E10.2 The computed velocities and times needed for bubbles of different sizes to rise.

	A_i	B_i	C_i	D_i
Mn	23,600	85.49	−23.92	2.191
Ni	−4552	−165.9	51.135	−4.476

Solution:

The ratio of the flux of nickel, J_{Ni}, and manganese, J_{Mn}, is obtained from Eq. 10.14:

$$J_{Ni}/J_{Mn} = (p_{Ni}/p_{Mn}) \times (M_{Mn}/M_{Ni}) \qquad \text{(E10.3)}$$

where p_i is the equilibrium vapor pressure of element i over the liquid alloy and M_i is the atomic weight of element i. The equilibrium vapor pressures of element i over the alloy, p_i can be obtained by multiplying the equilibrium vapor pressures by the respective activities. Thus, $p_{Ni} = p^0_{Ni} a_{Ni}$ where p_{Ni} is the equilibrium vapor pressure over the alloy, p^0_{Ni} is the equilibrium vapor pressure over pure Ni, and a_{Ni} is the activity of Ni in the alloy. Since the temperatures in the fusion zone are commonly much higher than the melting point, the alloy may be treated as an ideal solution, i.e., a_{Ni} can be assumed to be equal to its mole fraction. The mole fractions of all elements can be obtained by considering any given weight of the alloy, calculating the number of moles of each element, and calculating the mole fractions of each element. From the alloy composition given, the weight percentages of Fe, Ni, Cr, and Mn add up to 99.3%. The remaining 0.7% of the alloy constituents is not known. The mole fractions of Fe, Cr, Ni, and Mn can be approximately calculated as: 0.689, 0.191, 0.045, and 0.076, respectively. So, for a given temperature, the flux ratio expressed by Eq. E10.3 can be calculated.

The vaporization occurs from all locations on the surface of the fusion zone but the equilibrium vapor pressures are highly sensitive to temperature. As a result, most of the vapors originate from the highest temperature region of the fusion zone, i.e., from directly under the heat source. The computed values of J_{Ni}/J_{Mn} are plotted as a function of temperature in Figure E10.4 below. For a flux ratio of J_{Ni}/J_{Mn} = 0.05, the temperature is about 2980 K. This value is approximate because vapors originate from all locations of the surface, although the highest temperature regions contribute most toward the vaporization.

Worked out example 10.11   (Continued)

Figure E10.4   A plot of the ratio of the computed rates of vaporization of nickel and manganese as a function of temperature.

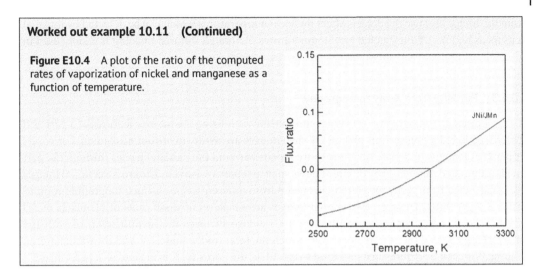

10.6   Characterization methods

Although high-quality parts are made using additive manufacturing, parts need to be adequately tested for flaws such as pores, cracks, inclusions, residual stress, and distortion without damaging them to ensure quality. Residual stress and distortion have been covered in Chapter 11 in detail and here we discuss the nondestructive testing of parts for pores, cracks, and inclusions to ensure their fitness for service. Nondestructive tests are also used to determine the evolution of defects in parts due to service where parts are exposed to high temperatures, cyclic loading, or corrosive environments. Reliable detection of defects by NDT depends on the size, geometry, and location of the defect, as well as the complexity, density, and surface finish of the part.

10.6.1   Porosity using Archimedes' principle

Nondestructive methods are widely used for measuring the porosity of additively manufactured parts. The porosity of the part is determined by measuring its weight in both air and another liquid such as water. First, the density of the part, ρ, is accurately computed from the measured values of masses of the part in air, m, and in a liquid, m_f, and the density of fluid ρ_f as:

$$\rho = m\rho_f/(m - m_f) \tag{10.15}$$

Finally, the measured density can be compared with the theoretical density of the material of the part to estimate the porosity of the part. These estimations provide a simple but overall value of porosity. They do not provide the size distribution or the specific location of the porosity. An appropriate imaging technique such as microscopy of the surface can provide both the size distribution and the location of the pores on flat surfaces. The distribution of pores of different sizes and shapes can be determined in three dimensions using X-ray tomography.

10.6.2   Infrared imaging

Infrared photographs of the part surface containing a defect show the contrast in thermal radiation between the defect and the surrounding areas. The shape and the contour outline can be observed from the image because the defects affect the local temperatures which can be detected by infrared sensors as regions of unexpected local temperatures. Although infrared imaging is widely used in research, it cannot detect small micron size pores.

10.6.3 Porosity from X-ray tomography

X-ray Computed Tomography (CT) is a technique for visualizing the interior features of a part as if it were sectioned along various planes. CT slice images are made up of volume elements or voxels or picture elements or pixels. An X-ray source rotates around the part and passes through the part and is attenuated differently by different flaws such as a pore or a crack. The cross-sectional images are constructed from X-ray measurements taken from different angles using tomographic computational algorithms. These cross-sectional images are made up of small units of 2D pixels or 3D voxels. The measurement of porosity from a CT image involves the identification of both the material regions and the void regions from which the total void volume, Vv, and the total material volume, Vm, can be determined and used to calculate the porosity, P, expressed as a percentage:

$$P = 100 \times Vv/(Vm + Vv) \tag{10.16}$$

The separation of a CT volume into regions of dense material and voids requires a segmentation algorithm. Small voids in metallic materials are challenging to find by CT because of their poor grey value contrast or low contrast-to-noise ratio. Appropriate algorithms are required to improve accuracy and reduce uncertainty in measurements. Figure 10.17 shows the effectiveness of using an appropriate segmentation algorithm to enhance the accuracy of CT in measuring the percentage of porosity.

10.6.4 Surface roughness measurement using profilometers

The surface roughness of AM parts is commonly determined by measuring the average surface heights and depths. The roughness may be determined by a mechanical device called a contact profilometer. Typically a diamond stylus is in contact with the sample surface. The stage containing the sample travels laterally as the tip of the stylus moves vertically to measure the roughness. In some cases, the stylus moves and the sample remains stationary. The measurement accuracy depends on the probe radius, stylus force, and scan length. The physical contact between the sample and the stylus in some cases may damage the surface of the specimen. The roughness may also be measured using an optical probe. There are several variations of this technique all aimed to reconstruct the surface topology in a noncontact manner. For example, an optical image of the surface can be obtained using optical interference, using a confocal aperture, focus, and phase detection.

Figure 10.17 Comparison between the X-ray computed tomography raw image (left) and the processed image (right) [29]. The background of the left image was removed to improve image clarity showing 3% porosity in IN718. The figure is taken from an open-access article [29] under the terms and conditions of the Creative Commons Attribution (CC BY) license.

10.6.5 Microscopy

Both optical and scanning electron microscopy are perhaps the most widely used laboratory techniques for characterizing defects in printed parts. Although specimens need to be prepared for examination in most cases, they reveal details of surface conditions quickly and inexpensively. Starting with the characterization of powder size distribution to powder morphology to the details of grain structure and phases in the microstructure, microscopy has been an indispensable tool in additive manufacturing laboratories. Different forms of porosity and lack of fusion defects and types of cracking are commonly characterized by optical and scanning electron microscopy. Similarly, microcopy remains an indispensable tool for characterizing morphological characterization of solidification structure in additively made parts.

10.6.6 Ultrasonic defect detection

Ultrasonic testing is a nondestructive technique that makes use of high-frequency sound waves to find cracks, pores, and inclusions in parts. A probe emits ultrasonic waves to the surface of a part. A couplant material, usually a liquid is placed between the probe and the part surface to bridge the impedance mismatch between the air and the solids. As the sound waves are transmitted inside the component, they encounter flaws and the reflected waves are identified based on the time difference between the reflected and transmitted signals. Apart from the measurement using a stationary probe, the probe is also moved in different patterns in some tests to cover various areas. The size of the flaws and their locations can be determined from the test. It is a relatively simple test and any metallic material can be tested at a low cost.

Ultrasonic testing works best with smooth surfaces. So, the rough surfaces need to be machined for the best results. Moreover, pores smaller than 1 mm, small nonmetallic inclusions, and tiny cracks near the surface are often missed. Somewhat smaller pores, as small as 0.2 mm in diameter, can be detected using multiple transmitters and detectors called pashed array ultrasonic equipment. The measurement efficiency depends on the part size and geometry, the frequency of the sound wave, the direction of the incoming sound wave, and the type of the transducer. The need to use the liquid couplant limits the temperature of the part and precludes real-time application of the technique. In contrast, pulsed lasers may be used to produce ultrasonic waves using short laser pulses which may be used in the detection of flaws in laser ultrasonic testing. A laser interferometer is used to determine the defects. However, this technique also requires smooth surfaces, and most ultrasonic nondestructive testing is conducted using conventional probes.

10.6.7 X-ray and neutron radiography

Transmission of radiations such as X-rays and neutrons through a medium depends on the properties of the medium according to the Beer-Lambert law:

$$I = I_0(1\text{-}e^{-\beta l}) \tag{10.17}$$

where I is the transmitted flux of radiation, I_0 is the incident flux, β is the absorptivity of radiation, and l is the length of the path. Eq. 10.17 was proposed for a single wavelength of radiation. However, the radiations used in commercial X-ray and neutron radiation equipment are in a band of wavelengths. The absorption coefficients vary with wavelength and a single effective value of the absorption coefficient is used in the calculations. Radiographic techniques take advantage of the difference in the transmission of radiation between a single-phase material and one that contains discontinuities such as porosity, lack of fusion, and inclusions. A test specimen is placed between the radiation source and a detector and an image of the transmitted radiation is recorded either on photographic paper or recorded electronically by a suitable detector. Absorptivity depends on the density and atomic weight of the medium. Materials with higher density and atomic weight absorb more radiation. In contrast, a void space such as a pore or a lack of fusion void absorbs less radiation and shows up as a bright spot in the transmitted radiation because of its higher intensity. Radiographic methods are excellent for finding internal flaws within the inspected part's volume. The technique can be used to examine the effect of HIPing on the reduction of porosity or cracks. The X-ray source's voltage, current, and exposure time are adjusted to penetrate a material based on its density and atomic weight. However, the detection of cracks is dependent on their orientation relative to the direction of transmission of the beam and as a result, some cracks are not detectable by this method. In addition, there is a limitation of the size of the test piece that could be investigated because of the limits of the depth of penetration. It is also difficult to determine small pores in large specimens. X-ray radiography has been used for flaw detection in welds using portable units and a rich knowledge base is available for its application in welding. Radiographic applications in additive manufacturing are still evolving and more work is needed for the optimization of the technique for additive manufacturing.

X-rays and neutrons interact differently with materials. X-rays interact with electrons in atoms, while neutrons interact with the nucleus. As a result, some materials absorb X-rays more intensely than neutrons and vice versa. Neutron radiography can show many light materials within dense materials and they are often used to probe composite materials and ceramic parts. The neutrons for radiography are produced in a nuclear reactor. This dependence severely limits the commercial use of this technique and makes the process expensive.

10.7 Defect mitigation

Defects such as porosity, cracks, residual stress, distortion, and surface roughness are common in additively manufactured parts. Depending on the end use, defects in parts need to be removed to improve their properties and serviceability. Defect mitigation by post-processing often involves hot isostatic pressing (HIP), heat treatment, and surface modifications.

10.7.1 Post-process machining/surface treatment

Parts are built on a base plate and the build needs to be separated from the plate after manufacturing. This is often achieved using a pressurized gas nozzle or using an appropriate saw. After the part is separated from the support structure and base plate, the surface becomes rough, and depending on the application and the alloy, an appropriate surface treatment is used. Also, additive manufacturing of some parts requires support structures. The removal of the support structure also makes the surface uneven. Surface treatments are required for the parts to comply with the specifications. In addition, surface roughness, waviness, cracks, porosity, and oxide formation are also removed by a surface modification technique to achieve improved mechanical and chemical properties. These include machining, mechanical and chemical polishing, grinding, and electropolishing.

10.7.2   Hot isostatic pressing

In hot isostatic pressing (HIP), a part is exposed to a high-pressure gas such as argon, at a high temperature below the solidus temperature of the alloy to promote plastic flow and increase density. The pressure is typically in the range of 100 to 200 MPa to enhance plastic flow and atom/vacancy diffusion to reduce porosity and internal cracks. The chamber is filled with inert gases such as argon, helium, or in some cases nitrogen, depending on the alloy. The physical processes that take place during processing include plastic flow and creep. Plastic flow tends to reduce pore fraction as the applied hydrostatic pressure exceeds the reduced yield point of the elevated temperature. Creep deformation often involves various defects, particularly dislocations or grain boundaries. Factors such as dislocation density, grain size and shape, and temperature affect grain boundary diffusion commonly known as Coble Creep. The temperature increases the creep rate because of the faster rates of diffusion resulting from the Arrhenius-type temperature dependence of diffusion. Cobble creep is often the main mechanism in parts with fine grains and at low temperatures since the activation energy for grain boundary diffusion is low. Diffusion may also occur within the interior of the grains rather than along the boundaries, and this type of creep is called Nabarro-Herring creep and contributes to the plastic flow. When the temperature is relatively high and grains are large, this form of creep typically dominates over Coble creep. Both diffusional creeps occur at low applied stresses. In metallic materials, the creep that occurs at high applied stresses approaching yield stress involves the motion of dislocations resulting in plastic deformation.

HIP reduces the volume percentage of porosity and voids. However, microstructural changes, particularly changes in the grain structure, also occur because of the exposure of parts to elevated temperatures. Mechanical properties may be affected by both changes in porosity and microstructural features. The specific contribution of each of these factors depends on the alloy and the conditions of the HIP.

10.7.3   Real-time control

Real-time reliable measurement of melt-pool dimensions including the track width, and layer characteristics such as thickness and surface morphology, and the temperature field has the potential to detect defect formation and take remedial action by adjusting process variables. In-situ process signals are valuable data for quality improvement. However, additive manufacturing is a relatively new field and currently, there is a lack of a unified approach to process control for quality improvement. Most manufacturers adjust process parameters based on prior experience. A discussion of sensing and control is available in Chapter 6 where the available sensors, such as optical and infrared cameras, photodiodes, and pyrometers, are discussed. Their applications in monitoring powder flow, melt pool and seam tracking, distortion control, and the availability of commercial equipment are also discussed in the chapter.

10.7.4   Modeling

Trial-and-error adjustment of multiple process variables to print largely defect-free parts is slow, expensive, and often unsuccessful. Emerging digital tools, mechanistic modeling based on the laws of physics, and machine learning using data gathered from various sources facilitate solving complex problems such as the evolution of defects. The quantitative relations indicated in the chapter provide functional relationships among some of the factors that can be used to influence the prevention of defects before the parts are printed. However, most rigorous mechanistic models are computationally intensive and cannot be run in real-time. They may be used offline for planning when models are available. Examples of such modeling include models of heat, mass, and momentum transfer for reducing the lack of fusion defects, powder scale models for reducing porosity and surface roughness, models for the prevention of solidification cracking, and thermo-mechanical modeling for reducing residual stress and distortion. More details of such mechanistic modeling are available in Chapter 12.

In contrast with mechanistic modeling, machine learning programs are rapid because they do not require the solution of complex equations that represent many simultaneous physical processes such as heat transfer and fluid flow. Also, machine learning programs are easy to build because of the availability of many open-source, well-tested, algorithms [30]. They have been used for process parameter optimization, sensing and control, improvements of part attributes, and mitigating defects in laboratories. While commercial pieces of equipment such as pyrometers, various types of cameras, and photodiodes are becoming increasingly available with additive manufacturing machines as discussed in Chapter 6 on sensing, control, and qualifications, commercial applications of machine learning methods for defect mitigation are in their initial stages.

Various types of machine learning models have been used [30] for defect mitigation in laboratory tests. They include artificial neural networks for defect recognition, fusion zone geometry prediction, and compensation of thermal distortion. The process involves connections between input and output variables via an activation function. Classification programs such as decision trees have been used for the reduction of surface roughness, porosity, and residual stress. These programs progressively classify a group of variables based on certain rules and the root of the tree often displays the most important variable for the reduction of defects. Support vector machines that are used for classification and regression can split data into groups based on features. These programs have been used for the reduction of surface roughness and for monitoring composition. K-nearest neighbor models have been used for the reduction of porosity and compliance with dimensional accuracy. These models separate data based on the attributes of the nearest neighbor data. Random forest programs that consist of multiple decision trees have been used for the reduction of surface roughness, cracking, and porosity. A selection of open-source machine learning programs for the mitigation of defects is listed in Table 10.2.

Table 10.2 Open-source machine learning algorithms used in defect mitigation.

Open source programs	Description and features	Applications in defect reduction
Weka	Used for classification, clustering, and regression. Available from https://www.cs.waikato.ac.nz/ml/weka/citing.html	Image-based classification of defects, porosity reduction
Scikit learning	Used for classification, clustering, and regression. Available from https://scikit-learn.org/stable	Dimensional accuracy, distortion control
TensorFlow	Used for neural network and data flow programming. Available from https://www.tensorflow.org	Dimensional accuracy and defect detection
Keras	Neural network library, runs on multiple platforms. Available from https://keras.io	Dimensional accuracy, distortion control
Theano	Used for efficient processing of multidimensional arrays. Available from http://www.deeplearning.net/software/theano	Defect detection, control of part weight

10.8 Summary

Additively manufactured components often suffer from many types of defects such as cracking, porosity, and other voids, surface roughness and waviness, loss of alloying elements, residual stress, distortion, and delamination. These defects affect the properties and serviceability of parts. Prevention of defects is a major challenge because they require optimization of the many variants of the manufacturing process, alloys, the large number of process variables, and their wide range of values. The high cost of feedstock and machines makes the traditional trial-and-error optimization of process variables to avoid defects expensive and time-consuming. Significant progress is being made in understanding the mechanism of the formation of these defects. Better mechanistic understanding and process control will result in improved quality consistency and cost competitiveness. In many cases, post-processing is used to reduce defects in additively manufactured materials. The mechanism of formation of common defects, their detection, characterization, and mitigation strategies are reviewed in this chapter.

Takeaways

What are the common defects in additively manufactured parts?

- Properties and serviceability of additively manufactured parts are often affected by cracking, porosity, and other types of voids, surface roughness and waviness, loss of alloying elements, residual stress, distortion, and delamination.

Cracking

- The solidification cracking or hot cracking occurs when the fusion zone has insufficient strength to withstand the contraction stresses generated in the final stage of solidification. Tensile stress due to shrinkage, solidification temperature range of the alloy, shape of the temperature versus solid fraction curve, and the solidification morphology affect solidification cracking.
- Liquation cracking is caused by localized melting of low melting regions at a grain boundary under the influence of thermal strains. The important factors for this type of cracking include the alloy composition, the temperature range of solidification, solidification shrinkage, thermal contraction during cooling, and the presence of intermetallic compounds and precipitates.
- Ductility dip cracking (DDC) is a solid-state intergranular cracking in the temperature range where the ductility of many alloys shows a marked drop. The large size of grains, a certain type of precipitates at the grain boundary, temperature, and restraint are important factors.

Porosity and voids

- An alloy in the fusion zone containing dissolved gases acquired from either the consumable or the ambient gas upon cooling may reject the gas as tiny spherical gas pores. The important factors include the gas source and the solubility difference between the liquid and solid alloy.
- The lack of fusion porosity originates from an insufficient overlap of adjacent tracks or layers leaving a void in the part. The dimensions of the fusion zone, the thermophysical properties of the alloy, and the scanning speed affect the lack of fusion voids.
- The keyhole instability-induced porosity originates from the long and narrow keyholes. This type of defect depends on the process parameters and the thermophysical properties of the alloy.

(Continued)

Takeaways   (Continued)

Surface roughness

- Surface roughness is caused by spattering, uneven melting of the powder and wire feed-stock, surface porosity, steps due to the construction of a curved 3D object, and unevenness and surface marks due to the removal of the support structure after parts are made. AM process selection, process parameters, and alloy type affect the roughness of the surface.
- Balling on the surface is caused by the long fusion zone at high scanning speeds. Depending on the processing conditions and alloy properties it can break up into small islands or balls.

Alloying element loss due to vaporization

- At high temperatures prevailing at the surface of the fusion zone, volatile alloying elements are selectively vaporized resulting in the chemical composition of the part being different from that of the powder or wire from which it is built.
- The alloy composition and the process parameters affect the composition change.

Characterization of defects

- Depending on the type of defect, determination of part density, infrared imaging, X-ray tomography, profilometry, microscopy, ultrasonic testing, and X-ray and neutron radiography are used to characterize the defects.

Mitigation of defects

- Machining, mechanical and chemical polishing, grinding, and electropolishing are used to reduce surface roughness. Hot isostatic pressing is used to reduce porosity and voids.

Appendix – Meanings of a selection of technical terms

Atomization: It is a process to create metal powders from a stream of molten metal into droplets, which solidify into powder particles.

Buckingham pi theorem: This theorem helps to form dimensionless groups or pi terms from the independent variables to describe the behavior of a system. By eliminating the units, the number of independent variables in a problem can be reduced.

Fusion zone: A volume containing molten alloys.

Heterogeneous nucleation: Nucleation of a new phase on a preexisting surface.

Keyhole: When a high power density laser or electron beam impinges on a metal surface, the intense energy instantly vaporizes the metal forming a narrow cavity filled with metal vapors directly beneath the beam. The cavity is known as the keyhole.

Laves phases: They are intermetallic compounds with the stoichiometry AB_2, where A and B are metals and the crystal structure is cubic or hexagonal.

Recoil pressure: During the interaction of a high energy beam such as a laser or electron beam with metal the intense vaporization of the metal exerts a pressure on the workpiece known as the recoil pressure.

Segregation of alloying elements: During the solidification of alloys, alloying elements do not partition equally between the solid and the liquid. The liquid is often enriched in alloying elements and this phenomenon is called segregation.

Shot peening: Shot peening is a surface treatment process in which a material is bombarded with small hard spheres to impart compressive stress.

Vapor pressure: The pressure exerted by a vapor in thermodynamic equilibrium with its liquid or solid phase is known as vapor pressure.

Weldability: Weldability is the ability of an alloy to produce a strong and reliable weld joint i.e., welded without cracking or forming other defects. Some materials, such as low-carbon steels, are easy to weld, while other alloys, such as certain high-carbon steels, are difficult to weld.

Practice problems

1) Is gas porosity harmful to additively manufactured parts? Which mechanical property is adversely affected by gas porosity?
2) What features of keyhole porosity are different from the gas porosity in microscopic sections?
3) How does the lack of fusion porosity appear different from the gas and keyhole porosity?
4) What steps do you recommend to avoid keyhole instability-induced porosity?
5) What are the differences between solidification cracking and other types of cracking commonly observed in additively manufactured parts?
6) Discuss the factors responsible for the solidification cracking of additively manufactured parts.
7) What are the ways of avoiding solidification cracking in additively manufactured parts?
8) What factors are responsible for liquation cracking? How can liquation cracking be prevented?
9) What is ductility dip cracking and how can it be prevented during additive manufacturing?
10) What are the common causes of surface roughness in additively made metallic parts? How is the surface roughness problem mitigated?
11) Powder bed fusion, directed energy deposition and wire-arc directed energy deposition are three commonly used additive manufacturing processes for printing metallic parts. Which of these processes would likely result in parts with the most and least surface roughness?
12) How are rough surfaces characterized? What are the remedies for surface roughness problems?
13) The chemical composition of additively manufactured parts is sometimes different from that of the feedstock materials used to make them. What alloying elements are likely to be depleted during the printing of a Ti6Al4V alloy and why?
14) Five and six thousand series aluminum alloys are often used in the automotive industry. What common defects are of concern in the printing of these alloys?
15) What are the different methods of characterizing porosity in printed metallic parts?
16) What non-destructive techniques are commonly used for characterizing the distribution of pores and voids in additively made parts?
17) What post-processing step is commonly used for improving the density of additively made metallic parts? Discuss the underlying mechanism of improvement.
18) The morphology of the solidification structure is often cited as a cause for the solidification cracking of printed metallic parts. What are the possible remedies?

References

1 Carter, L.N., Martin, C., Withers, P.J. and Attallah, M.M., 2014. The influence of the laser scan strategy on grain structure and cracking behaviour in SLM powder-bed fabricated nickel superalloy. *Journal of Alloys and Compounds*, 615, pp.338–347.

2 Harrison, N.J., Todd, I. and Mumtaz, K., 2015. Reduction of micro-cracking in nickel superalloys processed by selective laser melting: A fundamental alloy design approach. *Acta Materialia*, 94, pp.59–68.

3 Martin, J.H., Yahata, B.D., Hundley, J.M., Mayer, J.A., Schaedler, T.A. and Pollock, T.M., 2017. 3D printing of high-strength aluminium alloys. *Nature*, 549(7672), pp.365–369.

4 Wei, H.L., Mazumder, J. and DebRoy, T., 2015. Evolution of solidification texture during additive manufacturing. *Scientific Reports*, 5(1), article no.16446.

5 Kou, S., 2021. Predicting susceptibility to solidification cracking and liquation cracking by CALPHAD. *Metals*, 11(9), article no.1442.

6 Ransley, C.E., 1948. The solubility of hydrogen in liquid and solid aluminium. *Journal of the Institute of Metals*, 74, pp.599–620.

7 Xu, S.X., Wu, S.S., Mao, Y.W. and An, P., 2006. Establishment of hydrogen measurement system for magnesium alloy melt. *Transaction of the Nonferrous Metal Society of China*, 6, pp.1677–1680.

8 Bird, R.B., Stewart, W.E. and Lightfoot, E.N., 2006. *Transport Phenomena*, 2nd Edition. John Wiley & Sons, Inc.

9 Gan, Z., Kafka, O.L., Parab, N., Zhao, C., Fang, L., Heinonen, O., Sun, T. and Liu, W.K., 2021. Universal scaling laws of keyhole stability and porosity in 3D printing of metals. *Nature Communications*, 12(1), article no.2379.

10 Hann, D.B., Iammi, J. and Folkes, J., 2011. A simple methodology for predicting laser-weld properties from material and laser parameters. *Journal of Physics D: Applied Physics*, 44(44), article no.445401.

11 King, W.E., Barth, H.D., Castillo, V.M., Gallegos, G.F., Gibbs, J.W., Hahn, D.E., Kamath, C. and Rubenchik, A.M., 2014. Observation of keyhole-mode laser melting in laser powder-bed fusion additive manufacturing. *Journal of Materials Processing Technology*, 214(12), pp.2915–2925.

12 Plateau, J.A.F., 1873. Raleigh instability Statique expérimentale et théorique des liquides soumis aux seules forces moléculaires, Gauthier-Villars Publisher, Paris.

13 Tang, M., Pistorius, P.C. and Beuth, J.L., 2017. Prediction of lack-of-fusion porosity for powder bed fusion. *Additive Manufacturing*, 14, pp.39–48.

14 Mukherjee, T. and DebRoy, T., 2018. Mitigation of lack of fusion defects in powder bed fusion additive manufacturing. *Journal of Manufacturing Processes*, 36, pp.442–449.

15 Zhou, X., Wang, D., Liu, X., Zhang, D., Qu, S., Ma, J., London, G., Shen, Z. and Liu, W., 2015. 3D-imaging of selective laser melting defects in a Co–Cr–Mo alloy by synchrotron radiation micro-CT. *Acta Materialia*, 98, pp.1–16.

16 Zagade, P.R., Gautham, B.P., De, A. and DebRoy, T., 2023. Scaling analysis for rapid estimation of lack of fusion porosity in laser powder bed fusion. *Science and Technology of Welding and Joining*, pp.1–9. https://doi.org/10.1080/13621718.2022.2164830.

17 Qiu, C., Panwisawas, C., Ward, M., Basoalto, H.C., Brooks, J.W. and Attallah, M.M., 2015. On the role of melt flow into the surface structure and porosity development during selective laser melting. *Acta Materialia*, 96, pp.72–79.

18 Strano, G., Hao, L., Everson, R.M. and Evans, K.E., 2013. Surface roughness analysis, modelling and prediction in selective laser melting. *Journal of Materials Processing Technology*, 213(4), pp.589–597.

19 Rahmati, S. and Vahabli, E., 2015. Evaluation of analytical modeling for improvement of surface roughness of FDM test part using measurement results. *The International Journal of Advanced Manufacturing Technology*, 79(5), pp.823–829.

20 Du, Y., Mukherjee, T., Finch, N., De, A. and DebRoy, T., 2022. High-throughput screening of surface roughness during additive manufacturing. *Journal of Manufacturing Processes*, 81, pp.65–77.

21 Li, R., Liu, J., Shi, Y., Wang, L. and Jiang, W., 2012. Balling behavior of stainless steel and nickel powder during selective laser melting process. *The International Journal of Advanced Manufacturing Technology*, 59(9), pp.1025–1035.

22 Long, J., Wang, M., Zhao, W., Zhang, X., Wei, Y. and Ou, W., 2022. High-power wire arc additive manufacturing of stainless steel with active heat management. *Science and Technology of Welding and Joining*, 27(4), pp.256–264.

23 de Moraes Coelho, D., dos Santos Paes, L.E., Guarato, A.Z., de Araújo, D.B., Scotti, F.M. and Vilarinho, L.O., 2022. A low-cost methodology for quality inspection of metal additive manufactured parts. *Journal of the Brazilian Society of Mechanical Sciences and Engineering*, 44(7), article no.293.

24 Taminger, K., 2010. Electron beam additive manufacturing: State-of-the-Technology, challenges & opportunities. In *Direct digital manufacturing workshop*.

25 Liu, T., Yang, L.J., Wei, H.L., Qiu, W.C. and DebRoy, T., 2017. Composition change of stainless steels during keyhole mode laser welding. *Welding Journal*, 96, pp.258s–270s.

26 Mukherjee, T., Zuback, J.S., De, A. and DebRoy, T., 2016. Printability of alloys for additive manufacturing. *Scientific Reports*, 6(1), article no.19717.

27 Khan, P.A.A. and Debroy, T., 1984. Alloying element vaporization and weld pool temperature during laser welding of AlSl 202 stainless steel. *Metallurgical Transactions B*, 15, pp.641-644.

28 He, X., DebRoy, T. and Fuerschbach, P.W., 2003. Alloying element vaporization during laser spot welding of stainless steel. *Journal of Physics D: Applied Physics*, 36(23), article no.3079.

29 Lifton, J. and Liu, T. 2021. An adaptive thresholding algorithm for porosity measurement of additively manufactured metal test samples via X-ray computed tomography. *Additive Manufacturing*, 39, article no.101899.

30 DebRoy, T., Mukherjee, T., Wei, H.L., Elmer, J.W. and Milewski, J.O., 2021. Metallurgy, mechanistic models and machine learning in metal printing. *Nature Reviews Materials*, 6(1), pp.48–68.

11

Residual Stresses and Distortion

Learning objectives

After reading this chapter the reader should be able to do the following:

1) Understand different mechanisms of evolution of residual stresses and distortion during additive manufacturing.
2) Select an appropriate technique to measure residual stresses and distortion.
3) Compare different alloys and additive manufacturing processing conditions to evaluate their relative susceptibilities to distortion.
4) Appreciate the use of numerical simulation to compute residual stresses and distortion.
5) Understand the effects of different process variables and printing strategies on residual stresses and distortion in additively manufactured components.
6) Realize the detrimental effects of residual stresses and distortion on part quality.
7) Select methods to control residual stresses and distortion in additively manufactured parts.

CONTENTS

Theory and Practice of Additive Manufacturing, First Edition. Tuhin Mukherjee and Tarasankar DebRoy.
© 2024 John Wiley & Sons, Inc. Published 2024 by John Wiley & Sons, Inc.

11.1   Introduction

Residual stresses and distortion are mostly thermally induced phenomena. If a part is heated, it expands. Similarly, it contracts upon cooling. However, if the expansion or contraction is hindered by some external restriction, internal stress is generated in the material. Under certain conditions, even if the external restriction to hinder the expansion or contraction is removed, the material may not gain back its original shape and size. Therefore, a permanent deformation may remain in the material which is called distortion. The corresponding internal stress that remains inside the material is called residual stress. Common examples of residual stresses and distortion can be found in many instances starting from the glaciers in Antarctica to the high school physics labs. The temperature difference between day and night results in repeated expansion and contraction in the ice layers of a glacier. The expansion and contraction are restricted by the solid rocks beneath the glacier and cause high internal stresses inside the ice layers. High stresses create long cracks in ice layers and catastrophic failure of the glacier. Two metals in a bimetallic strip used in the high school physics lab expand and contract differently under the same temperature change. Expansion of one strip is hindered by the other which distorts the bimetallic strip in a bow shape. The evolution of residual stresses and distortion is common in any manufacturing process involving temperature variations such as casting, welding, and hot metal forming.

In AM, residual stresses and distortion originate because of the spatially nonuniform temperature variation, solidification of the deposit, temperature-dependent elastic and plastic properties of the alloy, and geometric constraints such as fixtures and clamps. In addition, solid-state phase transformation during cooling such as austenite to ferrite transformation in steels may result in internal stresses in the component. Accumulation of high residual stresses in AM parts may result in premature brittle fracture under low applied force and defects such as cracking, warping, and delamination. In addition, high residual stresses significantly degrade the fatigue properties of the component. Distortion causes dimensional inaccuracy, which in extreme cases may lead to part rejection. Therefore, an accurate estimation of residual stresses and distortion is needed to reduce or control them. The well-known measurement techniques of residual stresses include hole drilling, curvature method, X-Ray and Neutron diffraction, ultrasonic and magnetic methods, and indentation testing. These measurement techniques are often time-consuming and expensive and do not provide the temporal evolution of the residual stresses and distortion. Therefore, well-tested numerical models are currently being used to calculate the residual stresses and distortion in AM parts. In AM industry there are several techniques to control residual stresses and distortion. They include both pre-process methods that carefully control the AM process to reduce residual stresses and distortion and post-processing to eliminate the accumulated stresses and distortion.

In this chapter, several key physical factors that contribute to the evolution of residual stresses and distortion are discussed. Commonly employed measurement techniques of residual stresses and distortion in AM parts are described along with their relative advantages and disadvantages.

Back-of-the-envelope calculations to quickly compare susceptibility to distortion for different AM processes, process parameters, and alloys are introduced. With the advent of high-speed computers and state-of-the-art algorithms, more realistic and accurate calculations of residual stresses and distortion can be performed numerically, which are also examined in this chapter. Variations in residual stresses and distortion for different AM processes, process parameters, and printing strategies are also discussed. Finally, several techniques for reducing or controlling residual stresses and distortion in AM parts are highlighted.

11.2 Origin of residual stresses and distortion

In AM, residual stresses and distortion are affected by the spatially variable, transient temperature field, part geometry and fixtures, and temperature-dependent mechanical properties of the alloy. There are three key physical factors responsible for the origin of residual stresses and distortion in AM, (1) spatial gradient of temperature in the part, (2) elastic and plastic behavior of the alloy used, and (3) solid-state phase transformation, if present. The accumulated strain during the process can be represented as [1]:

$$\alpha \Delta T + \varepsilon_E + \varepsilon_P + \varepsilon_O = 0 \tag{11.1}$$

where the first term in the left-hand side of the equation indicates the strain accumulation due to a change in temperature (ΔT). The symbol α is the coefficient of thermal expansion. The second, third, and fourth terms represent elastic strain, plastic strain, and strain accumulated due to solid-state phase transformation. The following subsections discuss different key factors that contribute to the accumulation of residual stresses and distortion in AM parts.

11.2.1 Thermal effects

In AM, residual stresses and distortion depend on the spatially variable, transient temperature field as well as temperature-dependent thermophysical and mechanical properties. AM and fusion welding share many of the same physical phenomena, especially the temperature effects governing the formation of residual stresses and distortion. Therefore, the classic bar-frame problem, commonly used to illustrate the origin of residual stresses in fusion welding [2], is adapted here to explain the evolution of residual stresses in AM.

Inset in Figure 11.1 shows the middle bar and the frame, both made of the same material. In the context of AM, the frame and the bar are analogous to the substrate and the deposit, respectively. Temperature is changed only in the middle bar while the temperature of the frame is kept constant at room temperature. When the middle bar is heated, compressive stresses arise in it because its expansion is resisted by the frame. The magnitude of the compressive stress increases along line AB as the temperature of the middle bar increases. At point B the yield strength in compression is reached. As the temperature increases further, the stress in the middle bar is equal to the yield strength, which decreases with increasing temperature as shown by curve BC. The heating is stopped when the temperature of the middle bar reaches the maximum temperature (point C). As the temperature of the middle bar decreases, the stress drops along the line CD. At some point, the stress changes from compressive to tensile and eventually reaches the yield strength in tension (point D). As the temperature of the middle bar decreases further, stress is limited by the tensile yield strength. Therefore, the stress increases as the temperature

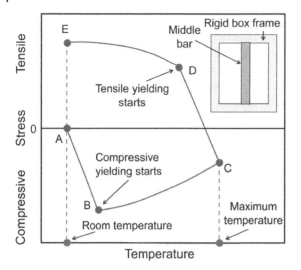

Figure 11.1 Effect of temperature variation on the evolution of residual stresses explained using a bar-frame problem commonly used in fusion welding. The setup of the bar and frame is shown in the inset. *Source:* T. Mukherjee and T. DebRoy.

decreases, as shown by curve DE, until the room temperature is reached at point E. Therefore, residual tensile stress equal to the stress at point E is accumulated in the middle bar. To balance the system, residual compressive stresses of half the value of the tensile stress at the middle bar, are set up on each side of the frame. Following this bar-frame problem, in AM, a deposit of a single track on substrate results in tensile residual stress in the deposit and compressive residual stress in the substrate near the deposit.

There are three important factors related to the effects of temperature on residual stresses and distortion. First, the temperature gradient is an important factor contributing to the accumulation of residual stresses. In the aforementioned example of the bar-frame problem, if both the bar and the frame are uniformly heated and cooled, they would expand and contract freely. Therefore, there would not be any accumulated stress. The stress is accumulated due to nonuniform expansion and contraction due to the temperature difference between the bar and the frame. Second, with a few hundred Kelvin changes in temperature, the thermal strain can exceed the elastic strain limit, resulting in the generation of plastic strain. This plastic strain results in the accumulation of tensile or compressive residual stresses when the component cools down to room temperature. Third, the elastic modulus of the alloy affects the slope of the lines AB and CD in Figure 11.1. In the elastic region, the amount of elastic strain accumulated is governed by the temperature gradient and the coefficient of thermal expansion. For a given increment in the elastic strain, the stress accumulated in the component is determined by the elastic modulus. The worked out example 11.1 illustrates the stress in a bar that is heated under constraints.

Worked out example 11.1

Consider a long bar fixed by its two ends. The bar is slowly heated up to 1200 K and then gradually cooled down to room temperature. What will be the residual stress accumulated in the bar? Assume the stress field is 1D along the axis of the bar. In addition, ignore any bending.

Useful data: Assume that the coefficient of thermal expansion or contraction (α): 19.7×10^{-6}/K, elastic modulus (E): 196.3 GPa are independent of temperature, the magnitude of tensile or compressive stress for yielding, $\sigma_y(T)$: $156 - 1.5 \times 10^{-2}$ T (in MPa), the relation between the true stress (in Pa) and temperature: $F(T) = -0.693\ T^3 + 1524.6\ T^2 - 11.6 \times 10^5\ T + 53.89 \times 10^7$, where "T" is the temperature in K.

Worked out example 11.1   (Continued)

Solution:

The calculation of residual stresses is described step-by-step below.

1) When temperature increases, the compressive stress accumulated $= \alpha \Delta TE$, where ΔT is the change in temperature.
2) At temperature $T = Ta$, the compressive stress accumulated will be equal to the compressive stress needed for yielding $(-\sigma_y)$. Therefore, α (Ta − 300) E $= -\sigma_y$(Ta) where 300 K is the room temperature.
3) For further increase in temperature, the stress will be controlled by true stress versus temperature relation. The maximum temperature attained is Tm = 1200 K. Therefore, at Tm, the compressive stress will be equal to $-\sigma_m$ = F(Tm) where F is the function representing the relation between the true stress and temperature.
4) During cooling from Tm, the elastic stress will be released, and the stress will become tensile until it reaches the tensile stress needed for yielding (σ_y). If it occurs at a temperature Tb, then we can write, $-\sigma_m + \alpha$ (Tm − Tb) E = σ_y(Tb) or − F(Tm) + α (Tm − Tb) E = σ_y(Tb). Using the values, Tb = 437.5 K.
5) As the temperature drops further, the stress will be controlled by the true stress vs. temperature relation. And the true stress accumulated during cooling down from Tb to room temperature (300 K) is the residual stress because this stress cannot be relaxed when constraints are removed. Therefore, the residual stress is equal to F(300) − F(Tb) = 44.2 MPa.

11.2.2   Effects of plasticity and flow stresses

The alloys commonly used in AM have a coefficient of thermal expansion above 1×10^{-5} K^{-1}. With a few hundred Kelvins change in temperature, the thermal strain can exceed the elastic limit and the component deforms plastically. In the plastic region, the true stress vs. true strain relation which governs the accumulation of residual stresses depends both on the temperature and strain rate. For example, Figure 11.2 shows that the true stress vs. true strain relation of a Ti-6Al-4V part can vary significantly with both temperature and strain rate. At higher temperatures, materials become softer and easier to deform plastically. Therefore, lower stresses are required to achieve the same strain at higher temperatures as shown in Figure 11.2 (a). Since temperature varies both spatially and with time during AM, the plastic behavior of the component shows a significant variation at different regions of the component which contributes to the accumulation of a spatially nonuniform residual stress field. A high strain rate hinders the movement of dislocations and increases the hardening of the alloy. Therefore, higher stresses are required to achieve the same strain at a higher strain rate as shown in Figure 11.2 (b). However, in AM, the strain rate is significantly low, approximately 0.001 to 0.01/s. Alloy plasticity is a crucial parameter contributing to the residual stresses. Particularly, plastic strain originated due to the hardening behavior of an alloy above the yield strength is an important factor.

11.2.3   Solid-state phase transformation

Most alloys exhibit solid-state phase transformation during cooling. These phase transformations are associated with a change in volume and accumulation of strain due to a change in the crystal structure. As a result, residual stresses are accumulated in the component. An example of this

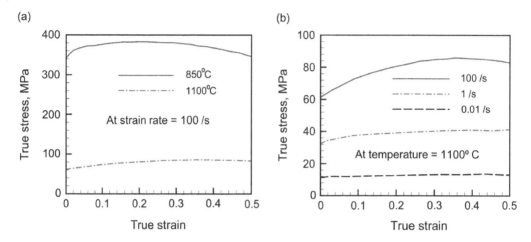

Figure 11.2 True stress vs. true strain relation of Ti-6Al-4V (a) at different temperatures for a constant strain rate of 100/s and (b) at different strain rates for a constant temperature of 1100°C. *Source:* Plots are made by T. Mukherjee and T.DebRoy based on data available in reference [3].

phenomenon is the freezing of water into ice inside the water supply pipelines in the winter. The freezing of water to form ice results in volume expansion because of the lower density of ice than water. An increase in volume due to ice formation results in high stress inside the pipeline which may cause the failure of the pipeline.

Table 11.1 summarizes how solid-state phase transformation causes the accumulation of residual stresses in commonly used alloys in AM. For example, during AM of steels, the change in the crystal structure from austenite to martensite due to rapid cooling generates high strain in the crystal lattice and results in residual stresses and distortion. Steels during cooling also undergo solid-state phase transformations [4] from austenite (FCC) to different types of ferrites (BCC). Figure 11.3 schematically shows the role of austenite to ferrite solid-state phase transformations in the development of residual stress in steels. After the deposition, when the temperature of the deposit is just below the solidus temperature, the expansion of the deposited material is restricted by the surrounding cooler material causing compressive stress. As the temperature decreases further, the material starts shrinking and experiencing tensile stress. Therefore, the stress in the austenite changes from compressive to tensile as cooling progresses. During the austenite to ferrite transformation, the volume of the material increases (see worked out example 11.2) resulting in an accumulation of compressive stress because the increasing volume is restricted by the surrounding material. Therefore, the stress changes from tensile to compressive. After the austenite-to-ferrite transformation is over and the temperature decreases further, the material starts shrinking and experiencing tensile stress.

Table 11.1 Solid-state phase transformations during cooling of different alloys affecting residual stresses.

Alloy	Transformation	Cause of residual stresses
Steel	Austenite to Ferrite	Volume change
Steel	Austenite to Martensite	Uniaxial dilatation and shear
Ti-6Al-4V	Beta to Alpha phase	Volume change
Cobalt	Martensite formation	Uniaxial dilatation and shear

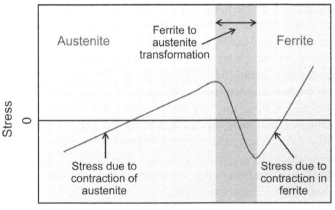

Figure 11.3 Schematic representation of the role of austenite to ferrite solid-state phase transformations in the development of residual stress in steels. *Source:* T. Mukherjee and T. DebRoy.

Worked out example 11.2

Additively manufactured steel parts undergo a solid-state phase transformation from austenite to ferrite between 800°C and 500°C. How much volumetric strain will be accumulated due to this phase change? How significant is the strain accumulation due to phase transformation compared to the thermal strain during cooing from the solidus to room temperature? Useful data: lattice parameter of austenite (FCC–iron): 3.591 A, lattice parameter of ferrite (BCC–iron): 2.863 A, average volumetric thermal expansion coefficient of steel: 48×10^{-6}/°C, solidus temperature: 1400 °C.

Solution:

Let us assume that there are N atoms in the part. Therefore, for austenite (FCC–iron), there are N/4 unit cells and for ferrite (BCC–iron), there are N/2 unit cells. Therefore, the volumetric strain due to volume change during the phase transformation = $((N/4) \times 3.591^3 - (N/2) \times 2.863^3)/((N/4) \times 3.591^3) = -0.0136$. The value is close to those reported in references [5, 6]. A negative value indicates that the volume of ferrite is higher than that of austenite resulting in a compressive strain due to an increase in volume. The volumetric strain during cooling from the solidus to room temperature = $(1400 - 25) \times 48 \times 10^{-6} = 0.066$. The cooling results in a decrease in volume. The strain due to cooling is significantly higher in magnitude than the strain due to volume change that accompanies the phase transformation.

Residual stresses due to the solid-state phase transformation in the welding of steels are generally minimized by a careful selection of filler materials that minimize the volumetric change due to phase transformation. In AM, prealloyed powders with the desired composition to control the volume change during the phase transformation are emerging.

11.2.4 Temporal evolution of residual stresses and distortion

In AM, the temporal evolution of residual stresses and distortion depends on the spatially nonuniform and transient temperature field. The distribution of temperature and corresponding stress evolution transverse to the deposit line is shown in Figure 11.4 for four locations along the deposit, A, B, C, and D. Along section A–A, which lies just ahead of the heat source, the temperature change due to the approaching heat source is almost zero. Therefore, the thermally induced stresses are also almost zero. Along section B–B, the temperature distribution is very steep since this section passes through the heat source. Stresses in this region, just at the center of the heat source, are also almost zero because molten metal cannot support a load. However, the stresses in the heat-affected region on either side of the heat source are compressive because thermal expansion in these regions is restrained by the surrounding metal at lower temperatures having higher strength. The compressive stresses in the heat-affected zone are balanced by tensile stresses accumulated away from the heat-affected zone. Behind the heat source, at section C–C, cooling has started, and the temperature distribution is less steep. In this section, the deposit and the heat-affected zone have cooled considerably. As they cool, tensile stresses develop because of the shrinkage. These tensile stresses are again balanced by compressive stresses away from the deposit to maintain mechanical equilibrium. Far behind the heat source, along section D–D, the temperature has dropped back to room temperature. High residual tensile stresses originate in the deposit and heat-affected zones, with compressive stresses in the substrate away from the deposit to balance the tensile stresses.

Figure 11.4 shows the evolution of stresses along the transverse directions to the deposit. Figure 11.5 indicates the stress evolution along the deposition (or scanning) direction during DED-L. The evolution of the stresses depends on the transient temperature distribution, especially during the cooling of the deposit. Figure 11.5 (a) shows the temperature distribution along line AB (Figure 11.5 (c)) during the cooling of the build. Figure 11.5 (b) represents the corresponding longitudinal stress profile developed along line AB. At t = 0 s, i.e., just after the laser beam traverses the entire deposit length and is switched off, the peak temperature of the deposit along line AB is about 950 K. This high temperature softens the material locally. Therefore, the magnitude of the stress is relatively low as it is limited by the yield strength at 950 K. The stress field evolves as the deposit continues

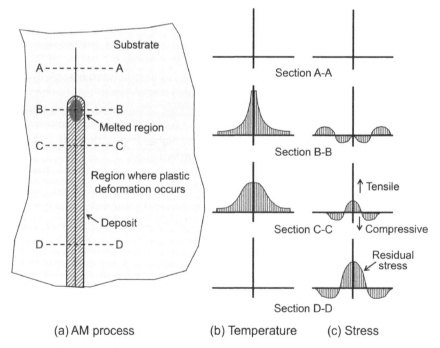

(a) AM process   (b) Temperature   (c) Stress

Figure 11.4 Schematic representation of (a) AM process, (b) temperature variations, and (c) corresponding stress fields while depositing a layer using AM. *Source:* T. Mukherjee and T. DebRoy.

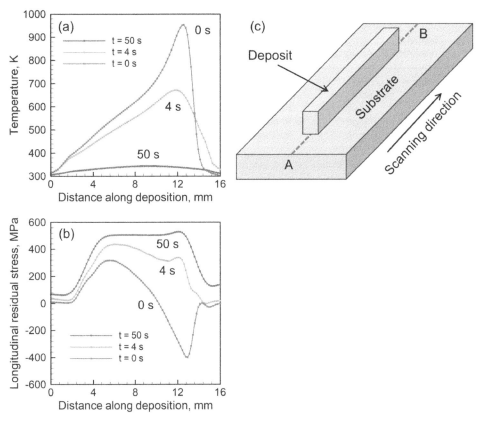

Figure 11.5 Variation in (a) temperature and (b) corresponding evolution of residual stresses during DED-L of Inconel 718 [7]. Here "t" represents time in seconds after the deposition ends and cooling starts. (c) Schematic of the deposit and substrate. AB is the line along the scanning direction, on the top of the substrate at the deposit-substrate interface, along which plots are made in figures (a) and (b). Here, the length of AB = 16 mm (A = 0 mm and B = 16 mm). The deposit is 12 mm long. Figures (a) and (b) are taken from [7] with permission from Elsevier. *Source* for figure (c): T. Mukherjee and T. DebRoy.

to cool down further. After 50 seconds, the deposit almost cools down to room temperature and the longitudinal stress along line AB is highly tensile, as shown in Figure 11.5 (b). If there is no additional layer deposited, the stress field at the end of cooling is the residual stress field in the part.

11.3 Measurement techniques

There are many well-accepted methods for measuring residual stresses and distortion. Table 11.2 summarizes several measurement techniques, their important attributes, and advantages and disadvantages in measuring residual stresses and distortion in AM parts.

11.3.1 Hole drilling and curvature methods

Hole drilling is one of the most popular methods that rely on monitoring the strain accumulated in the part based on which residual stresses are estimated. Generally, strain is measured using one

Table 11.2 Summary of various residual stress measurement techniques and their attributes.

Technique	Spatial resolution	Accuracy	Advantages	Disadvantages
Hole drilling	~50 μm	±50 MPa, poor accuracy at larger depths from the surface	• Can measure three principal components of stress. • Rapid and inexpensive.	• Destructive testing; holes damage the part surface. • Accurate measurement requires precise and careful positioning of strain gauges.
Curvature method	~5% of the specimen thickness	Limited by minimum measurable curvature	• Can measure stress variations along the depth of the part. • Less expensive experimental setup.	• Difficult to measure all three principal components of stress. • Needs to be used incrementally to uniquely determine stress.
X-ray diffraction	2–50 μm	±50 MPa, accuracy decreases with increasing depth from the surface	• Nondestructive testing method, good for expensive AM parts. • Can measure stress components in multiple directions	• Requires expensive and specialized equipment. • Measurement is time-consuming and requires special training. • Stress-free lattice spacing is often required which is often difficult to obtain.
Neutron diffraction	2–50 μm	±50×10^{-6} for the measured strain	• Nondestructive testing method, good for expensive AM parts. • Can measure all stress components with good accuracy and precision. • Can provide stress maps at different sections of a part.	• Requires expensive and specialized equipment. • Measurement is time-consuming and requires special training. • Stress-free lattice spacing required for stress calculation is often difficult to obtain. • Large grains can affect the diffraction pattern and cause measurement errors.
Ultrasonic and magnetic methods	1–5 mm	~10%	• Nondestructive testing method, good for expensive parts. • Portable equipment and easy to test • Rapid and less expensive.	• Difficult to measure all stress components. • Ultrasonic or magnetic wave velocities can depend on microstructural inhomogeneities and cause difficulties in separating the effects of multiaxial stresses.
Indentation testing	~150 μm	Accuracy depends on indenter geometry and applied force	• Easy to test • Rapid and less expensive.	• In destructive testing, a deep indentation may damage the part surface. • Stress is calculated using an empirical formula from the indenter measurement.

or multiple strain gauges fitted on the component. A small circular hole is drilled on the surface of a part and the stresses in the areas around the hole are partially relaxed. The residual stresses that exist in the drilled area are determined by measuring the strain in the areas around the hole and computing the stresses using a stress-strain relation that assumes the elastic behavior of the alloy.

Figure 11.6 schematically shows the commonly used hole drilling methods. According to the ASTM standards, there are three main methods of hole drilling, type A, B, and C (Figure 11.6 (a-c)), depending on the relative positions of the strain gauges around the drilled hole. Type A is the most commonly used method where three strain gauges are glued around the drilled hole (Figure 11.6 (a)). The angle between stain gauges 1 and 3 is 90°. An angle of 135° is used between gauges 2 and 3 as shown in Figure 11.6 (a). The surface of the AM part where stresses are measured is cleaned and then three stain gauges are glued on the surface. Then a small blind hole whose diameter and depth vary between 1 to 3 mm is drilled. For very thin parts, a through-hole is often drilled. Because of the hole drilling, stresses around the hole are partially relaxed and the resultant strain is measured by the three strain gauges. The principal stress components (σ_1 and σ_2) on the surface are calculated as [8],

$$\sigma_1 = \frac{E}{4A}(\varepsilon_1 + \varepsilon_3) - \frac{E}{4B}\sqrt{(\varepsilon_1 - \varepsilon_3)^2 + (2\varepsilon_2 - \varepsilon_1 - \varepsilon_3)^2} \tag{11.2}$$

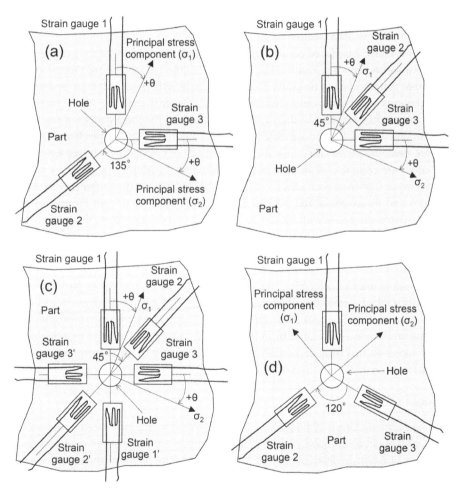

Figure 11.6 Schematic representation of different hole drilling methods (a) Type A, (b) Type B, (c) Type C, and (d) using an arrangement where the three gauges are separated by 120°. *Source:* T. Mukherjee and T. DebRoy.

$$\sigma_2 = \frac{E}{4A}(\varepsilon_1 + \varepsilon_3) + \frac{E}{4B}\sqrt{(\varepsilon_1 - \varepsilon_3)^2 + (2\varepsilon_2 - \varepsilon_1 - \varepsilon_3)^2} \tag{11.3}$$

where E is the modulus of elasticity (Pa), ε_1, ε_2, and ε_3 are the strains measured by the strain gauges 1, 2, and 3 respectively, and A and B are two constants whose values depend on the hole type (blind or through) and the depth and diameter of the hole. ASTM provides values of these two constants for both hole types for a wide range of hole diameters and depths [9]. In addition, manufacturers of the measurement equipment supply a chart with these values. The angle (θ) between the principal stress components and the strain gauges (Figure 11.6 (a)) can be estimated as,

$$\tan(2\theta) = \frac{2\varepsilon_2 - \varepsilon_1 - \varepsilon_3}{\varepsilon_3 - \varepsilon_1} \tag{11.4}$$

A positive value of θ represents a clockwise direction (Figure 11.6 (a)). The type A hole drilling method is commonly used to measure residual stresses in AM parts. To measure principal stress components along the length and width directions, the measurement is done on a horizontal plane. Measurements are taken at a vertical plane to measure the stress components along the depth direction and length or width direction.

In the Type B method, gauge 2 is placed between gauges 1 and 3. The angle between gauge 2 and the other two gauges is 45° as shown in Figure 11.6 (b). Type B method is used where there is some obstacle, and the three gauges need to be glued on the same side. When very high accuracy and precision in measurements are needed, three pairs of strain gauges are used in the Type C method. The type C method is a modified version of the Type B method where the three gauges of the Type B method have a pair glued diametrically opposite to the hole (Figure 11.6 (c)). Types B and C hole drilling methods are used in special situations and the methods of calculating stresses from the measured strain values are complex [8] and are not provided here. In addition, sometimes, three strain gauges are placed at 120° from each other and then a hole is drilled at the center (Figure 11.6 (d)). The magnitudes and directions of the principal stresses are calculated from the measured strain changes in the three gauges. For all methods, measured values depend on the size of the drilled hole, the positions of the gauges and the angle between the gauges, and the direction of principal stress components. Worked out example 11.3 explains the calculation of residual stress by hole drilling method.

The curvature method is used to determine the stresses within the layers. The deposition of a layer can induce stresses which cause the substrate to curve. Generally, the bottom surface of the substrate assumes a concave shape accumulating tensile stress. Curvature is measured using strain gauges. The resultant stresses are calculated from the curvature using a formula of bending of a beam. The changes in curvature during deposition allow the calculation of the corresponding variations in stress as a function of deposit thickness.

Worked out example 11.3

Type A hole drilling method using a blind hole of 2 mm diameter and 1 mm depth is used to measure residual stresses in an additively manufactured component. Three strain gauges 1, 2, and 3 (refer to Figure 11.6 (a)) are glued on the top surface (horizontal plane) of the component to measure the residual stress components along the length and width directions. The strains recorded at the three strain gauges 1, 2, and 3 (refer to Figure 11.6 (a)) are −0.0005, 0.0009, and 0.0006, respectively. Determine the principal stresses along the length and width directions of

---

Worked out example 11.3 (Continued)

the component. Find out the directions of these principal stress components with respect to strain gauges. What will happen if the 1st and 3rd strain gauges are glued along the direction of principal stresses? Elastic modulus of the material: 150 GPa. From the ASTM standard chart of hole drilling [9], the constants A and B for a blind hole of 2 mm diameter and 1 mm depth are found to be 0.173 and 0.37, respectively.

Solution:

Eqs. 11.2 and 11.3 can be used to calculate the principal stress components. For example, σ_1 is calculated using Eq. 11.2 as:

$$\sigma_1 = \frac{150 \times 10^9}{4 \times 0.173}(-0.0005 + 0.0006) -$$

$$\frac{150 \times 10^9}{4 \times 0.37}\sqrt{(-0.0005 - 0.0006)^2 + (2 \times 0.0009 + 0.0005 - 0.0006)^2} = -184 \times 10^6 \text{ Pa}.$$

Similarly, σ_2 is calculated using Eq. 11.3. The computed values of principal stress components are $\sigma_1 = -184$ MPa and $\sigma_2 = 227$ MPa. Therefore, σ_1 is compressive and σ_2 is tensile.

The direction of principal stresses with respect to the strain gauges can be found from the angle between the stress components and the strain gauges (θ) using Eq. 11.4. The computed value of θ is 28.6°. The positive angle indicates that the stress components shift clockwise from the strain gauges.

If the first and third strain gauges are glued along the direction of principal stresses, the value of θ is 0°. Therefore, the strain measured at the second strain gauge will be the average of the first and third strain gauges.

11.3.2 Diffraction techniques

When a polycrystalline material accumulates residual stresses, elastic strains are manifested in the crystal lattice of the individual grains. Diffraction techniques estimate the elastic strains in materials by measuring the strain on lattice planes. The strain on the lattice planes is measured from the interatomic spacing which is displaced under stress. As shown in Figure 11.7 (a), X-ray or neutron beams generated from a high energy source fall on the surface of the part. The beams reflected from different lattices or atomic planes are captured using a film or diffractometer. The phase differences between the reflected beams from different atomic planes are used to estimate the lattice spacing or interatomic spacing. The elastic strain is calculated from the difference between the measured lattice spacing and the lattice spacing of a stress-free material. Residual stress is then calculated from the elastic strain using the elastic constants of the material. The detailed methods are described below.

Figure 11.7 (b) schematically shows that incident X-ray beams are reflected from the lattice planes. There is a phase difference of $2d\sin\theta$ between the beams reflected from the two consecutive lattice planes, where "d" is the lattice spacing and "θ" is the angle of incidence of the X-ray beam (Figure 11.7 (b)).

$$\lambda = 2d\sin\theta \tag{11.5}$$

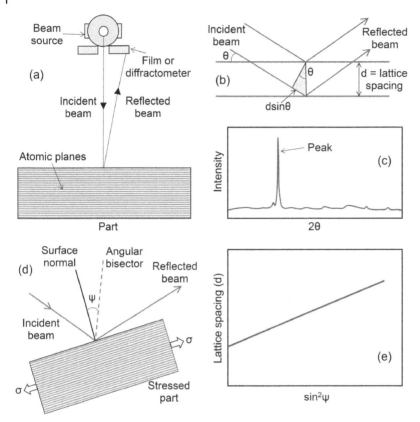

Figure 11.7 (a) Schematic representation of the diffraction method for measuring residual stresses. (b) Schematic representation of the principle of X-ray diffraction. (c) Typical X-ray diffraction result showing intensity versus 2θ plot. (d) Explanation of stress measurement using "Two-Angle Technique." (e) Lattice spacing versus $\sin^2\psi$ plot showing a linear relationship. *Source:* T. Mukherjee and T. DebRoy.

where λ is the wavelength and 2θ value is obtained from the diffraction result where peak intensity is observed (Figure 11.7 (c)). Therefore, for an X-ray beam of a known wavelength, X-ray diffraction results can be useful to predict the lattice spacing. For a part with residual stresses, the lattice spacing will be different from that of a stress-free part. Therefore, the strain due to the change in lattice spacing can be written as $\Delta d/d = -\cot\theta\,\Delta\theta$ where $\Delta\theta$ is the change in the angle in X-ray diffraction measurement. This measurement technique requires a strain-free lattice spacing of the material used which is often difficult to obtain. If strain-free lattice spacing is available, the strain due to the change in lattice spacing can be used to calculate stress.

To avoid the need for strain-free lattice spacing, the Society of Automotive Engineers (SAE) developed a method called the "Two-Angle Technique" for measuring residual stresses. In this method (Figure 11.7 (d)), the stress (σ) can be measured in a particular direction. The part should be placed inside the equipment in such a way that the stress component along that particular direction and incident and reflected beams exist in the same plane (Figure 11.7 (d)). The part is rotated about an axis that is perpendicular to that plane. The angle of rotation (ψ) is defined by the angle between the angular bisector of the incident and reflected beams and the normal direction of the part surface. However, often, the part is kept fixed, and the X-ray beam collimator is rotated at different angles. At different values of the angle (ψ), the values of θ are measured using X-ray diffraction (Figure 11.7 (c)) and corresponding values of the lattice spacing (d) are calculated using Eq. 11.5. From the results, a plot of lattice spacing (d) vs. $sin^2\psi$ is made (Figure 11.7 (e)). The plot is usually a straight line. Therefore, the plot can be made using two or more values of the angle ψ. For this reason, the method is called the "Two-Angle Technique." Commonly, two values of ψ are taken as 0° and 45°. However, for a good fitting, data at multiple angles are desirable.

The method assumes a plane strain condition where the stress component along the normal to the part surface is neglected. Under this condition, the strain due to a change in lattice spacing can be written as [10],

$$\frac{d-d_0}{d_0} = \left[\frac{1+v}{E}\sigma sin^2\psi\right] - \left[\frac{v}{E}(\sigma_1+\sigma_2)\right] \tag{11.6}$$

where d is the lattice spacing of the stresses part, d_0 is the lattice spacing of the stress-free part, E is the modulus of elasticity, v is the Poisson's ratio, and σ_1 and σ_2 are the principal stress components. Eq. 11.6 can be rewritten as,

$$d = \left[\frac{1+v}{E}\sigma d_0\right]sin^2\psi + d_0\left[1-\frac{v}{E}(\sigma_1+\sigma_2)\right] \tag{11.7}$$

Therefore, in d versus $sin^2\psi$ plot (Figure 11.7 (e)),

$$\text{slope} = \left[\frac{1+v}{E}\sigma d_0\right] \tag{11.8}$$

and y-intercept

$$= d_0\left[1-\frac{v}{E}(\sigma_1+\sigma_2)\right] \tag{11.9}$$

Since $E \gg (\sigma_1+\sigma_2)$, the y-intercept is almost the same as the strain-free lattice spacing (d_0). Therefore, the strain-free lattice spacing can be directly obtained from the d versus $sin^2\psi$ plot and there is no need to obtain the strain-free lattice spacing. The stress can be calculated by putting the value of strain-free lattice spacing (d_0) in Eq. 11.8 as,

$$\sigma = \frac{slope \times E}{d_0 \times (1+v)} \tag{11.10}$$

If the slope of the d versus $sin^2\psi$ plot is positive like what is shown in Figure 11.7 (e), the stress is tensile. It indicates that the lattice spacing of the stressed part is higher than that of a stress-free part. Compressive stress is observed if the slope of the plot is negative. Worked out example 11.4 illustrates the calculation method.

Diffraction techniques are nondestructive techniques suitable for measuring stresses inexpensive AM parts. However, challenges arise from the heterogeneous chemical and metallurgical properties of the part. Large grains can hinder the diffraction of the beam and result in a poor signal. Microstructural heterogeneity can cause variations in stress-free lattice spacing so that signals detected may be due to the changes in microstructure and not changes in residual stresses. Careful work with a sufficiently long detection time and accurate prediction of stress-free lattice spacing can enable a good-quality estimation of residual stresses.

Worked out example 11.4

The X-ray diffraction method with an X-ray beam of 1.54 Angstrom wavelength is used to measure residual stresses in an AM part. The lattice spacing versus $\sin^2\psi$ plot is given in Figure E11.1 below. Find out the stress. Useful data: Elastic modulus of the material: 70 GPa, Poisson's ratio: 0.3.

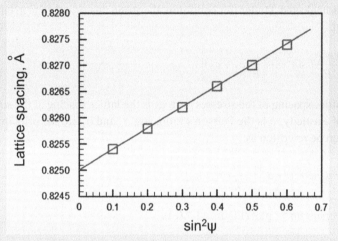

Figure E11.1 Lattice spacing versus $\sin^2\psi$ plot.

Solution:

The slope of the plot is positive which indicates tensile stress. The slope is estimated as 4×10^{-13} m. The intercept which is equal to the strain-free lattice spacing is 0.825 Angstroms or 8.25×10^{-11} m. Therefore, the stress can be calculated using Eq. 11.10 as 261 MPa.

11.3.3 Magnetic and ultrasonic techniques

When a material accumulates residual stresses, the magnetic properties of the part are significantly altered. Specifically, accumulated stresses result in anisotropy in the magnetic properties of the component. This stress-induced magnetic anisotropy leads to the rotation of an externally applied magnetic field away from its original direction. A sensor coil is used to monitor these small rotations in the plane of the component surface. The principal stress component is parallel to the direction of the applied magnetic field if no rotation is monitored by the sensor coil. Therefore, by changing the direction of the applied magnetic field the principal stress directions and their magnitudes can be estimated. The magnetic method is a nondestructive residual stress measurement technique that is also inexpensive and portable. However, this method is also sensitive to the variations in microstructure which must be accounted for during the calibration of the measurements.

In the ultrasonic method, an external ultrasonic wave is applied to the component having residual stresses. Accumulated residual stresses can change the speed of the ultrasonic wave through the material. Residual stresses are calculated from the differences in the ultrasonic speed measured from various directions. An acoustoelastic coefficient necessary for the calculation of stresses from the change in ultrasonic velocity is estimated from a calibration test using a stress-free component. Maximum sensitivity in the measurement is obtained when the ultrasonic wave propagates parallel to the principal component of stress. However, ultrasonic wave velocities can also depend on microstructural inhomogeneities which must be accounted for during the calibration of the measurements. However, low cost and nondestructive nature make this method suitable for measuring residual stresses in AM parts. Worked out example 11.5 illustrates the use of non-destructive testing for determining residual stresses.

11.3.4 Indentation testing

The indentation method is used to measure the response of a material to the applied stress and strain by a sharp indenter. The indenter is forced into the component surface as shown in Figure 11.8 (a), and both the variation in force and displacement of the indenter inside the component are monitored during the process. Residual stresses are calculated on the surface at the location where indentation is made, by using an empirical formula using the data obtained from the force vs. displacement curve. The indentation load required to reach a given depth depends on the residual stress values. Under the compressive residual stresses, the material around the indenter is squeezed by the stress. Therefore, a higher load is required to reach the same indentation depth than that for a stress-free material. In contrast, under the tensile residual stress, the material is relaxed by tension and a smaller load is necessary to reach the same indentation depth than that for a stress-free material. Generally, a Vickers indenter is used for the measurement. It is recommended to limit the indentation depth to 150 μm to avoid unnecessary indentation damage to the part surface.

11.3.5 Measurement techniques of distortion

Measurement of distortion is relatively straightforward than measuring residual stresses. Distortion in AM parts is usually measured both during the process and after the fabrication of the entire component. During the process, distortion can be measured by placing strain gauges on the substrate and monitoring the changes in strain. In addition, more sophisticated devices such as a laser displacement sensor can be used to monitor distortion during AM processes. As shown in Figure 11.8 (b), this device uses a laser beam to optically monitor the deflection of the substrate from its original position during the deposition of multiple layers. Although this device provides more accurate measurements than strain gauges, it is expensive and requires precise calibration. Distortion in the entire part after the deposition process can also be measured using different techniques. The most commonly used method involves using an optical or mechanical probe that measures the surface dimensions. By comparing the measured dimensions with the 3D design of the part, distortion is estimated. Instead of a probe, a 3D optical scanner is also used which scans the printed part from different directions and creates a 3D image of the part. The scanned 3D image is then compared with the 3D design to estimate the distortion.

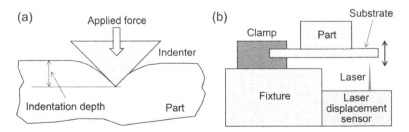

Figure 11.8 (a) A schematic representation of the indentation method for residual stress measurement. (b) Schematic representation of distortion measurement using laser displacement sensor. *Source:* T. Mukherjee and T. DebRoy.

The experimental measurement of the 3D distribution of residual stresses and distortion in an intricate AM part is complex and time-consuming. In addition, most of the methods to measure residual stresses provide the results at the end of the process. Therefore, it is not possible to estimate the temporal evolution of residual stresses during the fabrication process. Therefore, the evolution of residual stresses and distortion during the printing process is often calculated using commercial software. However, a reliable calculation requires rigorous experimental validation of the calculated results.

Worked out example 11.5

Which stress measurement technique should be used for a very expensive, single crystal turbine blade made using PBF-L?

Solution:

Nondestructive residual stress measurement techniques such as X-ray diffraction, neutron diffraction, and ultrasonic methods may be considered for expensive parts. Of these, neutron diffraction requires access to a beam line from a nuclear reactor which is not generally available. Ultrasound and X-ray diffraction [11] can be considered.

11.4   Analytical calculations of strain

Residual stresses and distortion in AM components exhibit significant spatial variation in 3D. Therefore, simple analytical calculations cannot provide the 3D distributions of residual stresses and distortion. However, simple, easy-to-use back-of-the-envelope calculations exist for the comparison of relative susceptibilities of different alloys to distortion under different processing conditions. A dimensionless strain parameter (ε^*) which quantitively represents the susceptibility to distortion in an AM part can be represented as [12]:

$$\varepsilon^* = \frac{\beta \Delta T}{EI} \frac{t}{F\sqrt{\rho}} H^{3/2} \tag{11.11}$$

where
ΔT = Temperature difference, K, represented by the difference between the peak temperature and the room temperature or preheat temperature.
β = Volumetric thermal expansion coefficient, /K.
E = Young's modulus of alloy, MPa.
I = Area moment of inertia of the substrate, mm^4. This is represented as $(1/12) \times$ width $\times$ (depth)^3.
EI = Flexural rigidity of the substrate, N-mm^2.
t = Total deposition time, s.
F = Fourier number calculated as thermal diffusivity of alloy/(pool depth $\times$ scanning speed).
The thermal diffusivity of an alloy is equal to thermal conductivity/(density $\times$ specific heat).
ρ = Density of alloy, kg/mm^3.
H = Heat input per unit length of deposit, J/mm, represented as (fraction of energy absorbed $\times$ power)/scanning speed.
Eq. 11.11 consists of important process parameters and alloy properties that affect distortion in AM parts. Peak temperature and pool depth to estimate the temperature difference (ΔT) and

Fourier number (F) in Eq. 11.11 can be calculated either by using a mechanistic model of heat transfer or by the back of the envelope calculations described in Chapter 7.

In AM, the evolution of the strain field depends on many complex thermomechanical phenomena. Therefore, calculation using Eq. 11.11 cannot provide the absolute value of thermal strain during AM. This equation is limited to the consideration of only elastic strain whereas the part may deform plastically. In addition, volumetric strain due to phase transformation and creep at high temperatures during AM is also not considered. Furthermore, this simple equation does not include variations in mechanical properties such as Young's modulus and thermal expansion coefficient with temperature. However, there are two prime benefits of this equation. First, it can provide an understanding of the effects of important process parameters and alloy properties on distortion. For example, it indicates that a sizable distortion may result from a large temperature change (ΔT), long deposition time (t), and high rates of heat input per unit length (H). In contrast, terms in the denominator of Eq. 11.11 indicate factors that can reduce distortion. For example, a high flexural rigidity (EI) of the structure resists distortion. A high Fourier number (F) which indicates rapid heat transfer relative to heat accumulation reduces the peak temperature and thus, the distortion. Second, Eq. 11.11 provides a usable scale to compare the thermal distortion for different alloys. For alloys that are highly susceptible to distortion, appropriate AM variables like laser power, layer thickness, and scanning speed need to be adjusted to reduce distortion. Applications of the dimensionless strain parameter to estimate distortion are explained in worked outs examples 11.6 and 11.7.

Worked out example 11.6

Two stainless steel 316 parts are made using a 15 mm/s scanning speed. For parts 1 and 2, the laser powers are 600 and 500 W, respectively. All other processing conditions are kept constant. Which of the two parts will experience more distortion? Use the appropriate thermophysical properties from Chapter 7.

Solution:

Susceptibility to thermal distortion during AM can be expressed by the dimensionless strain parameter which is calculated using Eq 11.11. The peak temperature and pool depth to estimate the temperature difference between the peak temperature and the ambient temperature (ΔT) and Fourier number (F) in Eq. 11.11 are analytically calculated as explained in Chapter 7 (see Section 7.4.3). For example, the peak temperature is calculated using Eq. 7.6 in Chapter 7 as:

$$T = T_0 + \frac{\eta P / V}{2 \pi k_s t}$$

where T_0 is the room temperature (298 K), η is fraction of energy absorbed (0.9), P is laser power (provided in the problem), V is the scanning speed (15 mm/s or 0.015 m/s), k_s is thermal conductivity (28 W/mK), and t is a very small time in second. Fourier number is calculated using Eq. 7.8 in chapter 7 as:

$$F = \frac{\alpha}{Vd}$$

where α, V, and d refer to the thermal diffusivity of the alloy (6.36×10^{-6} m^2/s), scanning speed (15 mm/s or 0.015 m/s), and fusion zone depth (in m), respectively.

Table E11.1 below shows the calculations of the strain parameter for the two parts. Higher laser power for part 1 results in a higher peak temperature and bigger pool both of which are responsible for larger distortion in part 1.

(Continued)

Worked out example 11.6   (Continued)

Table E11.1 Calculations of the strain parameter for the two parts.

Calculations of terms in Eq. 11.11	Parameters	Part 1	Part 2
Heat input (H)	Laser power (W)	600	500
	Laser scanning speed (mm/s)	15	15
	Fraction of energy absorbed	0.9	0.9
	Heat input (J/mm)	36	30
Temperature gradient (ΔT)	Peak temperature (K)	2345	2004
	Room temperature (K)	298	298
	Temperature gradient (K)	2047	1706
Volumetric thermal expansion coefficient (β)	Volumetric thermal expansion coefficient (/K)	5.85×10^{-5}	5.85×10^{-5}
Flexural rigidity (EI)	Young's modulus (GPa)	205.7	205.7
	Substrate width (mm)	10.0	10.0
	Substrate thickness (mm)	4.0	4.0
	Flexural rigidity (N-mm^2)	1.1×10^7	1.1×10^7
Fourier number (F)	Thermal conductivity (W/m K)	28	28
	Specific heat (J/kg K)	611	611
	Density (Kg/m^3)	7200	7200
	Thermal diffusivity (m^2/s)	6.36×10^{-6}	6.36×10^{-6}
	Pool depth (mm)	0.99	0.72
	Fourier number	0.429	0.589
Total time (t)	Total time (s)	1.0	1.0
Strain parameter (ε^*)	Strain parameter	**2.05×10^{-3}**	**0.94×10^{-3}**

Worked out example 11.7

Parts of stainless steel 316 and Ti-6Al-4V are manufactured by directed energy deposition using 600 W laser power and 15 mm/s scanning speed. All other processing conditions are kept constant. Which of the two parts will experience more distortion and why? Use the appropriate thermophysical properties from Chapter 7.

Solution:

Susceptibility to thermal distortion during AM can be expressed by the dimensionless strain parameter which is calculated using Eq 11.11. The peak temperature and pool depth necessary to estimate the temperature difference between the peak temperature and the ambient temperature (ΔT) and Fourier number (F) in Eq. 11.11 are analytically calculated as explained in Chapter 7 and as illustrated in worked out example 11.6. Table E11.2 below shows the calculations of the strain parameter for the two parts. Ti-6Al-4V exhibits a bigger molten pool owing to its low density. The big pool shrinks more during solidification and makes Ti-6Al-4V more susceptible to distortion.

Worked out example 11.7   (Continued)

Table E11.2   Calculations of the strain parameter for the two parts.

Calculations of terms in Eq. 11.11	Parameters	SS 316	Ti-6Al-4V
Heat input (H)	Laser power (W)	600	600
	Laser scanning speed (mm/s)	15	15
	Fraction of energy absorbed	0.9	0.9
	Heat input (J/mm)	36	36
Temperature gradient (ΔT)	Estimated peak temperature (K)	2345	2208
	Room temperature (K)	298	298
	Temperature gradient (K)	2047	1910
Volumetric thermal expansion coefficient (β)	Volumetric thermal expansion coefficient (/K)	5.85×10^{-5}	7.74×10^{-5}
Flexural rigidity (EI)	Young's modulus (GPa)	205.7	110.0
	Substrate width (mm)	10.0	10.0
	Substrate thickness (mm)	4.0	4.0
	Flexural rigidity (N-mm^2)	1.1×10^7	0.59×10^7
Fourier number (F)	Thermal conductivity (W/m K)	28	30
	Specific heat (J/kg K)	611	539
	Density (Kg/m^3)	7200	4000
	Thermal diffusivity (m^2/s)	6.36×10^{-6}	13.9×10^{-6}
	Pool depth (mm)	0.99	1.2
	Fourier number	0.429	0.77
Total time (t)	Total time (s)	1.0	1.0
Strain parameter (ε^*)	Strain parameter	$\mathbf{2.05 \times 10^{-3}}$	$\mathbf{3.52 \times 10^{-3}}$

11.5   Numerical simulation of residual stresses and distortion

Although the back of the envelope calculations provided in this chapter are helpful to compare distortion for different process conditions and alloys, they are not capable of estimating the complex evolution of 3D residual stresses and distortion for different AM processes. It needs a solution of 3D, transient constitutive equations of stresses and strains. These equations are second-order partial differential equations and may only be solved numerically. Several numerical models are being developed worldwide including commercial packages that implement finite element analysis (FEA) to numerically solve these equations and simulate residual stresses and distortion in AM. The commercial codes include 3DSim (http://3dsim.com), ESI (http://www.esigroup.com), Additive Works (https://additive.works), Abaqus (http://www.3ds.com), and Ansys (http://www.ansys.com).

Numerical modeling starts with generating the geometry models for the computational domain consisting of the part, and the substrate. The computational domain is subsequently divided into small cells where temperatures and mechanical properties are required. Property data are needed over a wide range of temperatures from room temperature to the solidus temperature of the alloy.

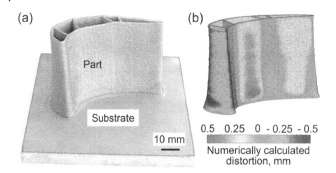

Figure 11.9 (a) An additively manufactured turbine blade made with stainless steel 316 using DED-L [13]. (b) 3D distribution of the distortion in the part numerically calculated using commercial software Simufact. Positive values of distortion indicate an increase in dimension compared with the designed value and negative values represent a shrinkage of part from the designed value [13]. The figures are taken from [13] with permission from Taylor & Francis.

Data for alloy plasticity at high temperatures and phase transformation-induced plasticity are also needed. The positions of fixtures are generally applied as boundary conditions. The constitutive equations are solved iteratively using small time steps. Calculations of the temperature field can be done separately at the beginning and using the computed values of temperatures residual stresses are calculated subsequently. However, simultaneous calculations of both temperature field and residual stresses and distortion can also be done using commercial packages. Figure 11.9 shows an additively manufactured turbine blade made with stainless steel 316 using DED-L for which 3D distribution of the distortion is calculated using the commercial software Simufact. Positive values of distortion indicate an increase in size from the designed value. The negative values represent a shrinkage of the part from the values specified in the design.

Although being increasingly popular in research and development, numerical models are still not widely used in practical applications because of the following reasons. These calculations are generally unsuitable for real-time applications because they require extensive computer time. Second, the calculations are complex and often require extensive training to use them and decades of experience to develop them. Finally, the numerical models currently available are not adequately standardized for different AM processes, process variables and alloy combinations and require further development.

11.6 Residual stresses and distortion in different AM processes

The evolution of residual stresses and distortion depends on 3D transient temperature fields, the geometry of the part, and temperature-dependent alloy properties. For a particular alloy, the heat transfer pattern and thus the temperature field vary widely for different AM processes as described in Chapter 7. In addition, the range of processing conditions such as heat source power, scanning speed, layer thickness, and deposit geometry vary significantly for different AM processes [14]. Therefore, there are significant variations in the resultant residual stresses and distortion for different AM processes. Numerical models are often used to explain those differences. For example, Figures 11.10 (a-c) show the numerically computed longitudinal stress (x-component, i.e. along the scanning direction) distribution in stainless steel 316 components printed using the three AM processes. In all three cases, the substrate dimensions and deposit height are the same. However, the same height of the deposit is made using different numbers of layers for three AM processes. For example, PBF-L requires the maximum number of layers since PBF-L commonly uses very thin layers. In addition, different processes use strikingly different heat inputs for printing components. The fusion zone in DED-GMA is the widest because of the highest heat input of this process

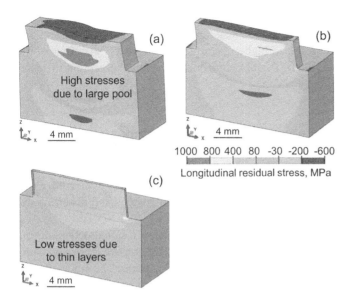

Figure 11.10 Longitudinal residual stress distribution in a SS 316 deposit printed using (a) DED-GMA (b) DED-L and (c) PBF-L. All parts are 16 mm long, 4 mm high, and built on a 20 mm long, 10 mm wide, and 10 mm thick substrate. The three parts are printed using 4, 5, and 16 layers for DED-GMA, DED-L, and PBF-L, respectively. Heat source powers for DED-GMA, DED-L, and PBF-L are 2130 W, 1500 W, and 110 W, respectively. Scanning speeds for DED-GMA, DED-L, and PBF-L are 10 mm/s, 10.6 mm/s, and 100 mm/s respectively. The scanning direction is along the positive x-axis. Half of the solution domain is shown because of the symmetry with respect to the XZ plane. The figures are generated using models developed by T. Mukherjee and T. DebRoy.

Figure 11.11 Strain parameters while depositing single layer SS 316 deposits using the three AM processes. The same substrate dimensions (20 mm long, 10 mm wide, and 10 mm thick) are taken for all three cases for consistency. Heat source powers for DED-GMA, DED-L, and PBF-L are 2130 W, 1500 W, and 110 W respectively. Scanning speeds for DED-GMA, DED-L, and PBF-L are 10 mm/s, 10.6 mm/s, and 100 mm/s respectively. *Source:* T. Mukherjee and T. DebRoy.

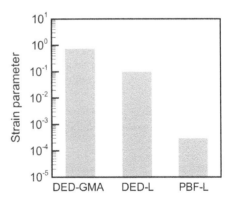

which forms the biggest molten pool among the three AM processes. In addition, DED-GMA components are printed using the thickest layers among the three processes. Thick and wide tracks accumulate high residual stresses during cooling and make the DED-GMA component the most susceptible to residual stresses among the three processes as shown in Figure 11.10 (a). Because of the smallest molten pool and thinnest layers in PBF-L, the part printed using this process exhibits the least residual stresses as shown in Figure 11.10 (c).

Back of the envelop calculations [14] using Eq. 11.11 can also be used to quantitatively compare the relative susceptibilities of DED-GMA, DED-L, and PBF-L parts to distortion. Figure 11.11 shows a comparison of the calculated strain parameters for stainless steel 316 components made by three AM processes. The molten pool in DED-GMA is significantly larger than those for DED-L and PBF-L respectively, primarily owing to the highest heat input used in DED-GMA. Bigger pools shrink more during solidification and result in higher distortion in AM parts. Shrinkage of the largest pool during solidification makes the DED-GMA components the most susceptible to distortion among the three AM processes.

11.7 Effects of process parameters and printing strategies

Residual stresses and distortion vary significantly with the important AM process parameters such as heat source power, scanning speed, substrate thickness, substrate preheat temperature and layer thickness. The effects of these process parameters on residual stresses and distortion are summarized in Table 11.3.

Figure 11.12 explains the effects of printing strategies on the evolution of residual stresses during the DED-GMA of Inconel 718. Three rectangular deposits of the same size are fabricated using three different printing strategies, long hatch, short hatch, and spiral patterns. The schematics for the three strategies are shown in Figure 11.12 (a-c). For the long hatch strategy, the scanning direction is along the longer side of the rectangular deposit. However, the scanning direction is reversed between two successive hatches. For the short hatch strategy, the scanning direction is along the shorter side of the rectangular deposit. Similar to the long hatch strategy, the scanning direction was also reversed between two successive hatches. In spiral deposition, the arc heat source follows a spiral path to deposit the rectangular component. Figure 11.12 (d-f) shows the

Table 11.3 Effects of important AM process parameters on residual stresses and distortion.

Parameters	Effect on residual stresses	Effect on distortion
Heat source power	An increase in heat source power results in a higher temperature that enhances the temperature gradient and the residual stresses.	Increases in heat source power result in a larger molten pool that shrinks more during solidification and distorts the part.
Scanning speed	An increase in scanning speed results in a lower temperature that reduces the temperature gradient and residual stresses.	Increases in scanning speed result in a smaller molten pool that shrinks less during solidification and reduces distortion in the part.
Substrate thickness	Thicker substrates have higher rigidity that can restrict the deformation of the component and accumulate high residual stresses.	Thick substrates can resist distortion due to their high rigidity. Therefore, parts made on thicker substrates exhibit less distortion.
Substrate preheat temperature	Stresses develop during cooling. At high preheat temperatures, the difference between the solidus temperature of the alloy and the preheat temperature is small resulting in lower residual stress.	High preheat temperatures result in a large molten pool that shrinks more during solidification and distorts the part.
Layer thickness	Thick layers accumulate high stresses during cooling. Therefore, parts made with thick layers exhibit high residual stresses.	Deposition of thick layers requires large molten pools that shrink more during solidification and distorts the part.

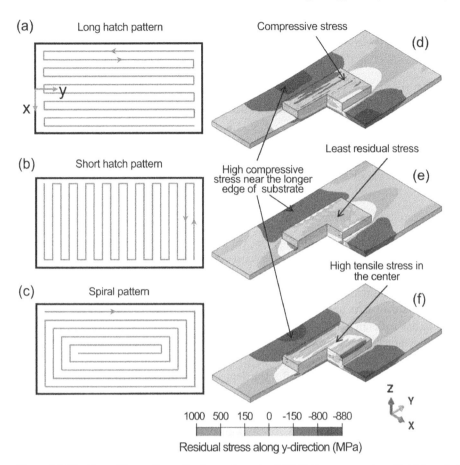

Figure 11.12 Deposition patterns for the rectangular DED-GMA components with (a) long hatch pattern, (b) short hatch pattern, and (c) spiral pattern. Residual stresses along the y-direction (along the longer side of the deposit) in Inconel 718 components deposited with (d) long hatch pattern, (e) short hatch pattern, and (f) spiral pattern when the deposits cooled down to room temperature. These results were obtained using a model developed by Q. Wu, T. Mukherjee, C. Liu, J. Lu, and T. DebRoy.

y-component of the residual stresses (along the longer side of the deposit) in parts made with three strategies. Cutaway isometric views are used to show the accumulations of residual stresses inside the component. The stress distributions in the substrate show high tensile stress near the shorter edge of the deposit and high compressive stress near the longer edge of the substrate. Y-direction stresses mainly originate from the expansion along this direction during heating and shrinkage during cooling. During the cooling time, all deposits shrink resulting in high tensile stresses near the shorter edge of the deposit. For the region near the longer edge of the substrate, compressive stresses originate due to the upward bending of shorter edges of the substrate. For the components fabricated with a long hatch pattern, compressive stresses are observed on the top surface of the deposits because the y-direction is the primary contraction direction for the deposit with a long hatch pattern. High tensile stresses can be found in the center of the deposit with a spiral pattern since the last hatch is deposited at that location. The deposit made using a short hatch pattern has the least residual stresses among the three strategies. This is because more hatches needed to fabricate the deposits using a short hatch pattern significantly alleviate the stresses due to the reheating effect. From the above analysis, it is evident that residual stress distribution can vary significantly depending on the scanning strategies. Therefore, residual stresses can be effectively minimized by carefully selecting the printing strategy [15].

11.8 Effects of residual stresses and distortion

Accumulation of high residual stresses and distortion in AM parts is undesirable because they affect the structural integrity and mechanical properties of the component and may cause defects. Detrimental effects of residual stresses and distortion in AM parts are discussed below.

11.8.1 Buckling, warping, and dimensional accuracy

Buckling is a common type of distortion observed in AM parts where the component distorts in a bow shape due to the non-uniformity in the residual stresses at the top and bottom of the component. Residual compressive stresses reduce the buckling strength of the part and the component may fail under external compressive load. Warping is defined as any unwanted distortion which may result in the failure of AM parts subjected to tensile, compressive, or torsional loading. Distortion in AM parts results in a deviation in geometry and dimensions specified in the design. Figure 11.13 (a) schematically shows dimensional inaccuracy where the dimensions of the distorted part significantly deviate from those of the target part. The arrow in Figure 11.13 (b) shows the distortion of the substrate (Inconel 718) after the deposition of blocks by PBF-L. Distortion of parts often leads to part rejection. Post-process machining is often employed to correct dimensional inaccuracy which, however, adds to the cost of an already expensive AM part.

11.8.2 Cracking and delamination

Stresses accumulated during solidification and cooling can cause cracks in AM parts even without any externally applied load. During AM, solidifying deposit contracts due to both solidification shrinkage and thermal contraction. However, the temperatures of the substrate and the previously deposited layers are lower than those of the depositing layer. Therefore, the contraction of the depositing layer is more than that of the lower layers and the substrate. Thus, the contraction of the solidifying layer is hindered by the substrate and the previously deposited layers. That

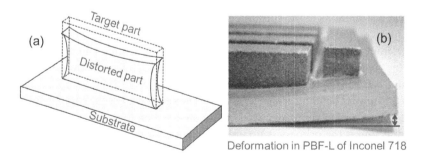

Deformation in PBF-L of Inconel 718

Figure 11.13 (a) Schematic representation of dimensional inaccuracy in AM parts. *Source:* T. Mukherjee and T. DebRoy. (b) Deformation in an Inconel 718 part fabricated using PBF-L [16]. Taken from [16] with permission from Elsevier.

generates tensile residual stress at the solidifying layer. If the magnitude of this tensile stress exceeds the tensile yield strength of the material, cracking may occur. Tensile residual stresses are transverse to the deposit in the heat-affected zone causing cracks perpendicular to the deposit along both the hatching and building directions. In the building direction, cracks may propagate over several layers. Therefore, in AM parts cracking can be very long spreading over several layers (Figure 11.14 (a)) or small with a maximum length equal to the layer thickness (Figure 11.14 (b)). Rapid cooling below the liquidus temperature of the alloy causes the formation of precipitates, carbide particles, and hard secondary phases. These hard particles often act as crack initiation sites under high tensile residual stresses.

Solidification cracking or hot tearing may occur in AM parts due to the accumulation of high solidification stress. A crack may initiate from the last remaining liquid in the inter-dendritic region under high solidification stress. Attempts have been made in the casting and fusion welding literature to quantify the solidification stress (σ_s) responsible for crack formation by using a modified Griffith's equation as [17]:

$$\sigma_s = \sqrt{\frac{2E\gamma}{\pi a}} \tag{11.12}$$

where E is the modulus of elasticity at a temperature close to the solidus temperature and γ is the specific surface energy. The variable "a" is half of the length of an internal crack which is related to the amount of remaining liquid during the last stage of solidification. Eq. 11.12 has been proven to be effective in predicting the susceptibility to crack formation due to the solidification stress as explained in worked out example 11.8.

High tensile residual stresses between two consecutive layers may exceed the tensile yield strength of the material and result in delamination which is the separation of two consecutive layers as shown in Figure 11.14 (c). In addition, high tensile residual stresses at the deposit-substrate interface may cause the detachment of the part from the substrate.

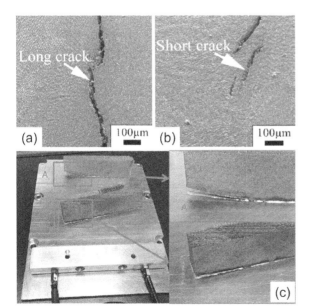

Figure 11.14 (a) Long crack and (b) short crack in a nickel-based superalloy part made using DED-L [18]. Figures (a, b) are taken from [18] with permission from Elsevier. (c) Delamination in AlSi10Mg parts fabricated using PBF-L [19]. Figure (c) is taken from an open-access article [19] under the terms and conditions of the Creative Commons Attribution (CC BY) license.

Worked out example 11.8

An aluminum alloy AA 6061 part is being made using the powder bed fusion-laser (PBF-L) using 200 W laser power and 100 mm/s scanning speed. Is this part susceptible to solidification cracking? Useful data: modulus of elasticity of AA 6061 near solidus temperature: 20 GPa, specific surface energy: 0.035 J/m^2, yield strength of AA 6061 near solidus temperature: 4 MPa, volume of the two-phase solid-liquid mushy zone at 200 W laser power and 100 mm/s scanning speed: 1.2 × 10^{-12} m^3. Assume that the crack length is equal to the radius of a sphere that can accommodate the residual liquid in the two-phase region at a fraction solid of 0.99.

Solution:

Eq. 11.12 can be used to predict the susceptibility to solidification cracking. Cracks may form if the solidification stress computed using Eq. 11.12 is higher than the local yield stress of the alloy. The value of "a" in Eq. 11.12 can be calculated from the amount of liquid remaining in the mushy zone during the last stages of solidification. It is well-known in the solidification cracking literature [20] that solidification cracking initiates at the solid fraction (f_s) for which the slope of the temperature versus (f_s)$^{1/2}$ plot is the maximum. All alloys have their characteristic temperature versus (f_s)$^{1/2}$ plot. For AA 6061, the maximum slope occurs at 0.99 solid fraction i.e. at (1 − 0.99) = 0.01 liquid fraction [20].

The volume of the remaining liquid can be calculated by multiplying 0.01 by the corresponding mushy zone volume. Therefore, the volume of the remaining liquid in the mushy zone that may initiate solidification cracking = 0.01 × 1.2 × 10^{-12} m^3.

The length scale "a" in Eq. 11.12 is approximately calculated as the cube root of the volume of the remaining liquid as 2.29 × 10^{-5} m. Since the remaining liquid may be distributed throughout the mushy zone, this length scale "a" can represent the largest possible flaw size inside the component.

Therefore, the solidification stress (σ_s) is calculated using Eq. 11.12 as:

$$\sigma_s = \sqrt{\frac{2 \times 20 \times 10^9 \times 0.035}{3.14 \times 2.29 \times 10^{-5}}} = 4412466.8 \text{ Pa} = 4.41 \text{ MPa}$$

Therefore, the computed value of solidification stress (4.41 MPa) is higher than the local yield stress of the alloy (4 MPa). Therefore, the part is susceptible to solidification cracking.

11.8.3 Fracture at low applied force

AM parts can exhibit premature brittle fracture at a low applied force due to the presence of high residual stresses. Fractures may occur at a stress level far below the yield strength of the material. Therefore, under certain conditions, the complete fracture of a part may happen even when the magnitude of the applied stress is considerably below the yield stress of the material. The fracture may initiate from a small flaw that would normally be stable if residual stresses were not present. When a flaw is present in a region that is either free of or contains compressive residual stresses, the fracture does not originate from the flaw. However, if a tip of a flaw lies in a region of high tensile residual stress, the residual stress would add to the applied stress and increase the intensity of the stress accumulation at the tip of the flaw. This increase in intensity may cause the flaw to initiate fracture. This eventually leads to a complete fracture of the component. However, if a fracture initiates locally at a region with high tensile residual stress which is surrounded by a region having compressive residual stress, a complete fracture of the part may be avoided. Therefore, the effect of residual stresses on brittle fracture depends on the magnitude of the external force, the presence of a flaw, and the nature of residual stresses. Often AM parts are post-processed using hot isostatic pressing to eliminate small flaws to prevent fracture.

11.8.4 Degradation of fatigue strength

Fatigue life which is the number of cycles required to fracture an AM part under a given fluctuating load depends on the residual stresses accumulated in the component. The accumulation of tensile residual stresses significantly degrades the fatigue strength. Fatigue crack may initiate from a small flaw in the component and propagate under high tensile residual stresses. Therefore, the component may fail before its specified fatigue strength. Figure 11.15, a plot based on independent literature data [21, 22], shows that the fatigue cracks in additively manufactured Ti-6Al-4V parts grow faster in the presence of a higher tensile residual stress in the component. The maximum tensile residual stress values in the figure are estimated from the reported stress distributions [21, 22]. The y-intercept of the curve is about 8×10^{-6} mm/cycle which corresponds to the crack

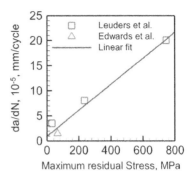

Figure 11.15 Fatigue crack growth rate per cycle (da/dN) for additively manufactured Ti-6Al-4V components as a function of maximum tensile residual stress. The plot is made by T. Mukherjee and T. DebRoy using the data obtained from the crack growth rate reported in independent literature by Leuders et al. [21] and Edwards et al. [22]. The stress intensity factor range is 10 MPa.m$^{1/2}$.

growth rate with very low residual stresses (such as the traditionally processed Ti-6Al-4V parts).

Propagation of fatigue cracks under cyclic loading may be reduced under compressive residual stresses. Therefore, residual stresses can increase or decrease the mean stress experienced by the component. Therefore, accumulated residual stresses must be accounted for in the prediction of the fatigue life of AM components. The relation between fatigue crack growth and residual stress is explained in worked out example 11.9.

Worked out example 11.9

Derive a formula to provide a quantitative relation between fatigue crack growth and residual stress.

Solution:

The fatigue crack growth rate per cycle, da/dN, is expressed as a function of the difference in the cyclic stress intensity factor ΔK. The relationship, known as the Paris law (see Chapter 9), is expressed as:

$$da/dN = C(\Delta K)^m$$

where a is related to the size of the crack, N is the number of cycles, ΔK is the stress intensity factor range, and C and m are empirical constants that depend on the composition and geometry of the material, temperature, and other factors. The fatigue crack growth will continue until the stress intensity factor is equal to the fracture toughness (K_{Ic}). At that point, the stress inside the component is equal to the maximum tensile residual stress that the part can withstand without failure. Therefore, we can write $K_{Ic} = Y\sigma\sqrt{\pi a}$ where Y is a geometric constant, σ is the maximum residual stress, and a is related to the size of the crack.

Therefore, $da/dN = C(Y\sigma\sqrt{\pi a})^m$

From the above expression, it is evident that the fatigue crack growth rate per cycle increases at higher residual stress. This relation is also evident from the data in Figure 11.15 where da/dN is shown to be proportional to the maximum residual stress.

11.9 Controlling residual stresses and distortion

Residual stresses and distortion in AM parts need to be controlled since they cause defects such as buckling, warping, cracking, and delamination. They lead to premature brittle fracture at low applied force and affect the fatigue strength. To control residual stresses and distortion effectively, it is essential to understand how they originate in AM parts depending on the process and process parameters used. Previous sections of this chapter provide this essential understanding. In this section, several techniques for reducing and controlling residual stresses and distortion in AM parts are discussed. These techniques are summarized in Table 11.4.

Table 11.4 Methods used for controlling residual stresses and distortion in AM.

Controlling methods		Features
Before and during printing	Adjusting process parameters	Important process parameters such as heat source power, scanning speed, substrate thickness, and layer thickness can be adjusted to control residual stresses and distortion. The use of mechanistic models and machine learning to predict the optimum process parameters is becoming increasingly popular.
	Adjusting scanning strategies	Scanning strategies and printing sequences significantly affect residual stresses and distortion (Section 11.7). A careful selection of scanning strategy and printing sequence can reduce residual stresses.
	Component design and fixture position	Residual stresses are developed during cooling when the shrinkage is hindered by the fixtures. Therefore, fixtures and their positions can be adjusted depending on the design of the part to reduce residual stresses.
	Preheating the substrate and/or powder bed	Residual stresses primarily develop during cooling. At a high preheat temperature, the difference between the solidus temperature of the alloy and the preheat temperature is small resulting in low residual stresses. Preheating is a commonly used technique to reduce stresses in welding.
	Adjusting feedstock composition	Pre-alloyed powders can be made with the desired chemical composition that results in a low-temperature phase transformation. That will avoid the accumulation of residual stresses due to solid-state phase transformation (Section 11.2.3) and can be used to control residual stresses.
	In-situ sensing and control of distortion	Part distortion can be monitored during the AM process using a high-speed camera, and mechanically or optically controlled displacement sensors that can send data to a computer. The computer can process the data from the sensor and send a feedback signal to the printing machine to adjust the process conditions to reduce distortion.
After printing post-processing	Post-process heat treatment	Post-process heat treatment consists of heating the as-fabricated parts to high temperatures below the solidus temperature of the alloy and holding them for some time while the stresses are relieved. However, this adds cost to an already expensive AM part.
	Shot peening	Shot peening involves striking a part surface with small metallic or ceramic particles with a force sufficient to create plastic deformation on the surface which imparts compressive stress on the surface.
	Machining to achieve correct dimensions	Distortion can cause dimensional inaccuracy that can be corrected by post-process machining. However, the machining allowance should be incorporated in the design of the part.

There are several contributing factors to residual stresses and distortion in AM. These factors are quite difficult to control in a complex component typically made by AM. All commonly used control techniques can be classified into two categories, before and during printing methods, and post-processing after printing. Methods applied before and during printing are primarily developed by experimental trials and experience. Post-process corrective methods (Table 11.4) are often used to correct a distorted part or reduce residual stresses accumulated in a component during the process. These post-process corrections are often expensive and time-consuming to implement. However, care should be taken to apply these control techniques since they may affect the properties of the component. With the advent of accurate computational models, new and creative techniques are emerging. It is now possible to use computational models to design new control methods and test the results with fewer experiments before implementing them.

11.10 Summary

In AM, residual stresses and distortion originate from the spatially nonuniform temperature variation, solidification of the deposit, temperature-dependent elastic and plastic properties of the alloy, geometric constraints such as fixtures and clamps, and solid-state phase transformations. A detailed understanding of these physical factors is needed to gain important insight into the evolution of residual stresses and distortion. While several well-known techniques are available for the measurement of residual stresses and distortion, numerical models are also being developed to calculate these quantities in AM parts. Understanding of the evolution of residual stresses and distortion in AM is still emerging and for most cases, the use of simple analytical equations or experiments is not enough because of the complexity of the process and the time and money needed to conduct experiments. In the last few years, numerical simulations are being increasingly used to simulate residual stresses in AM. However, further development of those models is needed to make them computationally efficient and applicable to different AM processes and alloys.

Takeaways

Origin of residual stresses and distortion

- Spatial change in the temperature field, temperature-dependent elastic and plastic behavior of alloy, and solid-state phase transformation are key contributors to residual stresses and distortion.
- The plastic properties of an alloy depend on both temperature and strain rate.
- In additive manufacturing, residual stresses and distortion evolve largely depending on the 3D, spatially nonuniform, transient temperature field.
- Change in crystal structure due to solid-state phase transformation results in a volumetric strain which causes internal stress accumulations in the material.

Measurement techniques

- Residual stresses can be measured by mechanical, diffraction, ultrasonic and magnetic methods, and indentation testing.
- Mechanical methods such as hole drilling and curvature method as well as indentation testing are destructive methods and may damage the part.
- Diffraction techniques are nondestructive methods with high spatial resolution and accuracy but are expensive and time-consuming.

(Continued)

Takeaways   (Continued)

- Ultrasonic and magnetic methods are nondestructive, inexpensive, and easy to use but with poor spatial resolution and accuracy.

Back of the envelop calculations

- Back of the envelop calculations using the dimensionless strain parameter can be useful for quantitative comparison of distortion susceptibility.
- Relative susceptibilities to distortion of different processes, process parameters, and alloy combinations can be evaluated using the dimensionless strain parameter.

Residual stresses and distortion in different additive manufacturing processes

- DED-GMA part exhibits the highest residual stresses compared to DED-L and PBF-L because of the deposition of thick layers with large pools that accumulate high stresses during cooling.
- Deposition of the thinnest layers and small molten pools makes the PBF-L part the least prone to residual stresses.
- A large molten pool in DED-GMA shrinks more during solidification and makes the DED-GMA process the most vulnerable to distortion.

Effects of process parameters and printing strategies

- Residual stresses increase with increasing heat source power, layer thickness, and substrate thickness. Rapid scanning and preheating the substrate reduce residual stresses.
- AM parts distort significantly when fabricated using high heat source power, slow scanning, thick layers, and substrates having low rigidity.
- Printing strategies with different scanning patterns significantly affect residual stresses and distortion and can be optimized to control them.

Effects of residual stresses and distortion

- High residual stresses in additively manufactured parts can cause premature failure under low applied force, and defects such as cracking, warping, and delamination.
- The fatigue properties of additively manufactured parts can be significantly affected by the accumulated residual stresses.
- Distortion in additively manufactured parts can result in dimensional inaccuracy, and in extreme cases, may lead to part rejection.

Reducing or controlling residual stresses and distortion

- There are several steps applied before and during printing as well as post-processing to reduce or control residual stresses and distortion in AM parts.
- Methods that are applied before and during printing may require trials. Post-processing adds an extra cost to an already expensive AM part.

Appendix – Meanings of a selection of technical terms

<u>Casting:</u> A manufacturing process where a liquid is poured into a mold and the material takes the shape of the mold after solidification.

Electron beam: A focused beam of electrons used as a heat source in additive manufacturing.

Fourier number: A dimensionless number representing the ratio of the rate of heat dissipation to the rate of heat storage.

Fusion welding: A manufacturing process that uses a heat source to melt two parts to form a sound joint after solidification.

Heat treatment: A process of heating a part in a furnace at a high temperature less than the melting of the material followed by a subsequent cooling process. The rate of cooling can vary depending on the requirement.

Laser beam: The acronym laser stands for "light amplification by stimulated emission of radiation." A focused beam of laser is often used in additive manufacturing as a heat source.

Mushy zone: A two-phase region containing both solid and liquid.

Neutron: A subatomic particle present in all atomic nuclei except hydrogen. Scattering or diffraction of neutrons is often used to determine atomic structure and residual stresses.

Peak temperature: Maximum temperature inside the molten pool.

Plane strain: A two-dimensional state of strain in which the shape change or the distortion of the material occurs on a single plane. A plane strain condition can be observed when the region of the strain accumulation is very small compared to the dimension of the entire component.

Scanning pattern: Pattern of the movement of the heat source to melt the material in additive manufacturing. Often referred to as the deposition pattern for the directed energy deposition process.

Temperature gradient: Spatial gradient of the temperature field during additive manufacturing. The temperature field is spatially nonuniform and is distributed in three dimensions.

Volumetric strain: Strain accumulated due to the volume change. It is expressed as (final volume – initial volume)/initial volume.

Practice problems

1) The thermal conductivity of alloy A is twice that of alloy B. If two identical deposits are made using the two alloys under the same processing conditions, which deposit will be more susceptible to distortion?

2) Explain quantitatively how a thick substrate can reduce the thermal distortion in AM parts.

3) Discuss quantitatively how substrate preheating can affect the distortion of AM parts.

4) A given part can be made using either long or short tracks. Which track length can reduce the residual stress in the part?

5) If the same process variables such as power and speed are used to make parts of identical geometry using Ti-6Al-4V and Inconel 718 which part will have lower residual stresses and why?

6) What factors are important in determining the relative susceptibilities of distortion in different additively manufactured metallic components?

7) How reliable are the simulated values of the residual stress and distortion? How is the accuracy of the computed residual stresses and distortions evaluated?

8) The selection of an appropriate material for additive manufacturing is a very important issue. Which factors should be considered in selecting a material to minimize residual stresses and distortion?

9) Discuss the factors that are important in selecting a method to measure residual stresses in additively manufactured parts.

References

1 DebRoy, T., Wei, H.L., Zuback, J.S., Mukherjee, T., Elmer, J.W., Milewski, J.O., Beese, A.M., Wilson-Heid, A.D., De, A. and Zhang, W., 2018. Additive manufacturing of metallic components–process, structure and properties. *Progress in Materials Science*, 92, pp.112–224.

2 American Welding Society, 2018. *Welding Handbook. Volume 1: Welding and Cutting Science and Technology*, 10th Edition. Miami, FL, USA.

3 Seshacharyulu, T., Medeiros, S.C., Frazier, W.G. and Prasad, Y.V.R.K., 2000. Hot working of commercial Ti–6Al–4V with an equiaxed α–β microstructure: materials modeling considerations. *Materials Science and Engineering: A*, 284(1–2), pp.184–194.

4 Withers, P.J. and Bhadeshia, H.K.D.H., 2001. Residual stress. Part 2–Nature and origins. *Materials Science and Technology*, 17(4), pp.366–375.

5 Cheon, J. and Na, S.J., 2017. Prediction of welding residual stress with real-time phase transformation by CFD thermal analysis. *International Journal of Mechanical Sciences*, 131, pp.37–51.

6 Piekarska, W., 2015. Modelling and analysis of phase transformations and stresses in laser welding process. *Archives of Metallurgy and Materials*, 60(4), pp.2833–2842.

7 Mukherjee, T., Zhang, W. and DebRoy, T., 2017. An improved prediction of residual stresses and distortion in additive manufacturing. *Computational Materials Science*, 126, pp.360–372.

8 TN, Tech Note, 1993. *Measurement of residual stresses by the Hole-Drilling* Strain Gage Method*.

9 American Society for Testing and Materials, 2013. *Standard test method for determining residual stresses by the Hole-drilling Strain-gage Method*. ASTM International.

10 Prevey, P.S., 1986. X-ray diffraction residual stress techniques. *ASM International, ASM Handbook*, 10, pp.380–392.

11 Deng, Y., Zhang, Y. and Zhou, Y., 2022. Measurement of residual stress in single-crystal SiC by X-ray diffraction method. *Chinese Journal of Theoretical and Applied Mechanics*, 54(1), pp.147–153.

12 Mukherjee, T., Zuback, J.S., De, A. and DebRoy, T., 2016. Printability of alloys for additive manufacturing. *Scientific Reports*, 6(1), pp.1–8.

13 Biegler, M., Elsner, B.A., Graf, B. and Rethmeier, M., 2020. Geometric distortion-compensation via transient numerical simulation for directed energy deposition additive manufacturing. *Science and Technology of Welding and Joining*, 25(6), pp.468–475.

14 Mukherjee, T. and DebRoy, T., 2019. Printability of 316 stainless steel. *Science and Technology of Welding and Joining*, 24(5), pp.412–419.

15 Wu, Q., Mukherjee, T., Liu, C., Lu, J. and DebRoy, T., 2019. Residual stresses and distortion in the patterned printing of titanium and nickel alloys. *Additive Manufacturing*, 29, article no.100808.

16 Prabhakar, P., Sames, W.J., Dehoff, R. and Babu, S.S., 2015. Computational modeling of residual stress formation during the electron beam melting process for Inconel 718. *Additive Manufacturing*, 7, pp.83–91.

17 Williams, J.A. and Singer, A.R.E., 1968. Deformation, strength, and fracture above the solidus temperature. *Journal of the Institute of Metals*, 96(1), pp.5–12.

18 Zhao, X., Lin, X., Chen, J., Xue, L. and Huang, W., 2009. The effect of hot isostatic pressing on crack healing, microstructure, mechanical properties of Rene88DT superalloy prepared by laser solid forming. *Materials Science and Engineering: A*, 504(1–2), pp.129–134.

19 Hehr, A., Norfolk, M., Kominsky, D., Boulanger, A., Davis, M. and Boulware,P., 2020. Smart build-plate for metal additive manufacturing processes. *Sensors*, 20(2), article no.360.

20 Soysal, T. and Kou, S., 2018. A simple test for assessing solidification cracking susceptibility and checking validity of susceptibility prediction. *Acta Materialia*, 143, pp.181–197.

21 Leuders, S., Thöne, M., Riemer, A., Niendorf, T., Tröster, T., Richard, H.A. and Maier, H.J., 2013. On the mechanical behaviour of titanium alloy TiAl6V4 manufactured by selective laser melting: fatigue resistance and crack growth performance. *International Journal of Fatigue*, 48, pp.300–307.

22 Edwards, P., O'conner, A. and Ramulu, M., 2013. Electron beam additive manufacturing of titanium components: properties and performance. *Journal of Manufacturing Science and Engineering*, 135(6), article no.061016.

12

Mechanistic Models, Machine Learning, and Digital Twins in Additive Manufacturing

Learning objectives

After reading this chapter the reader should be able to do the following:

1) Know several applications of mechanistic modeling in additive manufacturing.
2) Compare different models based on their applicability and computational efficiencies.
3) Select an appropriate machine learning algorithm to solve the common problems of additive manufacturing.
4) Recognize the common applications of machine learning in additive manufacturing.
5) Perform machine learning calculations using WEKA, an open-source package.
6) Appreciate the roles of digital twins for improving the standardization and part qualification in additive manufacturing.

CONTENTS

Theory and Practice of Additive Manufacturing, First Edition. Tuhin Mukherjee and Tarasankar DebRoy.
© 2024 John Wiley & Sons, Inc. Published 2024 by John Wiley & Sons, Inc.

12.1 Introduction

In additive manufacturing (AM), structurally sound and reliable parts are produced and certi-fied by trial and error for each part, material, and process variant. This is a time-consuming and expensive undertaking because of the need to experimentally determine the effects of many vari-ables. The small size of the fusion zone, the strong recirculating motion of the liquid alloy inside the molten pool, the movement of the heat source, and the rapid temperature variations make the accurate measurement of important variables challenging. In addition, experimental measure-ments are only practical on surfaces that are easily accessible and not in interior locations. In the last two decades, significant efforts have been made to quantitatively understand the role of the AM process parameters on product attributes using emerging digital tools such as mecha-nistic models, machine learning, and digital twins. The mechanistic models [1] [2] seek to gain an improved understanding of heat transfer, liquid metal flow, and mass transfer to calculate the important metallurgical variables that affect the microstructure, properties, and defect formation in components based on phenomenological understanding. When scientific understanding is missing but data are available, machine learning can provide quantitative relations between the process variables and product attributes. Digital twins can bring the models, machine learning, large volume of available data, experiments under one framework, create a digital replica of the AM machines, and perform rapid and cost-effective virtual tests. Therefore, it has the potential to reduce cost, improve quality, and accelerate part qualification. These digital tools will not be substitutes for experiments but will reduce the number of tests needed to print sound parts. The combined approach of experiments and the use of digital tools is critical for controlling micro-structure and properties and reducing common defects in AM. This chapter describes the applica-tions of mechanistic models, machine learning, and digital twin in additive manufacturing.

12.2 Mechanistic models

Figure 12.1 shows the important role of mechanistic models of AM in the overall understanding of the physical processes and product attributes. These models enable calculations of important var-iables such as temperature and velocity fields, cooling rates, and solidification parameters that are not easily measured during AM. These models also provide a phenomenological description of how microstructure, properties, and defects in an AM part evolve from process variables and ther-mophysical properties. Mechanistic models of AM are widely used because of their usefulness in predicting important trends between processing variables and computed part attributes. Table 12.1 summarizes many of the common mechanistic models, their features, and their applications in AM. Many of the physical processes need to be represented in multiple length scales, and in some cases in varying time scales. Most of the simulations require transient three-dimensional temper-ature fields. Considerable spread in the computational efficiency is achieved depending on the physical processes considered and the scale of calculations. In this subsection, we examine the progress made and the opportunities and challenges in the mechanistic modeling of various impor-tant aspects of metal printing.

12.2.1 Modeling of heat, mass, and momentum transfer

In additive manufacturing, a quantitative understanding of the formation of the deposit geometry, microstructure, properties, and defect formation starts with the simulation of transient tempera-ture field and the flow of liquid metal captured through the modeling of heat, mass, and momentum

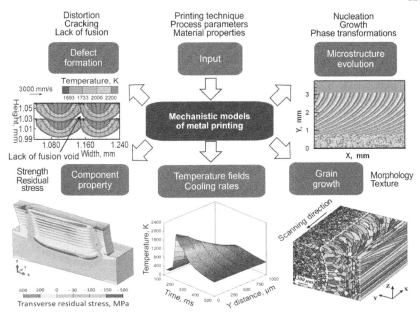

Figure 12.1 Applications of mechanistic models to understand the process and product attributes. *Source:* T. Mukherjee and T. DebRoy.

Table 12.1 Common mechanistic models for the simulation of additive manufacturing. The important features of the models, methodology, and applications are shown.

Purpose	Model	Features	Applications
Calculation of heat, mass, and momentum transfer	Part scale heat conduction model	The Fourier heat conduction equation is solved either analytically in 1D or 2D or numerically in 3D. Does not consider the effects of molten metal flow inside the pool and often provides inaccurate results.	Temperature fields, fusion zone geometry, cooling rates
	Part scale heat transfer & fluid flow	Solves 3D transient conservation equations of mass, momentum, and energy. Considers the effects of molten metal flow inside pool and therefore provides accurate temperature distribution and deposit geometry.	Temperature and velocity fields, fusion zone geometry, cooling rates, solidification parameters, lack of fusion
	Part scale volume of fluid and level set methods	Tracks the free surface of the molten pool. Computationally intensive. Accumulates errors in the computed variables.	3D deposit geometry, Temperature and velocity fields, cooling rates, solidification parameters
	Powder scale models	Involves free surface boundary conditions considering thermodynamics, surface tension, phase transitions, and wetting. Small time and length scales, computationally intensive. Lattice Boltzmann or Arbitrary Lagrangian-Eulerian.	Temperature and velocity fields, track geometry, lack of fusion, spatter, surface roughness

(Continued)

Table 12.1 (Continued)

Purpose	Model	Features	Applications
Microstructure, nucleation, and grain growth prediction	JMA-based models and TTT and CCT diagrams	Based on phase transformation kinetics during cooling. Widely used for simulating phase transformations in steels and common alloys with high computational efficiency.	Solid-state phase transformation kinetics
	Monte Carlo method	A probabilistic approach of grain orientation change. Provides grain size distribution with time. High computational efficiency.	Grain growth, solidification structure, texture
	Cellular automata	Simulates growth of grain and sub-grain structure during solidification. Medium accuracy and computational efficiency.	Solidification structure, grain growth, texture
	Phase-field model	Simulates microstructural features and properties by calculating an order parameter based on free energy that represents the state of the entire microstructure Computationally intensive	Nucleation, grain growth, evolution of phases, precipitate formation, solid-state phase transformation
Calculation of residual stresses and distortion	FEA-based thermomechanical models	Solves 3D constitutive equations considering elastic, plastic, and thermal behavior. Many existing software packages, easy to implement, can handle intricate geometries. Adaptive grid and inherent strain method are often used to increase calculation speed	Evolution of residual stresses, strains, distortion, delamination, warping

Notes: Time-temperature-transformation (TTT) diagrams provide the effects of time and temperature on the microstructure development at constant temperatures. The continuous cooling transformation (CCT) diagrams indicate the phase changes during cooling. The Johnson-Mehl-Avrami (JMA) equation provides the extent of phase transformations with time. The finite element analysis (FEA) is a numerical method to solve complex nonlinear equations.

transfer. These models need to simulate a series of physical processes such as energy absorption by feedstock and substrate materials, different modes of heat transfer, formation of the molten pool, liquid metal flow, vaporization of alloying elements from the pool surface, and solidification. The heat, mass, and momentum transfer calculations are typically based on the solutions of the equations of conservation of mass, momentum, and energy to obtain important variables such as the temperature-time history, fusion zone geometry, and the solidification growth rates.

There are generally four types of models [3-6] for heat, mass, and momentum transfer (Table 12.1). Heat conduction models solve Fourier's heat conduction equation to provide a 3D, transient temperature field and can be used for large and complex parts fabricated using multiple layers and hatches (Figure 12.2a). Heat transfer and fluid flow models can also be used in part scale to calculate 3D, transient temperature fields as well as velocities inside the molten pool (Figure 12.2b). Part scale volume of the fluid and level set methods are used to compute the temperature and velocity fields as well as the evolution of 3D deposit geometry by tracking the movement of the molten pool surface. Powder scale models can capture the interaction among the individual powder particles and the heat source and therefore are restricted for simulating very small time and length scales (Figure 12.2 c and d). However, since these models can capture

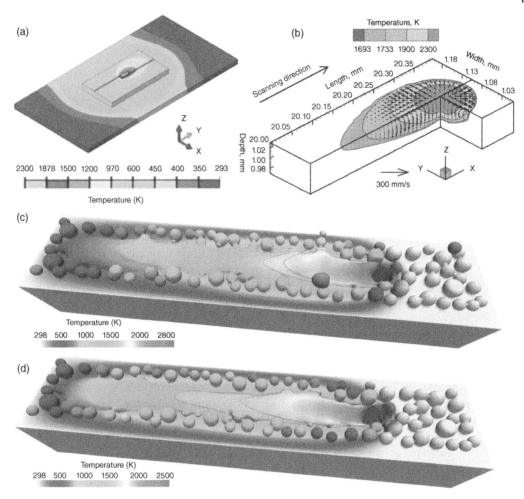

Figure 12.2 Calculation of heat, mass, and momentum transfer using (a) heat conduction model for DED-GMA of Ti-6Al-4V, (b) heat transfer and fluid flow model for PBF-L of SS 316, and powder scale model for PBF-L of (c) Ti-6Al-4V and (d) Hastealloy. Figures (a) and (b) are generated using models developed by T. DebRoy. Figures (c) and (d): Courtesy of Professor H.L. Wei from Nanjing University of Science and Technology, China.

physics on a micrometer scale, they can provide important insights into spatter formation, inter-particle heat transfer, and evolution of geometry by powder accumulation that cannot be obtained from the part scale models.

Selection of these models depends on the AM processes. For example, powder bed fusion processes are simulated using both the powder scale as well as part scale models. Powder scale models are often integrated with a separate model of powder spreading that can provide an accurate powder size distribution and packing density to the heat and fluid flow calculations. Similarly, a separate powder injection model is used along with the powder scale model of directed energy deposition processes. These models are useful for providing the powder flow velocity, amount of powder captured to form the molten pool, the temperature of the powders before reaching the substrate, and many other important parameters. The heat transfer and fluid flow calculations by solving the conservations equations of mass, momentum, and energy are similar for both powder and wire-based directed energy deposition processes. The 3D, transient temperature fields and deposit geometry can provide thermal cycles, cooling rates, and solidification parameters from which important information can be obtained to simulate microstructure and grain growth as described below.

12.2.2   Simulation of microstructure evolution and grain growth

Modeling microstructure by computing phase fractions of various constituents helps to understand the properties of components. Each alloy undergoes unique phase transformations during heating and cooling during the multilayer additive manufacturing process. Therefore, the modeling of microstructures is highly alloy specific for representing the various phase transformations involved in the evolution of microstructure. Similarly, the morphology, orientation, and dimension of the grains affect the mechanical and chemical properties of components. Therefore, modeling microstructure evolution and grain growth (Table 12.1) is important to understand and control the structure and properties of components.

In AM, modeling of phase transformations and scale of microstructure is done considering the thermal history and alloy composition. Reliable microstructure calculations are achieved using detailed kinetic information of phase transformation manifested in continuous cooling transformation (CCT) diagrams and relations of phase fractions with time such as the Johnson Mehl Avrami (JMA) equation. Although these models provide reliable results of phase fractions, they do not provide morphological information.

Phase-field simulations [7, 8] are used to predict microstructural features in small length scales. For example, phase-field simulations of the temporal growth of lamellar alpha-phase in Ti-6Al-4V are shown in Figure 12.3 (a-d). The challenges in using these models include representation of the physical processes such as nucleation, heating, and cooling considering fluid flow, and the prescription of energy density fields, especially at boundaries. Lack of quantitative comparisons of the evolution of phase fractions with experimental data and the computationally intensive nature of the 3D calculations at part scales add to the difficulties. Cellular automata models are used to simulate dendritic growth during the solidification process, solidification morphology, and grain texture.

Models of grain growth [9-11] based on the Monte Carlo method can simulate the transition of different grain morphologies such as columnar to equiaxed grains, the variation of grain growth directions under location-dependent solidification conditions, and the solid-state grain growth under multiple thermal cycles. For example, Figure 12.4 (a) shows the computed influence of nuclei density on the grain morphology during PBF-L of stainless steel 304. The amount of equiaxed grains increases with the nuclei density and columnar to equiaxed transition (CET) is observed where the nuclei density is high, and the growth of columnar grains is terminated by the equiaxed grains. Figure 12.4 (b) shows the grain growth process during solidification of an aluminum alloy computed using a Monte Carlo method-based grain growth model. For computing the temporal evolution and spatial distribution of grain size, the 3D, transient temperature fields, fusion zone geometries, and the local temperature gradient and solidification growth rate are obtained through modeling based on which growth of columnar grains are simulated in 3D. In AM, columnar grains grow epitaxially from the partially melted grains following the maximum heat flow directions at the solidification front with the movement of the molten pool. During AM, heat primarily flows along a direction which is perpendicular to the boundary of the molten pool (Figure 12.4 c-d). Computed results (Figure 12.4 e) show that the columnar grains grow along the maximum heat flow direction which is consistent with the corresponding experimental observations (Figure 12.4 f). This 3D grain growth model can uncover the mechanisms

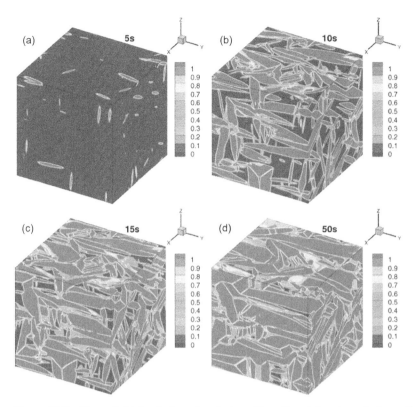

Figure 12.3 Phase-field simulation of evolution of lamellar alpha-phase during solidification of Ti-6Al-4V after (a) 5, (b) 10, (c) 15, and (d) 50 seconds of solidification [7]. The scale shows the fraction of alpha-phase. The figure is reprinted with permission from Springer Nature.

of grain structure evolution and provide comprehensive information on the morphology, dimension, and orientation of the grains and texture. Therefore, such models can also be used to understand microstructural inhomogeneity and anisotropy in mechanical properties [14, 15]. Worked out example 12.1 explains the calculation of solidification parameters to understand the grain morphology.

12.2.3 Models of defect formation

Mechanistic models to simulate common defects in AM, such as porosity, lack of fusion, surface roughness, balling, and cracking are emerging. In part-scale modeling, small features such as surface roughness, tiny gas pores, balling are not simulated. Powder scale models are suitable for simulating these defects since they can simulate very small length and time scales. Figure 12.5 (a-b) shows simulations of the void formation using a powder scale model. Pores of different shapes and sizes may be observed inside the deposited track due to the entrapment of gas during PBF and DED of commonly used AM alloys. However, in these models, the time step is often restricted to a few nanoseconds to maintain computation convergence at small grid spacing and high liquid metal flow velocities. Thus, these models often take a day or more in multiprocessor computers to simulate very small domains.

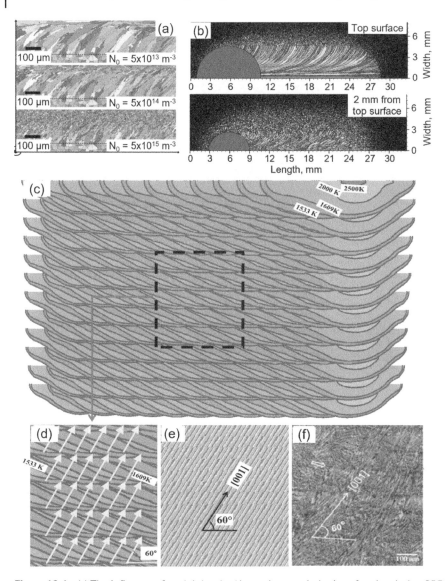

Figure 12.4 (a) The influence of nuclei density, N_0, on the morphologies of grains during PBF- L of SS 304 stainless steel shows that columnar to equiaxed transition is favored at high nuclei density. Figure courtesy: Professor Wenda Tan from University of Michigan, Ann Arbor. (b) Monte Carlo simulation of grain growth during gas tungsten arc welding of 1050A aluminum alloy shows that columnar grains may appear equiaxed at certain cross-sections because of the curvature of the columnar grains. The results are generated using a model developed by T. DebRoy and H.L. Wei. (c-f) Prediction of columnar grain growth along the maximum heat flow direction during DED-L of Inconel 718 [11]. Figures (c-f) are taken from an open-access article [11] under the terms and conditions of the Creative Commons Attribution (CC BY) license.

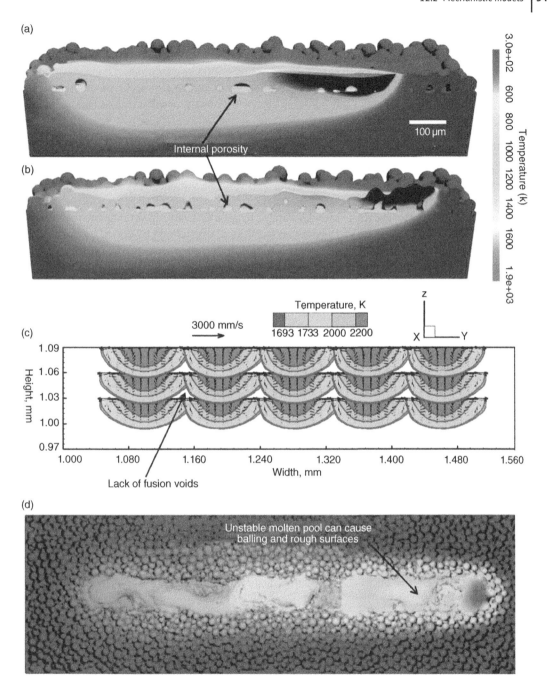

Figure 12.5 (a-b) Powder-scale modeling of defect formation such as voids in PBF-L of Ti-6Al-4V [12]. The figures are reprinted with permission from Elsevier. (c) Part-scale modeling of lack of fusion defects during PBF-L of stainless steel 316. The figure is generated using a model developed by T. Mukherjee and T. DebRoy. (d) Powder-scale modeling of fluctuation of the molten pool which may cause balling and rough surfaces [13]. The figure is reprinted with permission from Elsevier.

Worked out example 12.1

Long columnar grains are often observed in additively manufactured parts. Columnar grains are often undesirable because they may cause solidification cracking (see Chapter 10) and anisotropy in mechanical properties (see Chapter 9). In practice, process parameters are adjusted by trial and error to form equiaxed grains and the process is called columnar to equiaxed transition (CET). Explain how CET can be predicted using a mechanistic model.

Solution:

CET depends on the ratio of the temperature gradient (G) to solidification growth rate (R) which can be calculated using a heat transfer and fluid flow model [16]. Every alloy has a threshold value of G/R ratio above which columnar grains are formed. Heat transfer and fluid flow models are used to identify process conditions to get a G/R value lower than the threshold value to achieve CET. Figure E12.1 provides an example of predicting CET during fusion welding of aluminum alloy 6082. A heat transfer and fluid flow model was used to compute the variation in G/R along the width of the fusion zone. For aluminum alloy 6082, the threshold value of G/R is around 9 Ks/mm^2. G/R values higher than 9 Ks/mm^2 resulted in columnar grains near the edges of the fusion zone [16]. Equiaxed grains were found near the centerline of the fusion zone. The observation made based on the variations in G/R was consistent with the results from a Monte Carlo-based grain growth model and experiment.

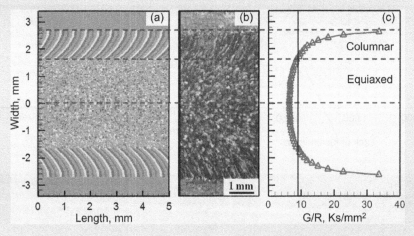

Figure E12.1 Observation of CET during fusion welding of aluminum alloy 6082 using (a) Monte Carlo-based grain growth model, (b) experiment, and (c) G/R ratio. *Source:* T. DebRoy and H.L. Wei.

Worked out example 12.2

Formation of the lack of fusion voids is a serious problem in additive manufacturing. Under the same processing conditions, all alloys are not equally susceptible to lack of fusion. Explain how a mechanistic model can be used to compare the susceptibility to lack of fusion of stainless steel 316 and Ti-6Al-4V under the same processing conditions.

Solution:

For the same processing conditions, lack of fusion is affected by the pool shape and size [17]. Wider and deeper pools result in a better fusional bonding among neighboring tracks and reduce lack of fusion. For the same processing conditions, pool shape and size vary for different alloys. A heat transfer and fluid flow model can calculate the pool shape and size for different

Worked out example 12.2 (Continued)

alloys. For example, Figure E12.2 below shows the transverse view of the molten pools for multitrack deposits of stainless steel 316 and Ti-6Al-4V. Because of its lower density, Ti-6Al-4V exhibits larger pool which reduces the intertrack lack of fusion. Thus, a mechanistic model of AM process can be used to predict lack of fusion before performing experiments.

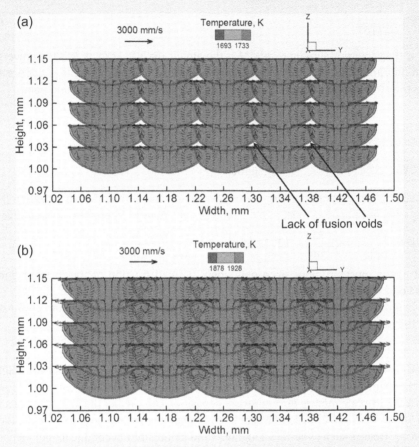

Figure E12.2 Calculation of lack of fusion using a mechanistic model of additive manufacturing for (a) stainless steel 316 and (b) Ti-6Al-4V at 60 W laser power and 1000 mm/s speed. *Source:* T. Mukherjee and T. DebRoy.

Worked out example 12.3

Keyholes are is often formed during additive manufacturing. Unstable keyholes may result in keyhole pores. Find an expression to calculate the minimum energy needed to sustain evaporation and form a keyhole during additive manufacturing of steels. Assume that the formation of keyhole in steels depends only on the evaporation of iron. Data needed for the calculations: Atomic mass of iron: 0.056 kg, boiling point of iron: 3134 K, and enthalpy of vaporization of iron: 354.1 kJ/gm mole.

(Continued)

Worked out example 12. 3 (Continued)

Solution:

The evaporative flux, J (mol/(m^{2}s)) from the molten pool can be estimated from the Langmuir equation as:

$$J = \frac{P}{\sqrt{2\pi MRT}} \tag{E12.1}$$

Here, P is the vapor pressure (in N/m^2), M is the atomic mass in kg/mol, R is the universal gas constant (in J/(mol K)), and T is the temperature (in K). At boiling point, the vapor pressure is equal to the atmospheric pressure (1 atm or 101,325 Pa). Therefore, at boiling point,

$$J = \frac{101,325 \left(\frac{N}{m^2}\right)}{\sqrt{2\pi \times 0.056 \text{ kg/mol} \times 8.314 \left(\frac{J}{molK}\right) \times 3134 \,(K)}} = 1058.5 \,\frac{mol}{m^2 s}$$

The product of the evaporative flux, J (mol/(m^{2}s)) and the enthalpy of vaporization (kJ/mol) provides the evaporative heat flux (kJ/(m^{2}s) or W/m^2). Therefore, evaporative heat

flux $= 1058.5 \,\frac{mol}{m^2 s} \times 354.1 \frac{kJ}{mol} = 0.375 \times 10^9 \,\frac{W}{m^2} = 0.375 \times 10^5 \,\frac{W}{cm^2}$

Incident power must be more since only about a small fraction of the incident energy is absorbed on a flat surface.

If we take 10% absorption, we get 0.375 million watts/cm^2 as the minimum power density needed to form a keyhole.

The lack of fusion defects originate due to improper fusional bonding among the neighboring tracks. If the shape and size of the deposited track can be accurately predicted, lack of fusion can be simulated. Part-scale models for multitrack deposition can predict the lack of fusions voids as shown in Figure 12.5 (c). Worked out example 12.2 explains a methodology to calculate the lack of fusion defects. Defects such as surface roughness and balling may occur due to molten pool instability, attachment of partially melted powders on the deposit, and improper melting and can be simulated using powder scale models (Figure 12.5d). Worked out example 12.3 explains a methodology to calculate the keyhole formation.

Cracking is a major defect in AM parts and is difficult to model because of the complexity involved in its evolution mechanism. Therefore, susceptibility to cracking is predicted by modeling those individual mechanisms but the evolution of cracking by considering all probable mechanisms simultaneously is a difficult undertaking. For example, massive cracking often occurs at the boundaries of the columnar grains whose formation can be simulated using a grain growth model. Similarly, a tensile strain that helps in the growth of a crack can be calculated using a thermomechanical model of AM process. Common defects in AM occur because of simultaneously occurring, multiple physical phenomena many of which are difficult to model. Therefore, models of defect formation vary significantly in their complexity, and in most cases, they have several simplified assumptions to make the calculations tractable.

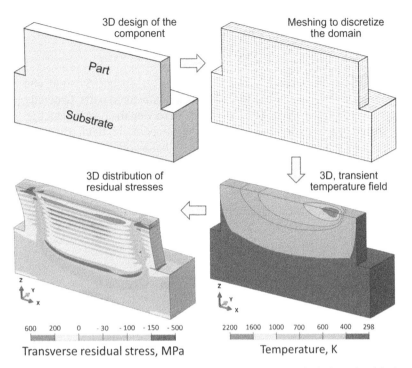

Figure 12.6 Schematic representation of the step-by-step calculation of residual stresses using a thermomechanical model. *Source:* T. Mukherjee and T. DebRoy.

12.2.4 Modeling residual stresses and distortion

Experimental prediction of the spatiotemporal evolution of residual stresses and distortion is challenging, but well-tested thermomechanical models [14] are widely used. Thermomechanical models compute residual stresses and distortion from the 3D, transient temperature fields using temperature-dependent mechanical properties of alloys. These models are primarily developed using commercial software and can be applied for large parts but are computationally intensive. They are mostly based on heat conduction models that ignore liquid metal flow which is often the main mechanism for heat transfer within the liquid pool. More accurate models that consider convective heat transfer arc emerging with improvements in computational software and hardware. Such models, at first, accurately calculate the 3D, transient temperature fields in the discretized solution domain by considering the effects of the convective flow of molten metal. Then, the geometry, meshing, and computed temperature fields are used to predict residual stresses and distortion (Figure 12.6). Where computationally intensive calculations are not practical, back-of-the-envelope analytical calculations provide a means to mitigate distortion. Worked out example 12.4 explains how to predict residual stresses in additively manufactured large components.

Worked out example 12.4

Large aircraft frames have been printed in laboratories. To increase the productivity, a large volume of material is deposited per unit time that often results in an accumulation of high residual stresses. Explain how to predict residual stresses in additively manufactured aircraft frames.

(Continued)

Worked out example 12.4 (Continued)

Solution:

Thermomechanical models using commercial packages such as Ansys, Abaqus, Sysweld, and Altair are often used to calculate residual stresses in additively manufactured parts. Generally, these models first calculate the 3D transient temperature fields by solving the heat conduction equation and stresses are computed subsequently from the temperature field using the mechanical properties of the material. For example, Figure E12.3 below provides an example of calculation of residual stresses in a 3-meter long large aircraft frame of aluminum alloy 2219 made using laser-wire-based additive manufacturing. However, calculation of residual stresses for such a large part is time-consuming. For this case, the simulation using Abaqus took 120 hours in a computer with Intel Xeon Gold 6140 CPU (2.3 GHz and 36 cores) and 64 GB RAM.

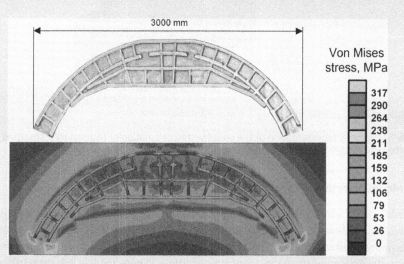

Figure E12.3 Calculation of residual stresses using Abaqus for a 3 meters long aircraft frame of aluminum alloy 2219 made using laser-wire-based additive manufacturing [18]. The figure is reprinted with permission from Elsevier.

12.2.5 Prediction of component properties

Predictions of mechanical properties such as yield strength, hardness, and toughness of the component are complex because these properties depend on microstructural features, solidification morphologies, precipitates, grain structure, texture, presence of defects, and many other factors. In addition, inhomogeneity in the microstructure in AM often results in anisotropy in properties. For example, the presence of long columnar grains along the vertical direction (build direction) decreases the tensile strength along that direction compared to that along the horizontal direction (scanning direction). Despite these difficulties, several models are available to predict mechanical properties based on several simplifying assumptions. For example, hardness can be predicted from the computed thermal cycle and isothermal heat treatment data (Figure 12.7). Accurate thermal cycles are calculated using a heat transfer and fluid flow model of AM process. Hardness is calculated using the temperature data in a JMA equation [15] where the alloy-specific parameters are predicted from an isothermal heat treatment experiment of that alloy. However, this approach

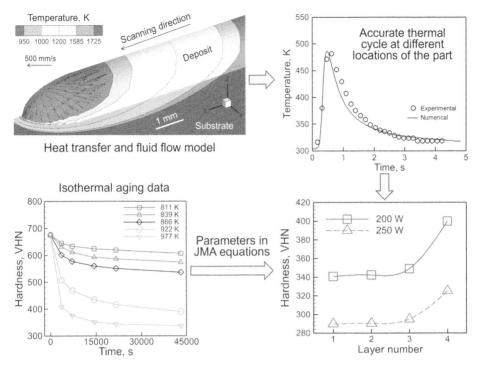

Figure 12.7 Schematic representation of the step-by-step calculation of hardness using a property prediction model. *Source:* T. Mukherjee and T. DebRoy.

can only be applied for predicting simple properties like hardness in simple alloy systems such as ferritic-martensitic steel or precipitation hardened nickel and aluminum alloys. A rigorous model for the prediction of mechanical properties by considering all factors contributing to the property is not yet available and needs further work.

12.2.6 Available computational tools

Several commercial computational tools are available for calculating temperature fields, deposit geometry, cooling rates, microstructural features, grain growth, the evolution of defects, and residual stresses and distortion. There are a few open-source packages, such as OpenFOAM that can calculate temperature fields and deposit geometry. Table 12.2 summarizes several commercially available computational tools used in AM simulations along with their salient features and applications in AM.

12.3 Machine learning

Additive manufacturing is still a developing field and the detailed mechanisms to explain various part attributes such as microstructures and properties are largely unknown. Machine learning can synthesize available data and discover hidden relations among process variables and product attributes and use the relations to make useful predictions without any phenomenological understanding. Machine learning extracts useful information and relations hidden within this data

Table 12.2 Available computational tools for mechanistic modeling of AM.

Computational tools	Features	Applications in additive manufacturing
Abaqus	Based on the finite element method For part scale simulations Contains a module for AM https://www.3ds.com/products-services/simulia/products/abaqus	Calculations of temperature field, deposit geometry, residual stresses, and distortion.
Ansys	Based on the finite volume method For part scale simulations https://www.ansys.com	Calculations of temperature field, deposit geometry, residual stresses, and distortion.
COMSOL Multiphysics	Based on the finite element method For part scale simulations https://www.comsol.com	Calculations of temperature and velocity fields, deposit geometry, residual stresses, distortion, and simple defects.
Sysweld	Based on the finite element method For part scale simulations Originally developed for welding https://www.esi-group.com/products/cfd-multiphysics	Calculations of temperature field, deposit geometry, residual stresses, and distortion.
Altair	Based on the finite element method For part scale simulations AM module for process optimization https://www.altair.com	Calculations of temperature field, deposit geometry, residual stresses, and distortion.
OpenFOAM	Ideal for powder scale modeling Open-source software https://openfoam.org	Temperature field, fusion zone geometry, powder size distribution, and powder packing in PBF. Modeling of powder flow in DED
Flow 3D	Powder scale modeling https://www.flow3d.com	Powder spreading, solidification, microstructural evolution, melt pool dynamics, and porosity formation

without any phenomenological guidance or explicit programming. The accuracy of the predictions improves with the quality and volume of data without developing any new algorithm. The availability of powerful open-source programs facilitates their applications for solving many complex problems in additive manufacturing that may appear intractable at first. There are several important problems in AM where a database of results can be used to train machine learning algorithms to help analyze results and serve as predictive tools. In this subsection, we discuss the common machine learning algorithms, indicate the availability of open-source algorithms and codes, describe important applications in AM, and provide worked out examples to show how machine learning calculations can be done using open-source codes.

12.3.1 Machine learning algorithms

There are two types of machine learning algorithms, supervised and unsupervised, both of which are used in additive manufacturing (Table 12.3). To implement supervised machine learning algorithms, a training dataset with both input data and the corresponding output data determines the

Table 12.3 Common machine learning algorithms used in additive manufacturing [2].

Algorithms	Description and features	Applications in additive manufacturing
Artificial neural networks	Layers of hidden nodes connect input and output variables. An activation function is used to connect nodes with each other. Errors in predictions are minimized by adjusting weights for each connection.	Defect recognition; geometry prediction; thermal deformation compensation; process parameter optimization; anomaly detection, quality monitoring; topology optimization.
Decision tree	Progressively classifies a group of variables based on rules and displays them as an upside-down tree. The root of the tree often displays the most important variable and the apex shows the least important one.	Surface roughness reduction; porosity prediction; dimensional variation; printing speed modeling; design considering residual stresses and support requirement.
Support vector machines (SVM)	Used for classification and regression, it can split each data into one group or other based on its location in feature space. The features of data fully determine its location and there is no stochastic element involved.	Defect detection; real-time composition monitoring; surface roughness; tensile strength prediction; construction of process maps; monitoring temperature field.
Bayesian networks (BN) or Bayesian classifiers (BC)	A statistical model that represents probabilistic relation between cause and effect. The conditional probabilities are computed using Bayes' theorem.	Quality inspection; fault diagnosis; thermal field prediction; porosity prediction; optimization of process parameters; fusion zone depth prediction.
K-nearest neighbor (K-NN)	Separates data into different classes based on the attributes or the class of the majority of the nearest neighbors. The number of nearest neighbors, k, is selected by trial and error.	Quality monitoring; printing speed monitoring; porosity prediction; dimensional variation; design of metamaterials.
Random forest	It consists of multiple decision trees, each with a classification. The forest gets a classification from the attributes of the greatest number of trees. For regression, it considers the average of the outputs of different trees.	Surface roughness determination; tensile strength prediction; reducing macro porosity and cracks; printing speed monitoring; minimize porosity by optimizing parameters.

hidden relationship through the use of an appropriate algorithm. Once trained, the algorithm can predict the outputs for specified input data. Supervised machine learning often requires some prior knowledge of the influential factors (e.g., process parameters) that affect a property (e.g., pool geometry), and the goal is to establish a quantitative relationship that can accurately predict a property for any set of input data. Commonly used supervised machine learning in AM applications include neural networks, support vector machines, Bayesian networks, random forest, k-nearest neighbor, and decision tree.

Unlike supervised machine learning, unsupervised machine learning identifies features or relationships within the data fed into the algorithm. This is useful in situations where causative factors are unknown but there is still a need to find order among sets of data. These features/relationships are useful for obtaining important correlations that are often qualitative. A common application of unsupervised machine learning is the classification of features or objects within a digital image or output signal from a sensor. This could be used to identify unique features from the output of in situ sensors, such as classifying various features of a powder bed surface. The important applications of machine learning algorithms in additive manufacturing are discussed below.

12.3.2 Common applications in additive manufacturing

The rapid adoption and application of machine learning in additive manufacturing have been driven by both the need to control the process and product attributes and the availability of powerful open-source codes. Recent applications ranged from process parameter optimization, sensing and control, to improving fusion zone attributes, tailoring microstructures, and mitigating defects. These applications are broadly classified into two categories. In regression-based problems, a mathematical correlation between the input and output variables is developed. Classification problems are used to divide the cases into two or more classes. The examples in this subsection illustrate the importance of machine learning in addressing both regression and classification problems in additive manufacturing. In addition, we provide worked out examples 12.5 to 12.13 that will help readers to know how to use the common regression-based and classification-based algorithms to solve important problems in additive manufacturing.

12.3.2.1 Regression problems

The most important regression problems in additive manufacturing include controlling part geometry, fusion zone dimensions, and controlling microstructure and properties. Part geometry often deviates from the design due to process instability and distortion and may result in part rejection in extreme cases. Machine learning is often used to control part geometry during AM. For example, measured data on deposit geometry are used to correlate with the processing conditions and alloy properties using a neural network. The trained neural network can be later used to predict part geometry for a new set of processing conditions without performing new experiments. Such a trained network can also be used during printing to control part geometry by in-situ adjustment of processing conditions. In addition, machine learning is used to reduce the geometric deviation of parts by optimizing the process parameters. The process parameter selection is an important factor in controlling part geometry, and the quality of the data-based optimization will be enhanced as more data are accumulated with time. The applications of machine learning for controlling part geometry improve compliance with the geometric specifications of the design ultimately facilitating part qualification.

Microstructural features such as grain size, distribution, orientation, and properties such as tensile strength, hardness, and fatigue life are used to correlate with the processing conditions using machine learning [19] [20]. Figure 12.8 provides an example of using machine learning to predict and control grain size. The figure shows that grain size and distributions were computed using a grain growth

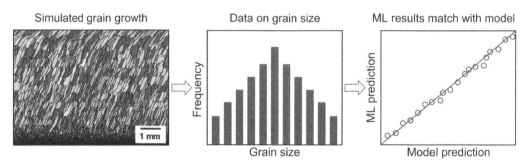

Figure 12.8 An example of the application of machine learning in microstructure control. From the results of the model, the grain growth is visualized in 2D and the data on the frequency versus grain size are extracted. These data are used to train a neural network that could then correctly calculate grain growth. *Source:* T. Mukherjee and T. DebRoy.

model. From the results of the model, the grain growth was visualized in 2D and the data on the frequency versus grain size were extracted. A neural network trained with the average grain size and process variables can rapidly predict the grain size for different sets of process variables. The predictions of the neural network matched well with those from the computationally intensive grain growth model. Thus, a neural network-based machine learning is often used to predict and control grain size.

Worked out example 12.5

In additive manufacturing, molten pool dimensions are influenced by process parameters such as heat source power, scanning speed, and heat source spot size. For example, Table T12.1 in Appendix A3 provides a set of data on the variations in molten pool width for various combinations of power, speed, and spot size. Use a suitable machine learning algorithm to provide a quantitative relation between the pool width and power, speed, and spot size which can be used to predict width for new processing conditions.

Solution:

Linear regression is a simple machine learning algorithm that can be used to provide such quantitative relation. The calculation can be performed using the regression tool in MS Excel. The regression tool can be found under the "Data Analysis" section of the "Data" tab. The data under columns P, S, and D need to be selected as "Input X Range" and the data under "Pool width" should be considered as "Input Y Range". Using the data provided, the relation is obtained as:

$$\text{Pool width} = 0.84 + 4.4 \times 10^{-4} P - 2.4 \times 10^{-4} S - 0.49\,D$$

where P is power in W, D is spot diameter in mm, and S is scanning speed in mm/s. The above correlation is valid for the range of values indicated in Table T12.1. The coefficient of power is positive because an increase in power results in a wider pool. In contrast, negative coefficients of speed and spot size indicate a diminishing effect of these two variables on pool width. The above equation can be used to predict pool width for a new set of power, speed, and spot size within their range in Table T12.1 in Appendix A3. Predicted values of pool width using the above equation for 27 cases (Table T12.1) are provided in Table T12.1 and plotted against the actual values of the pool width in Figure E12.4 below. The mean absolute error in predicting the pool width is 0.014 mm (Table T12.1).

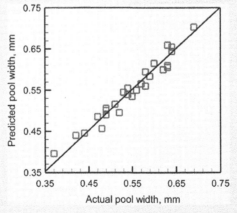

Figure E12.4 A plot showing the predicted values of pool width using a linear regression vs. the actual pool width. *Source:* T. Mukherjee and T. DebRoy.

Worked out example 12.6

Vickers hardness of additively manufactured parts of different aluminum alloys are given in Table E12.1 below along with their composition (in wt.%). Use a linear regression to provide a quantitative relation between the hardness and chemical composition which can be used to predict hardness for other aluminum alloys.

Solution:

Linear regression is a simple machine learning algorithm that can be used to provide quantitative relations. The calculation can be performed in MS Excel as explained in worked out example 12.5. Using the data provided, the relation is obtained as:

$$\text{VHN} = 103.4 - 2.2\,\text{Cu} - 19.8\,\text{Fe} - 0.07\,\text{Mg} + 31.9\,\text{Mn} + 0.8\,\text{Si} - 124.1\,\text{Ti} + 168.4\,\text{Zn}$$

The above correlation is valid for the range of values indicated in the above table. Predicted values of the hardness using the above equation for different aluminum alloys are plotted against the actual values of the hardness in Figure E12.5 below.

Table E12.1 Vickers hardness of aluminum alloy parts for various compositions (in wt.%).

Cu	Fe	Mg	Mn	Si	Ti	Zn	VHN
5.3	0.08	0.52	0.31	0.051	0.064	0	92
0	0.55	0.4	0.45	10	0	0.1	136
0.3	0.8	0.1	0.15	12	0	0.2	135
0.003	0.12	0	0	12.2	0	0	111
0.05	0.25	0.4	0.1	10	0.1	0.1	114
0.1	0.55	0.4	0.45	10	0	0.1	127
0.001	0.16	0.35	0.002	10.08	0.01	0.002	107
0.08	0.36	0	0	12.1	0	0	106
4.47	0	1.95	0.55	0	0	0	111

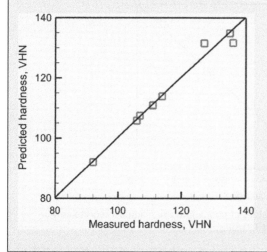

Figure E12.5 A plot showing the predicted values of hardness using a linear regression vs. the actual hardness. *Source:* T. Mukherjee and T. DebRoy.

Worked out example 12.7

In additive manufacturing, selective vaporization of alloying elements results in a composition change (see Chapter 10). Vaporization of elements is affected by the vapor pressure which is a function of temperature. Figure E12.6 (a) shows the measured vapor pressure of an element at different temperatures where a discontinuity is observed near the boiling point. Use nonlinear regression to derive an equation correlating the vapor pressure with temperature. Explain under fitting and over fitting using this example.

Solution:

Figure E12.6 (b) shows a nonlinear regression with power fitting. The R^2 value of 0.95 indicates the fitting is not optimum. Figure E12.6 (c) shows a nonlinear regression with cubic polynomial fitting. The R^2 value of 0.99 indicates a good fitting. However, the same data can be fitted using a polynomial of order 6 (Figure E12.6 (d)) which also gives a R^2 of 0.99. The fitting does not improve but the polynomial becomes more complex. Therefore, figure (c) shows the best fitting and figures (b) and (d) indicate under fitting and over fitting, respectively.

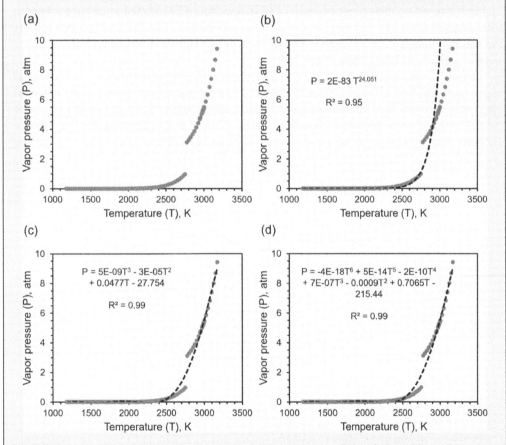

Figure E12.6 Vapor pressure vs. temperature data (a) data, (b) power fitting, fitting with polynomials of degree (c) three and (d) six. *Source:* T. Mukherjee and T. DebRoy.

Worked out example 12.8

Train a neural network for predicting molten pool width from power, speed, and spot diameter. Use the data in Table T12.1 in Appendix A3 which is also used in Worked out example 12.5 to predict pool width from power, speed, and spot diameter using a linear regression. Compare the results from the neural network with those from the linear regression used in Worked out example 12.5.

Solution:

An open-source machine learning package WEKA (see Appendix A2) can be used to train a neural network. In WEKA, neural network module is called "Multilayer Perceptron" which can be found under the "Classify" menu. The neural network parameters such as numbers of hidden layers, hidden nodes, hyperparameters, learning rates can be adjusted from the "Multilayer Perceptron" tab (see Figure 12.7 below). The calculation can be started by clicking on "Start" (see Figure 12.7 below).

The trained neural network can be used to predict pool width for a new set of power, speed, and spot size within their range in Table T12.1 in Appendix A3. Predicted values of pool width using neural network for 27 cases (Table T12.1) are provided in Table T12.1 and plotted against the actual values of the pool width in Figure E12.8 below. The mean absolute error in predicting the pool width is 0.012 mm (Table T12.1).

The mean absolute error (see Figure E12.9 below) is slightly less than the error in linear regression (0.014 mm). For examples like predicting pool width from power, speed, and spot size, neural network does not provide a significant advantage over linear regression. For these cases, the use of linear regression is easy and time-efficient and can provide accurate results. However, for complex, nonlinear problems neural network is a better choice than regression.

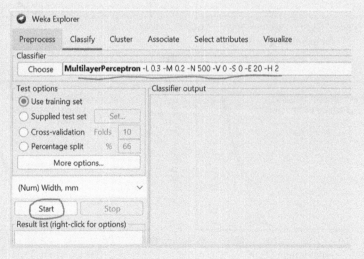

Figure E12.7 Running neural network in WEKA.

Worked out example 12.8 (Continued)

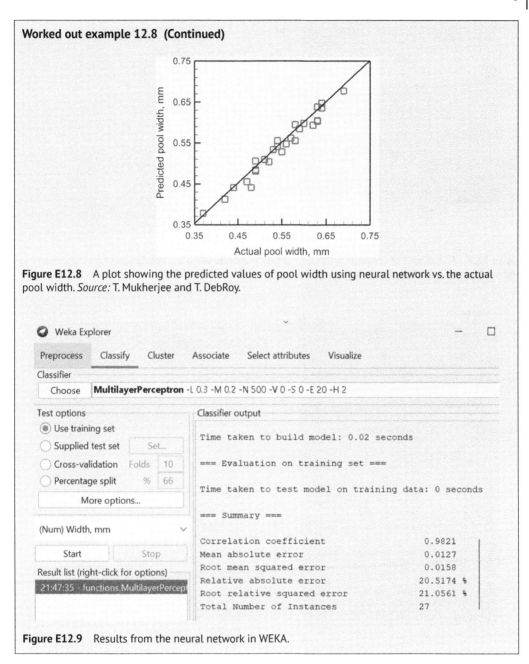

Figure E12.8 A plot showing the predicted values of pool width using neural network vs. the actual pool width. *Source:* T. Mukherjee and T. DebRoy.

Figure E12.9 Results from the neural network in WEKA.

12.3.2.2 Classification problems

The most important classification problem in additive manufacturing is to predict and control common defects such as balling, cracking, porosity, lack of fusion, distortion, and surface roughness. For example, a support vector machine-based machine learning was used to minimize the lack of fusion defects in Ti-6Al-4V parts fabricated using DED-L [21]. In that research, the temperature field during the DED-L process was monitored using an IR camera. A molten pool boundary was extracted from the temperature field by tracking the solidus temperature contour. A support vector machine was trained using the data extracted from the molten pool results. The trained

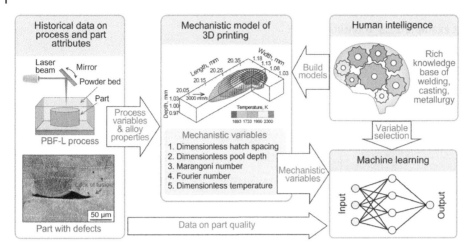

Figure 12.9 An example of controlling the lack of fusion defects using a combination of machine learning and mechanistic modeling. *Source:* M. Jiang, T. Mukherjee, Y. Du, and T. DebRoy.

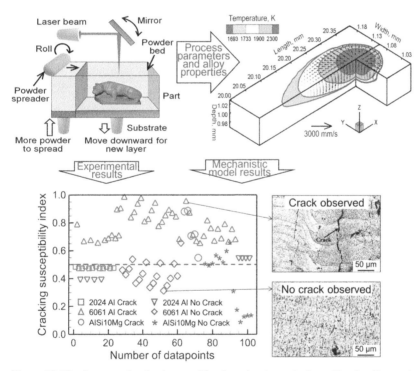

Figure 12.10 An example of using machine learning to control cracking in alloy parts made by PBF-L. *Source:* B. Mondal, T. Mukherjee, and T. DebRoy.

support vector machine was used to classify the processing conditions into two categories, normal and abnormal, based on the probability of lack of fusion defect formation. Defects were found in the components printed using the abnormal processing conditions responsible for the defect formation. Figure 12.9 also provides another example of controlling lack of fusion defects using a combination of machine learning and mechanistic modeling.

Figure 12.10 provides an example of using a machine learning algorithm to reduce cracking during PBF-L. Machine learning algorithms were trained using both the experimental data on crack formation as well as several computed variables that represent the mechanism of cracking. These variables were calculated using a heat transfer and fluid flow model. A cracking susceptibility index [22] was derived to provide a threshold value classifying the crack and no crack cases. The index could identify conditions for reducing cracking during PBF-L of a wide variety of engineering alloys. In addition, feature selection-based machine learning was used to provide the hierarchical influence of important variables on cracking. This hierarchy is helpful to guide engineers about which variables to adjust for controlling crack formation. Machine learning has provided useful guidelines to control the common AM defects, the physics behind the formation of which is often not known.

Worked out example 12.9

In additive manufacturing, discontinuity in the deposited track often causes balling defect and rough surfaces (see Chapter 10). Discontinuity in track occurs due to the improper melting of the feedstock materials which is affected by a poor selection of the heat source power, scanning speed, and heat source spot size. Table T12.2 in Appendix A3 provides a set of data on the continuity of the deposited tracks for various combinations of power, speed, and spot size. Use a logistic regression to classify the continuous and discontinuous tracks based on power, speed, and spot size.

Solution:
Logistic regression is used for classification problems. It correlates the probability of occurrence of a particular case with the input variables using a logarithmic equation. An open-source machine learning package WEKA (see Appendix A2) can be used for performing the analysis using a logistic regression. In WEKA, logistic regression module can be found under the "Classify" menu. The parameters for the logistic regression can be adjusted from the "Logistic" tab. The calculation can be started by clicking on "Start". The calculation using WEKA provides the coefficients for the logistic regression (see Figure E12.10 below) that can be used to express the probability of occurrence of class "C" as:

$\ln (p/(1 - p)) = 170.08 + 0.39 \times$ power $- 0.156 \times$ speed $- 390.11 \times$ spot diameter

Here, "p" denotes the probability of occurrence of class "C" (continuous track).

The negative coefficients for the speed and spot diameter indicate that high values of these two variables reduce the probability of continuous track formation. High speed results in a longer molten pool that may be unstable and break into small isolated pools to cause discontinuous tracks. Energy density is inversely proportional to the square of spot size. Large spot reduces the energy density, causes improper melting, and results in a discontinuous track.

The logistic regression equation can be used to determine whether the track will be continuous or not for a new set of power, speed, and spot size within their range in Table T12.2 in Appendix A3. The probability of continuous track "p" for all cases (Table T12.2) are plotted along with the actual experimental observations in Figure E12.11 below. The logistic regression predicts the occurrence of continuous or discontinuous tracks for all data accurately.

(Continued)

Worked out example 12.9 (Continued)

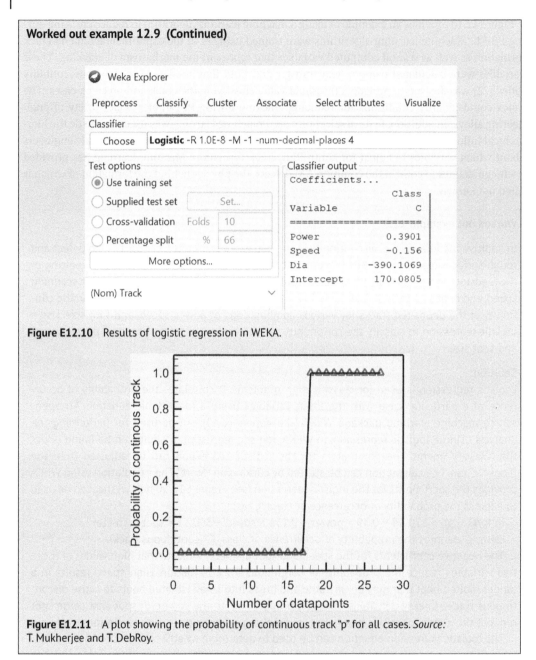

Figure E12.10 Results of logistic regression in WEKA.

Figure E12.11 A plot showing the probability of continuous track "p" for all cases. *Source:* T. Mukherjee and T. DebRoy.

Worked out example 12.10

In additive manufacturing, discontinuity in the deposited track often causes balling defect and rough surfaces (see Chapter 10). Discontinuity in track occurs due to the improper melting of the feedstock materials which is affected by a poor selection of the heat source power, scanning speed, and heat source spot size. Table T12.2 in Appendix A3 provides a set of data on the continuity of the deposited tracks for various combinations of power, speed, and spot size. Use a suitable machine learning algorithm to identify which variable among the three variables has the highest influence on the track continuity.

Worked out example 12.10 (Continued)

Solution:

Iterative Dichotomiser 3 algorithm is a machine learning algorithm that calculates an index called information gain to rank the input variables based on their relative influence on the output. Here, input variables are power, speed, and spot size. The output is binary, continuous or discontinuous. The calculation of information gain is based on a factor called entropy (S) which is defined by [23]:

$$S = -\sum_{i=1}^{n} P(x_i) \log_2 P(x_i)$$ (E12.2)

where the outputs $x_1, x_2, \ldots x_n$ have respective probabilities $P(x_1)$, $P(x_2), \ldots P(x_n)$. In this example, we have two types of outputs (n = 2) indicating the continuity and discontinuity, respectively. In the data set of 27 datapoints (Table T12.2), there are 17 cases with discontinuity and 10 cases where tracks are continuous. Therefore, from Eq. E12.2, the total entropy is

$$S = -\left(\frac{17}{27}\right)\log_2\left(\frac{17}{27}\right) - \left(\frac{10}{27}\right)\log_2\left(\frac{10}{27}\right) = 0.951$$

The information gain should be calculated on normalized values of the variables where normalized value = (value − minimum value)/(maximum value − minimum value). The normalized values of the variables are also provided in Table T12.2.

The information gain of a particular variable (V) is given by [23]

$$IG = S - \sum_{i=0}^{n} P(V) \times S(V)$$ (E12.3)

where $P(V)$ is the probability of the variable for a data set and $S(V)$ is the entropy of the variable. The variables can have values greater (indicated by suffix "high") or smaller (indicated by suffix "low") than a threshold value and accordingly we will have $P(V_{high})$, $P(V_{Low})$, $S(V_{high})$ and $S(V_{low})$. S is the entropy of the total data set before partition (Eq. E12.2).

For power, the calculation of information gain is explained as follows:

The threshold value for power is 0.5 (taken as the midvalue of the range of the normalized values). Using this threshold value for power, out of 27 total data points, 18 cases that have values equal to or higher than the threshold (V_{high}) and 9 cases that have values lower than the threshold (V_{Low}).

$$P(V_{high}) = \frac{18}{27} = 0.67 \text{ and } P(V_{Low}) = \frac{9}{27} = 0.33$$

Out of the 18 cases of V_{high}, 9 cases are continuous and the remaining 9 cases are discontinuous. Thus, entropy $S(V_{high})$ is

$$S(V_{high}) = -\left(\frac{9}{18}\right)\log_2\left(\frac{9}{18}\right) - \left(\frac{9}{18}\right)\log_2\left(\frac{9}{18}\right) = 1.0$$

Similarly, out of the 9 cases lower than the threshold value, 1 case is continuous and the remaining 8 cases are discontinuous. Thus, entropy $S(V_{Low})$ is

$$S(V_{low}) = -\left(\frac{1}{9}\right)\log_2\left(\frac{1}{9}\right) - \left(\frac{8}{9}\right)\log_2\left(\frac{8}{9}\right) = 0.503$$

(Continued)

Worked out example 12.10 (Continued)

The information gain for the variable power is thus calculated using Eq. E12.3 as

$$IG = S - P(V_{High}) \times S(V_{High}) - P(V_{Low}) \times S(V_{Low}) \qquad \text{(E12.4)}$$

$$= 0.951 - 0.67 \times 1.0 - 0.33 \times 0.503 = 0.116$$

Using the same procedure, the information gain for speed and spot size are calculated as 0.135 for both the variables. Therefore, speed and spot size are equally important and both are more important than power. High speed results in a longer molten pool that may be unstable and break into small isolated pools to cause discontinuous tracks. Energy density is inversely proportional to the square of spot size. Large spot reduces the energy density, causes improper melting, and results in a discontinuous track. This example explains how to evaluate the relative influence of variables using machine learning. Readers are encouraged to build their own data sets using the data from the literature and practice this exercise.

Worked out example 12.11

Information gain ratio is a commonly used index to evaluate the hierarchical importance of variables. Use information gain ratio to identify which variable among power, speed, and spot size in the worked out example 12.10 has the highest influence on the track continuity.

Solution:

Information gain ratio (IGR) is defined by [23]:

$$IGR = IG / S(V) \qquad \text{(E12.5)}$$

where IG is the information gain (worked out example 12.10) and $S(V)$ is the entropy of the variable. For the variable, power, $IG = 0.116$ (worked out example 12.10). Using this threshold value for power (0.5), out of 27 total data points, 18 cases that have values equal to or higher than the threshold and 9 cases that have values lower than the threshold. Therefore,

$$S(V) = -\left(\frac{18}{27}\right)\log_2\left(\frac{18}{27}\right) - \left(\frac{9}{27}\right)\log_2\left(\frac{9}{27}\right) = 0.918$$

The information gain ratio for the variable power is thus calculated using Eq. E12.5 as

$$IGR = 0.116 / 0.918 = 0.127$$

Using the same procedure, the information gain ratio for speed and spot size are calculated as 0.147 for both the variables. Therefore, speed and spot size are equally important and both are more important than power which is consistent with Worked out example 12.10.

Worked out example 12.12

Gini index is commonly used to evaluate the hierarchical importance of variables. Use Gini index to identify which variable among power, speed, and spot size in the Worked out example 12.10 has the highest influence on the track continuity. Compare the results with those in the Worked out examples 12.10 and 12.11.

Worked out example 12.12 (Continued)

Solution:

The threshold value for power is 0.5 (taken as the midvalue of the range of the normalized values). Using this threshold value for power, out of 27 total data points, 18 cases that have values equal to or higher than the threshold (V_{high}) and 9 cases that have values lower than the threshold (V_{Low}).

Gini index (G) is defined by [23]:

$$G = 1 - \sum_{i=1}^{n} (P(x_i))^2 \tag{E12.6}$$

where the outputs $x_1, x_2, \dots x_n$ have respective probabilities $P(x_1), P(x_2), \dots P(x_n)$. In this example, we have two types of outputs (n=2) indicating the continuity and discontinuity, respectively.

Out of the 18 cases of V_{high}, 9 cases are continuous and the remaining 9 cases are discontinuous. Thus, entropy $G(V_{high})$ is

$$G\left(V_{high}\right) = 1 - \left[\left(\frac{9}{18}\right)^2 + \left(\frac{9}{18}\right)^2 \right] = 0.5$$

Similarly, out of the 9 cases lower than the threshold value, 1 case is continuous and the remaining 8 cases are discontinuous. Thus, entropy $G(V_{Low})$ is

$$G(V_{low}) = 1 - \left[\left(\frac{1}{9}\right)^2 + \left(\frac{8}{9}\right)^2 \right] = 0.2$$

Therefore, overall Gini index for the variable power = $0.5 \times \frac{18}{27} + 0.2 \times \frac{9}{27} = 0.4$

Using the same procedure, the Gini index for speed and spot size are calculated as 0.38 for both the variables. The variable with the least Gini index is the most important variable [23]. Therefore, speed and spot size are equally important and both are more important than power which is consistent with Worked out examples 12.10 and 12.11.

Worked out example 12.13

Lack of fusion defects originate from the improper fusional bonding among the neighboring tracks and layers creating void space in the part. In powder bed fusion, for a given layer thickness and hatch spacing, fusion bonding among the neighboring tracks is affected by the molten pool width and depth. Table T12.3 in Appendix A3 provides a set of data on the lack of fusion defect formation for various combinations of molten pool width and depth. Use a suitable machine learning algorithm to provide a quantitative tool to predict the occurrence of lack of fusion defects from the molten pool dimensions.

Solution:

Support Vector Machines (SVM) can be used to provide a quantitative tool to predict a binary output. For this example, presence of defects and absence of defects represented by "Y" and "N," respectively, in Table T12.3 in Appendix A3. SVM can provide an equation of a hyperplane that delineates the two classes of data. For a new set of data, when inserted in the equation of the hyperplane, determines which side of the plane the data lies. SVM calculations can be done in an open-source package WEKA using the module SMO (Sequential Minimal Optimization) as shown in Figure E12.12.

(Continued)

Worked out example 12.13 (Continued)

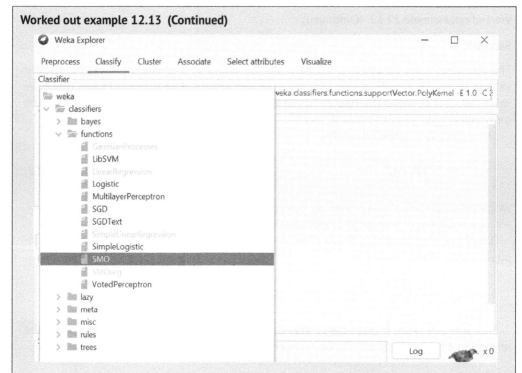

Figure E12.12 Use of SVM in WEKA by selecting "SMO" in the "Classify" module.

The SMO parameters can be changed from the SMO tab. The three changes are highlighted in Figure E12.13 below. A polynomial kernel is used for simplicity. Other complex kernels such as RBF kernel with an exponential function are also available.

The output of the calculations is the equation of the hyperplane (see Figure E12.14) as:
Hyperplane (Z) = 41.7663 − 0.0423 (Pool depth) − 0.042 (Pool width).

The plane is plotted in 3D in Figure E12.15 (a) where it intersects with Z = 0 plane along a line indicated by a dashed line. This intersecting line delineates the two classes of data. Figure E12.15 (b) shows the top view of Figure E12.15 (a) where the dashed intersecting line can correctly delineate all 80 data points. From Figure E12.15 (a), it is evident that one class of data lies above Z = 0 and the other class lies below Z = 0. The values of Z are calculated for all 80 data points (Table T12.3). It is shown that all Z values for defects are positive, and for the cases with no defects the Z values are negative.

The classification accuracy of SVM is 100% as evident from Figure E12.15 (b), the calculated values of Z in Table T12.3 and from the WEKA result (see Figure E12.16 below).

This problem provides a simple example of applying SVM where there are only two input variables. In addition, the equation of the hyperplane can be accurately represented by a linear function of the input variables. However, there may be problems where the output variable depends on more than two inputs. In such cases, the hyperplane is complex and cannot be plotted even in 3D. In addition, the equation of the hyperplane may be a complex nonlinear function of the input variables. For such problems, SVM cannot be applied in WEKA using the SMO module but another module "LibSVM" is suitable. Readers are encouraged to consult the WEKA manual to learn more about "LibSVM" and how to apply this module in WEKA.

Worked out example 12.13 (Continued)

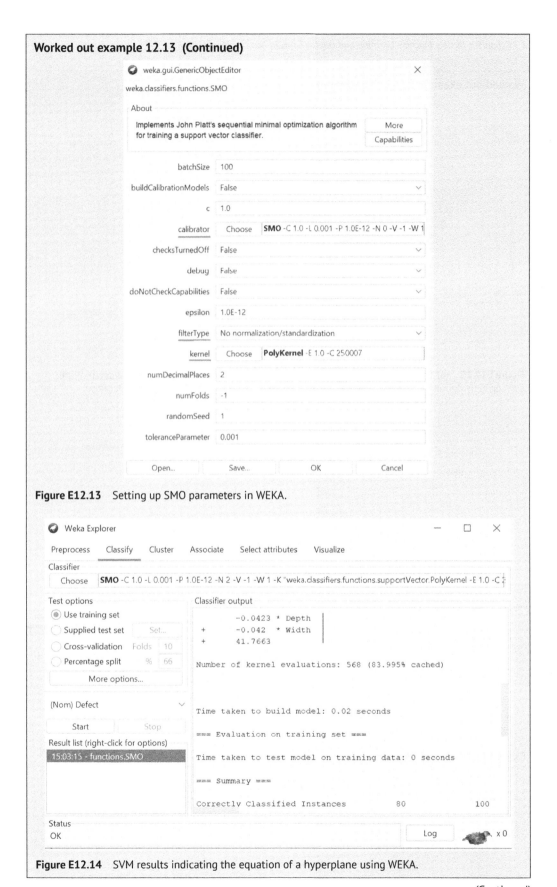

Figure E12.13 Setting up SMO parameters in WEKA.

Figure E12.14 SVM results indicating the equation of a hyperplane using WEKA.

(Continued)

Worked out example 12.13 (Continued)

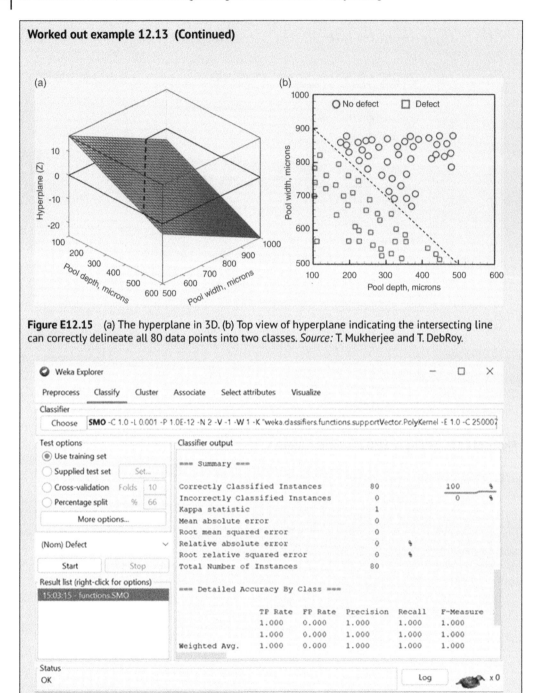

Figure E12.15 (a) The hyperplane in 3D. (b) Top view of hyperplane indicating the intersecting line can correctly delineate all 80 data points into two classes. *Source:* T. Mukherjee and T. DebRoy.

Figure E12.16 The results of SVM using WEKA.

12.3.3 Available resources

The growing applications of machine learning in AM have been facilitated by the availability of open-source programs. A list of commonly used, open-source codes is presented in Table 12.4. The table also describes the features of different algorithms and provides examples of their recent applications in AM. Classification algorithms such as decision tree, random forest, k-nearest neighbor are useful for data classifying problems such as the "detected" or "not-detected" pores in additively manufactured parts. These algorithms are also used for decision-making. Regression-based algorithms such as neural networks, Bayesian networks, support vector machines are used to correlate the inputs and outputs based on a function and can predict the values of the output variable for a set of input parameters. These open-source computer programs can be easily used because they are accompanied by extensive manuals and test cases. In this chapter, we provided several worked out examples to describe how to use an open-source code WEKA for applying machine learning in additive manufacturing.

12.3.4 Important considerations before using machine learning

Machine learning is widely used to address important issues in additive manufacturing. Machine learning can discover hidden relations among process variables and product attributes and use the relations to make useful predictions often without any phenomenological understanding. The availability of powerful open-source programs facilitates their applications for solving many

Table 12.4 Open-source codes for using machine learning in additive manufacturing [2].

Codes	Description and features	Applications in additive manufacturing
Weka	Written in Java, used for classification, clustering, regression; visualization, etc. available from https://www.cs.waikato.ac.nz/ml/weka/citing.html Online course available	Image classification-based defect detection; energy consumption in additive manufacturing; data classification; fault diagnosis; porosity reduction.
Scikit learn	Written in Python, Cython, C, C++, used for classification, clustering, regression, etc., available from https://scikit-learn.org/stable	Printing speed modeling; dimensional accuracy; temperature profile prediction; the relation between several microstructures; process monitoring; grain structure simulation.
TensorFlow	Written in Python, C++, CUDA, used for neural network and data flow programming, available from https://www.tensorflow.org	Dimensional accuracy; defect detection; mechanical behavior of structures; online part quality monitoring; melt pool images.
Keras	Cross-platform neural network library is written in Python; runs in multiple platforms, available from https://keras.io	Distortion prediction; thermal history prediction; dimensional accuracy; quality monitoring.
Theano	Written in Python for Windows, Linux, and Mac, available from http://www.deeplearning.net/software/theano	Defect detection; prediction of part weight and building time.
PyTorch	Written in Python and C++ for Windows and Linux environment, available from https://pytorch.org	Defect and anomaly detection, process monitoring.

complex problems in additive manufacturing that may appear intractable at first. However, there are several limitations to using machine learning in additive manufacturing as discussed below.

Understand the data: The success of a machine learning algorithm largely depends on the quality of the data. A data set may have outliers that need to be removed before using the dataset for machine learning. Sometimes data set can be unbalanced with significantly different volume of data under each classes. Unbalanced data should be oversampled or undersampled before usage. Users need to carefully evaluate the data to make sure that it does not have any contradictory trends. The quality of data also depends on the data collection methods. For example, the occurrence of defects in AM parts is a classification problem with two classes, parts with defects and parts where no defects are found. Even if the experimental characterization method does not identify any defects, defects may be present in the parts. Such experimental data significantly degrades the performance of machine learning.

Selection of input variables: Machine learning algorithms synthesize the data to establish correlations between the input and output variables. For example, the width of the deposited track in the DED-L of an alloy can be correlated with the process variables such as laser power, scanning speed, laser beam radius, preheat temperature, powder mass flow rate, and powder size. However, in many cases, it involves many input variables, and the selection of appropriate sets of input variables affects the accuracy of predictions. Correlation coefficients between each pair of variables are calculated. Only one variable is selected from the two highly correlated variables. This process is used to select the most relevant and inter-independent input variables. Often machine learning analysis needs a large volume of data which is time-consuming and expensive to generate. For example, recently, it has been shown that cracking in AM is affected by 12 process variables [22]. Based on the 2-factor design of experiments, it needs a minimum of 4096 experiments to study the effects of these variables. An alternative is to identify the variables that can capture the physics of the problem. Four physics-based variables that capture the combined effects of the 12 process variables were identified that required less volume of data to correctly predict cracking.

Selection of an algorithm: The machine learning algorithms are divided into two categories to address two types of problems. Regression-based problems require a mathematical correlation between the input and output variables and classification problems are used to divide the cases into two or multiple classes. Users need to understand the type of the problem and then select an appropriate algorithm. In addition, the selection of an algorithm should also consider the nature of the data, computational time, and accuracy. For example, support vector machines are good for classification problems with an unbalanced, skewed set of data because it does not depend on the entire set of the data. However, this algorithm is often computationally intensive. In contrast, a decision tree is a simpler, easy-to-use classification algorithm but often provides lower accuracy than support vector machines.

Identify the correct metric: The soundness of a machine learning analysis is evaluated by different metrics such as error metrics and coefficient of determination for regression type problems and accuracy, precision, sensitivity, and specificity for classification type problems. Let us consider an unbalanced dataset that has 95% of data for class "A" and the remaining 5% for class "B." Any classification algorithm will show 95% accuracy by predicting all class "A" cases correctly. Therefore, for this case, accuracy is not a good metric to evaluate the machine learning analysis. In contrast, sensitivity calculation can be used. Users need to identify the correct metric to evaluate a certain machine learning analysis.

Correlation does not mean causation: Two variables may show highly accurate fitting, but they may not have any scientific relation between them. For example, data may show that the productivity of AM processes is significantly enhanced when residual stresses accumulated are less. Residual stresses and productivity have no relation between them. Productivity is increased at a faster scanning speed which also reduces the residual stresses. For these cases, users need to understand the data, select the appropriate input and output variables, and interpret the results from the machine learning analysis.

12.4 Digital twins in additive manufacturing

A digital twin [24] is a virtual replica of a machine that has been successfully tested to replicate the behavior of the hardware. A digital twin can reduce the number of trial-and-error tests to obtain the desired set of product attributes and reduce the time required for part qualification. General Electric currently uses over 550,000 digital twins for a wide variety of applications ranging from jet engines to power turbines. In addition, NASA and the US Air Force have also utilized digital twins to increase the reliability and safety of vehicle designs. AM is an emerging technology and the digital twins of AM are still developing. Here, we discuss the most important building blocks of a digital twin of AM and their functions.

Various components of a digital twin and the interconnections among them are shown in Figure 12.11. It consists of mechanistic and statistical models, machine learning, big data analytics, and sensing and control. A mechanistic model can estimate the transient temperature field, fusion zone geometry, cooling rates, solidification parameters, microstructures, properties, and defects. Another important component of the digital twin is a sensing and control model. This model may interface with multiple sensors and monitoring systems for in-situ measurements of important

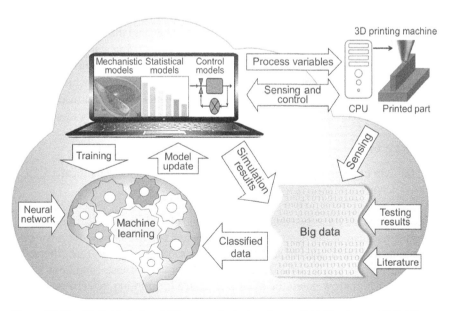

Figure 12.11 Digital twin in additive manufacturing. *Source:* T. Mukherjee and T. DebRoy.

variables. Both the mechanistic and control models can have errors due to uncertainties in input parameters, thermophysical data, and simplifying assumptions. Statistical models correct these errors and improve the accuracies of the model predictions. In addition, machine learning is an effective tool to model highly complex attributes of AM that are not well understood phenomenologically. This machine learning algorithm is first trained using a data set classified from the big data of AM and then validated and tested using a suitable data set independent of the data used for training.

Connecting the physical and virtual world of printing by creating a set of verifiable models and ultimately a digital twin will accelerate the qualification process, improve quality, reduce cost, and increase the market share of AM components. The advanced software and hardware capabilities and a rich knowledge base of metallurgy are crucial factors that will make building and utilization of digital twin in AM a realistic venture. Worked out example 12.14 explains how a digital twin of additive manufacturing can accelerate product qualification.

12.5 Summary

Mechanistic models of AM are immensely useful for predicting various metallurgical attributes such as temperature field, cooling rates, solidification parameters as well as deposit geometry, microstructure, properties, and defects. However, they require an understanding of underlying physical mechanisms and the complex relationship among process, structure, and properties that are not always available. In addition, mechanistic models are often very complex and require significant computational resources and user skills. Furthermore, they cannot be used in real-time. These complexities are addressed, almost always, by modeling the most important physical processes and ignoring the less important ones. These simplifying assumptions compromise fidelity, the extent of which is checked by comparing model predictions with experimental results. In addition, the task is often leveraged using the experience of building models of fusion welding and metallurgy. When physics-based understanding is lacking, mechanistic models cannot be developed. In such cases, data-driven machine learning can be used to control the process and product attributes. However, they often need a large volume of reliable data that may not be available. Machine learning models are being developed that can be trained using a small set of data and will be of significant interest to solve AM problems. A combination of machine learning, models, experiments, and big data in AM can create a digital twin of 3D printing which has the potential to address many issues in AM. These powerful emerging digital tools can provide important insights about the process as well as products that are not obtainable by any other means.

Worked out example 12.14

Explain how a digital twin of additive manufacturing can accelerate product qualification.

Solution:

Currently, part qualifications are done by experimental trial and error. This process is time-consuming and expensive because of the need for many experiments to explore the large range of AM process parameters. A digital twin of additive manufacturing can virtually test different sets of process parameters and predict whether defect will form under those conditions (see Figure E12.17 below). It can narrow down the parameter space for printing defect-free parts and reduce the volume of experiments need to be performed. Thus, the digital twins will have the ability to significantly expedite part qualification.

Worked out example 12.14 (Continued)

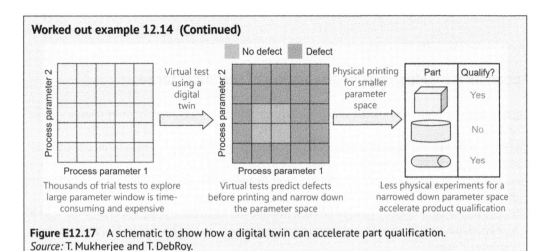

Figure E12.17 A schematic to show how a digital twin can accelerate part qualification.
Source: T. Mukherjee and T. DebRoy.

Takeaways

Mechanistic models

- Mechanistic models are used in additive manufacturing for calculating temperature and velocity fields, deposit geometry, thermal cycles, cooling rates, solidification parameters, microstructure, grain growth, mechanical properties, defects, and residual stresses.
- Models can be of diverse time and length scales. For example, powder scale models can simulate powder-heat source interactions but are computationally expensive and are restricted to a very small length scale. In contrast, part scale models can simulate additive manufacturing of large real-life components.
- Models are developed by considering the most important physical processes and ignoring the less important ones. These assumptions compromise fidelity, the extent of which is checked by comparing model predictions with experimental results.
- The available commercial packages for modeling additive manufacturing include Ansys, Abaqus, Comsol, Altair, and Sysweld.

Machine learning

- When the scientific understanding is missing but data are available, machine learning can provide quantitative relations between the process variables and product attributes.
- Commonly used machine learning algorithms in additive manufacturing include neural network, Bayesian network, support vector machines, random forest, k-nearest neighbor, and decision tree.
- Open-source machine learning codes that are used in additive manufacturing are Weka, Scikit Learn, Keras, Tensor Flow, and Theano.
- Machine learning is used in additive manufacturing for process parameter optimization, part geometry control, tailoring microstructure and properties, reducing defects, and sensing and control.

Digital twin

- A digital twin of additive manufacturing hardware consists of mechanistic and statistical models, machine learning, big data analytics, and sensing and control.
- Digital twins can virtually test different sets of process parameters, minimize the volume of expensive and time-consuming experiments, reduce cost, and accelerate part qualification.

Appendix A1   Meanings of a selection of technical terms

Big data analysis: Methods to store, analyze, interpret, and extract information from a large volume of data.

CAD: The acronym of CAD stands for computer-aided design. CAD represents a three-dimensional design of a part.

Fusion zone: The region of a part in additive manufacturing and fusion welding that is melted by the heat source.

Infrared: Wavelength of electromagnetic radiation greater than that of the red end of the visible light spectrum but less than that of microwaves. Infrared radiation has a wavelength from about 800 nm to 1 mm.

Laser beam: The acronym laser stands for "'light amplification by stimulated emission of radiation." A focused beam of laser is often used in additive manufacturing as a heat source.

Phase transformation: Transformation from one state of material to another. Common examples include liquid to solid transformation and austenite to ferrite transformation in steels.

Plasma: An electrically neutral gas consisting of electrons, excited atoms, and positively charged ions. A plasma arc is often used as a heat source in additive manufacturing.

Temperature gradient: Indicates the extent of spatial variation of temperature in a part and affects the cooling rates, solidification parameters, and grain growth.

Appendix A2   How to use the open-source machine learning package WEKA

WEKA is an open-source machine learning package provided by The University of Waikato in New Zealand. This software can be freely downloaded from https://www.cs.waikato.ac.nz/ml/weka for Windows, Mac, and Linux systems. Extensive manuals for using this package are also available on this website. Below are the steps for using this package for machine learning analysis.

1) Common format of the data file that can be imported to WEKA is .csv format (Figure A12.1). To generate the data file, the data can be put in an MS Excel sheet using the format shown below. The first row contains the names of the input and output variables. Here, "Depth" and "Width" are two input variables and the output variable is "Defect." The Excel sheet can be saved as a .csv file.

	A	B	C
1	Depth	Width	Defect
2	222.977	858.6332	N
3	244.4501	863.8948	N
4	105.0028	783.2523	Y
5	162.1635	733.9386	Y
6	260.9733	865.9713	N

Figure A12.1   CSV data file format for WEKA.

2) Open WEKA installed on your computer and click on "Explorer" As shown in Figure A12.2.

Figure A12.2 WEKA graphical user interface.

3) Select "Open file" and browse the .csv file from your computer to import the file in WEKA as show in Figure A12.3.

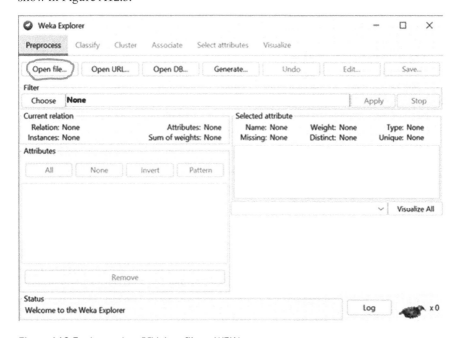

Figure A12.3 Importing CSV data file to WEKA.

4) The dataset is imported to WEKA and three variables can be correctly seen in Figure A12.4. How to use a data set for neural network and support vector machine calculations are explained in the worked out examples.

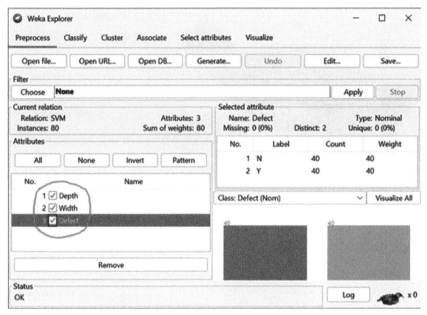

Figure A12.4   Data imported to WEKA.

Appendix A3   Data for worked out examples on machine learning

Table T12.1   Data used in Worked out examples 12.5 and 12.8. There are 27 data points in total.

Power, W	Speed, mm/s	Spot diameter, mm	Width, mm	Width from linear regression, mm	Absolute error	Width from NN, mm	Absolute error
100	500	0.3	0.60	0.615	0.015	0.598	0.002
200	500	0.3	0.63	0.659	0.029	0.638	0.008
300	500	0.3	0.69	0.704	0.014	0.677	0.013
100	750	0.3	0.54	0.555	0.015	0.555	0.015
200	750	0.3	0.62	0.599	0.021	0.593	0.027
300	750	0.3	0.64	0.644	0.004	0.636	0.004
100	1000	0.3	0.52	0.495	0.025	0.504	0.016
200	1000	0.3	0.54	0.539	0.001	0.541	0.001
300	1000	0.3	0.59	0.584	0.006	0.584	0.006
100	500	0.4	0.57	0.565	0.005	0.562	0.008
200	500	0.4	0.63	0.610	0.020	0.603	0.027

Table T12.1 (Continued)

Power, W	Speed, mm/s	Spot diameter, mm	Width, mm	Width from linear regression, mm	Absolute error	Width from NN, mm	Absolute error
300	500	0.4	0.64	0.654	0.014	0.647	0.007
100	750	0.4	0.49	0.505	0.015	0.506	0.016
200	750	0.4	0.56	0.550	0.010	0.547	0.013
300	750	0.4	0.58	0.594	0.014	0.594	0.014
100	1000	0.4	0.44	0.445	0.005	0.441	0.001
200	1000	0.4	0.49	0.490	0.000	0.481	0.009
300	1000	0.4	0.55	0.534	0.016	0.528	0.022
100	500	0.5	0.51	0.516	0.006	0.509	0.001
200	500	0.5	0.58	0.560	0.020	0.555	0.025
300	500	0.5	0.63	0.605	0.025	0.605	0.025
100	750	0.5	0.48	0.456	0.024	0.441	0.039
200	750	0.5	0.49	0.500	0.010	0.484	0.006
300	750	0.5	0.53	0.545	0.015	0.534	0.004
100	1000	0.5	0.37	0.396	0.026	0.378	0.008
200	1000	0.5	0.42	0.440	0.020	0.412	0.008
300	1000	0.5	0.47	0.485	0.015	0.455	0.015
Mean absolute error					0.014		0.012

Table T12.2 Data used in Worked out examples 12.9–12.12. In output columns, "C" represents continuous tracks and "D" denotes discontinuous tracks. There are 27 data points in total. Normalized value = (variable value – minimum value)/(maximum value – minimum value).

Actual values				Normalized values			
Power W	Speed mm/s	Spot size mm	Output	Power	Speed	Spot size	Output
100	500	0.3	C	0	0	0	C
200	500	0.3	C	0.5	0	0	C
300	500	0.3	C	1	0	0	C
100	750	0.3	D	0	0.5	0	D
200	750	0.3	C	0.5	0.5	0	C
300	750	0.3	C	1	0.5	0	C
100	1000	0.3	D	0	1	0	D
200	1000	0.3	D	0.5	1	0	D
300	1000	0.3	C	1	1	0	C
100	500	0.4	D	0	0	0.5	D
200	500	0.4	C	0.5	0	0.5	C

(Continued)

Table T12.2 (Continued)

Actual values				Normalized values			
Power W	Speed mm/s	Spot size mm	Output	Power	Speed	Spot size	Output
300	500	0.4	C	1	0	0.5	C
100	750	0.4	D	0	0.5	0.5	D
200	750	0.4	D	0.5	0.5	0.5	D
300	750	0.4	C	1	0.5	0.5	C
100	1000	0.4	D	0	1	0.5	D
200	1000	0.4	D	0.5	1	0.5	D
300	1000	0.4	D	1	1	0.5	D
100	500	0.5	D	0	0	1	D
200	500	0.5	D	0.5	0	1	D
300	500	0.5	C	1	0	1	C
100	750	0.5	D	0	0.5	1	D
200	750	0.5	D	0.5	0.5	1	D
300	750	0.5	D	1	0.5	1	D
100	1000	0.5	D	0	1	1	D
200	1000	0.5	D	0.5	1	1	D
300	1000	0.5	D	1	1	1	D

Table T12.3 Data used in (Worked out example 12.13). There are 80 data points in total. 40 cases have the lack of fusion defects and no lack of fusion defects are in the remaining 40 cases.

Depth, microns	Width, microns	Lack of fusion defect? Y: Yes N: No	Hyperplane (Z)
223.0	858.6	N	−3.73
244.5	863.9	N	−4.86
105.0	783.3	Y	4.43
162.2	733.9	Y	4.08
261.0	866.0	N	−5.64
251.2	819.0	N	−3.26
234.6	836.2	N	−3.28
280.0	844.6	N	−5.55
303.3	801.8	N	−4.74
331.3	846.5	N	−7.80
303.1	869.0	N	−7.55
351.1	861.4	N	−9.26
134.9	722.3	Y	5.72
136.5	761.8	Y	4.00
118.9	821.6	Y	2.23

Table T12.3 (Continued)

Depth, microns	Width, microns	Lack of fusion defect? Y: Yes N: No	Hyperplane (Z)
103.6	701.1	Y	7.94
105.1	740.5	Y	6.22
208.5	710.3	Y	3.11
196.8	761.6	Y	1.46
170.3	794.8	Y	1.18
375.1	846.3	N	−9.65
389.9	867.6	N	−11.17
366.7	876.3	N	−10.55
418.0	871.8	N	−12.53
224.1	745.5	Y	0.98
245.0	658.9	Y	3.73
195.4	674.0	Y	5.19
183.7	700.8	Y	4.56
447.7	878.1	N	−14.05
468.4	854.6	N	−13.94
484.9	878.0	N	−15.62
440.3	852.5	N	−12.67
461.9	817.2	N	−12.10
164.1	645.3	Y	7.72
276.4	649.2	Y	2.81
327.8	603.1	Y	2.57
324.6	566.8	Y	4.23
355.2	566.7	Y	2.94
479.3	826.8	N	−13.23
481.0	786.2	N	−11.60
425.6	810.9	N	−10.30
438.0	822.6	N	−11.31
368.6	810.1	N	−7.85
334.6	823.0	N	−6.96
287.7	762.3	N	−2.42
249.6	780.6	N	−1.58
226.4	805.2	N	−1.63
198.3	829.9	N	−1.48
175.0	858.8	N	−1.71
191.5	876.9	N	−3.16
352.7	587.0	Y	2.20
415.6	547.3	Y	1.20
437.9	530.1	Y	0.98
448.7	514.1	Y	1.20
314.5	648.0	Y	1.25

(Continued)

Table T12.3 (Continued)

Depth, microns	Width, microns	Lack of fusion defect? Y: Yes N: No	Hyperplane (Z)
254.8	689.8	Y	2.01
257.6	593.7	Y	5.93
283.1	629.9	Y	3.33
215.4	611.0	Y	6.99
197.3	568.3	Y	9.55
226.3	567.2	Y	8.37
225.3	599.2	Y	7.07
191.6	851.3	N	−2.09
316.7	752.6	N	−3.24
307.7	714.2	N	−1.24
322.6	693.9	N	−1.02
356.5	693.7	N	−2.45
370.6	670.2	N	−2.06
380.4	707.5	N	−4.04
340.7	731.2	N	−3.35
356.3	765.3	N	−5.45
263.5	556.4	Y	7.25
285.8	545.6	Y	6.76
288.4	525.3	Y	7.51
314.8	531.6	Y	6.12
342.9	517.6	Y	5.52
110.6	567.6	Y	13.25
90.6	608.2	Y	12.39
80.0	576.2	Y	14.18
96.4	595.4	Y	12.68

Practice problems

1) Between a powder scale model and a part scale heat conduction model, which one should be used for calculating the temperature field during powder bed fusion of a part with 500 mm × 300 mm × 100 mm dimensions and why?

2) Explain why powder scale models are so computationally intensive?

3) Why simulation of a part made with thinner layers is computationally more expensive?

4) Which microstructure model should be used to calculate the distribution of laves phases during DED-L of Inconel 718? Do a literature review to justify your answer.

5) Large brackets made by DED-GMA often suffer from buckling due to high thermal distortion. Explain how a mechanistic model can be used to reduce buckling in such parts.

6) Find an expression to calculate the minimum energy needed to form a keyhole during additive manufacturing of Inconel 718. Assume that the formation of keyhole in Inconel 718 depends only

on the evaporation of nickel. Data needed for the calculations: molecular weight of nickel: 59 g/g mol, boiling point of nickel: 3003 K, and enthalpy of vaporization of nickel: 6278 kJ/kg.

7) Among the following problems, identify which ones are the regression problems and which ones are classification problems: (a) identify and reduce the occurrence of gas porosities, (b) control the amount of laves phases in the printed Inconel 718 parts by adjusting process variables, (c) minimize the occurrence of delamination due to accumulation of high residual stresses, and (d) control the hardness of the printed SS 316 parts by adjusting the process variables.

8) Worked out example 12.5 provides an equation representing the pool width as a function of power, speed, and spot size. Use this relation to generate a process map for predicting pool width within 100–500 W power and 500–1000 mm/s speed at a constant spot size of 0.5 mm.

9) Cracking is one of the major problems in additive manufacturing. Cracking is affected by multiple AM variables. If data are available on crack formation, discuss how the relative influence of the causative variables on crack formation can be evaluated using machine learning?

10) Based on an Internet search, make a list of organizations that apply digital twins in additive manufacturing and discuss, where practical, their usage.

11) Chapter 10 discusses several back-of-the-envelope calculations for predicting defect formation in additive manufacturing. What are the advantages and disadvantages of these calculations with respect to the mechanistic models of defect prediction?

References

1 Wei, H.L., Mukherjee, T., Zhang, W., Zuback, J.S., Knapp, G.L., De, A. and DeBroy, T., 2021. Mechanistic models for additive manufacturing of metallic components. *Progress in Materials Science*, 116, article no.100703.

2 DebRoy, T., Mukherjee, T., Wei, H.L., Elmer, J.W. and Milewski, J.O., 2021. Metallurgy, mechanistic models and machine learning in metal printing. *Nature Reviews Materials*, 6(1), pp.48–68.

3 Wu, Q., Mukherjee, T., Liu, C., Lu, J. and DeBroy, T., 2019. Residual stresses and distortion in the patterned printing of titanium and nickel alloys. *Additive Manufacturing*, 29, article no.100808.

4 Mukherjee, T. and DeBroy, T., 2018. Mitigation of lack of fusion defects in powder bed fusion additive manufacturing. *Journal of Manufacturing Processes*, 36, pp.442–449.

5 Ibarra-Medina, J., Pinkerton, A.J., Vogel, M. and N'Dri, N., 2012. Transient modelling of laser deposited coatings. In *26th International Conference on Surface Modification Technologies, Valardocs*, India.

6 Khairallah, S.A., Anderson, A.T., Rubenchik, A. and King, W.E., 2016. Laser powder-bed fusion additive manufacturing: Physics of complex melt flow and formation mechanisms of pores, spatter, and denudation zones. *Acta Materialia*, 108, pp.36–45.

7 Radhakrishnan, B., Gorti, S. and Babu, S.S., 2016. Phase field simulations of autocatalytic formation of alpha lamellar colonies in Ti-6Al-4V. *Metallurgical and Materials Transactions A*, 47(12), pp.6577–6592.

8 Gong, X. and Chou, K., 2015. Phase-field modeling of microstructure evolution in electron beam additive manufacturing. *JOM*, 67(5), pp.1176–1182.

9 Li, X. and Tan, W., 2018. Numerical investigation of effects of nucleation mechanisms on grain structure in metal additive manufacturing. *Computational Materials Science*, 153, pp.159–169.

10 Wei, H.L., Elmer, J.W. and DeBroy, T., 2017. Three-dimensional modeling of grain structure evolution during welding of an aluminum alloy. *Acta Materialia*, 126, pp.413–425.

11 Wei, H.L., Mazumder, J. and DeBroy, T., 2015. Evolution of solidification texture during additive manufacturing. *Scientific Reports*, 5(1), article no. 16446.

12 Wei, H.L., Cao, Y., Liao, W.H. and Liu, T.T., 2020. Mechanisms on inter-track void formation and phase transformation during laser powder bed fusion of Ti-6Al-4V. *Additive Manufacturing*, 34, article no.101221.

13 Yan, W., Ge, W., Qian, Y., Lin, S., Zhou, B., Liu, W.K., Lin, F. and Wagner, G.J., 2017. Multi-physics modeling of single/multiple-track defect mechanisms in electron beam selective melting. *Acta Materialia*, 134, pp.324–333.

14 Mukherjee, T., Zhang, W. and DebRoy, T., 2017. An improved prediction of residual stresses and distortion in additive manufacturing. *Computational Materials Science*, 126, pp.360–372.

15 Mukherjee, T., DebRoy, T., Lienert, T.J., Maloy, S.A. and Hosemann, P., 2021. Spatial and temporal variation of hardness of a printed steel part. *Acta Materialia*, 209, article no.116775.

16 Wei, H.L., Elmer, J.W., and DebRoy, T., 2016. Origin of grain orientation during solidification of an aluminum alloy. *Acta Materialia*, 115, pp.123–131.

17 Mukherjee, T., Wei, H.L., De, A. and DebRoy, T., 2018. Heat and fluid flow in additive manufacturing–Part II: powder bed fusion of stainless steel, and titanium, nickel and aluminum base alloys. *Computational Materials Science*, 150, pp.369–380.

18 Li, R., Wang, G., Zhao, X., Dai, F., Huang, C., Zhang, M., Chen, X., Song, H. and Zhang, H., 2021. Effect of path strategy on residual stress and distortion in laser and cold metal transfer hybrid additive manufacturing. *Additive Manufacturing*, 46, article no.102203.

19 Popova, E., Rodgers, T.M., Gong, X., Cecen, A., Madison, J.D. and Kalidindi, S.R., 2017. Process-structure linkages using a data science approach: application to simulated additive manufacturing data. *Integrating Materials and Manufacturing Innovation*, 6(1), pp.54–68.

20 Miyazaki, S., Kusano, M., Bulgarevich, D.S., Kishimoto, S., Yumoto, A. and Watanabe, M., 2019. Image segmentation and analysis for microstructure and property evaluations on Ti–6Al–4V fabricated by selective laser melting. *Materials Transactions*, article no.MBW201806.

21 Khanzadeh, M., Chowdhury, S., Marufuzzaman, M., Tschopp, M.A. and Bian, L., 2018. Porosity prediction: supervised-learning of thermal history for direct laser deposition. *Journal of Manufacturing Systems*, 47, pp.69–82.

22 Mondal, B., Mukherjee, T. and DebRoy, T., 2022. Crack free metal printing using physics informed machine learning. *Acta Materialia*, 226, article no.117612.

23 Mitchell, T., 1997. *Machine Learning*, 1st Edition. McGraw-Hill.

24 Mukherjee, T. and DebRoy, T., 2019. A digital twin for rapid qualification of 3D printed metallic components. *Applied Materials Today*, 14, pp.59–65.

13

Safety, Sustainability, and Economic Issues in Additive Manufacturing

---

Learning objectives

After reading this chapter the reader should be able to do the following:

1) Understand the safety issues related to the handling of powders, the risks of fire and explosion, and the potential health hazards from the irradiation of laser and electron beams.
2) Know several techniques for creating and maintaining a safe and healthy workplace.
3) Appreciate the energy consumption and emission of greenhouse gases during additive manufacturing.
4) Recognize the need for new alloy development, reduction in materials wastage and energy consumption, and recycling materials for making additive manufacturing sustainable.
5) Understand the different costs associated with additive manufacturing processes and products.

CONTENTS

Theory and Practice of Additive Manufacturing, First Edition. Tuhin Mukherjee and Tarasankar DebRoy.
© 2024 John Wiley & Sons, Inc. Published 2024 by John Wiley & Sons, Inc.

13.1 Introduction

Additive manufacturing (AM) is a rapidly emerging technology used in a wide variety of industries including aerospace, automotive, healthcare, energy, and consumer products. AM, like all other manufacturing processes, is subject to several safety concerns. Fine metallic powders, for example, which are used as feedstock in AM, can cause fire and explosion if not stored properly. Fine powders can also affect the eyes and respiratory system and can cause asthma and other lung diseases. Heat sources like laser and electron beams need to be handled with special care since they can harm an operator's eyes. Therefore, creating a safe working environment in the AM business requires knowledge of fire safety, health risks, and safety protocol for using laser and electron beams.

Unlike traditional manufacturing processes, AM is uniquely able to produce near-net shape components in a single step without the need for assembling multiple components. This reduces materials wastage and contributes to sustainable manufacturing. However, AM requires expensive equipment and feedstocks. The cost of AM parts also includes the expenses related to the part design, equipment operation, and post-processing. Currently, the high cost of AM parts is justified primarily by the fabrication of specialized, complex parts that are difficult to produce using traditional manufacturing methods. An in-depth understanding of the economic issues of AM is needed to expand the market share of AM parts.

We cover several safety concerns in this chapter, including those relating to fire safety, health risks, safety around laser and electron beam radiations, and measures for establishing a secure working environment. Additionally, we list several sustainability concerns with additive manufacturing (AM), including energy use, carbon footprint, and the development of new materials, and we offer solutions to make AM a sustainable manufacturing technique for future applications. Finally, we discussed several AM-related economic challenges and solutions for boosting the market share of AM components.

13.2 Safety

The safety issues in additive manufacturing are currently addressed by the laws and regulations applied to the manufacturing industry. For example, in the US, AM shop floor health and safety issues are handled by the general scope of the Occupational Safety and Health Administration (OSHA). Safety issues in transporting and handling potentially hazardous materials such as powders are addressed by both OSHA and the US Department of Transportation (DoT) regulations. Environmental Protection Agency (EPA) regulations cover various environmental issues related to the generation, treatment, storage, and disposal of materials and waste. AM safety issues are also discussed in the ASTM F42 Subcommittee and the ISO/ASTM TC261 Working Group on environmental health and safety. In this section, we discuss the fire safety issues, potential health hazards related to AM, and impacts of laser and electron beam radiation during AM and protective measures taken against them. These discussions will assist engineers in creating a safe working environment in the additive manufacturing industry.

13.2.1 Fire safety

Metal or alloy powders used as feedstock in AM can be a potential source of fire or explosion [1]. These powders are small, nearly spherical with a diameter of less than 100 μm in most cases. Such small powders have a very high surface area-to-volume ratio and have the potential to catch fire in the air and explode under certain conditions. The susceptibility of powders to explosion is indicated by an explosion parameter represented by the pressure rise per unit time inside a dust explosion vessel during a controlled dust explosion experiment [2]. Finer powders with a higher surface area to volume ratio are more susceptible to explosion and catching fire (Figure 13.1).

Special care should be taken during unpacking, handling, transferring to the machine, and processing the powder feedstock. Some alloy powders are more vulnerable to explosion than others as indicated by the value of the explosion parameter (Table 13.1). There are several strict procedures for handling pure aluminum and aluminum alloy powders that are highly susceptible to fire and explosion. For example, the storage of 45 kg or more of aluminum alloy powders is

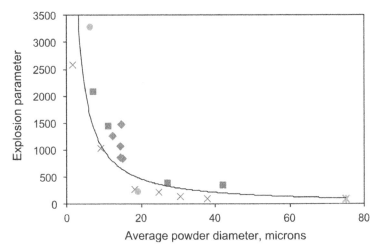

Figure 13.1 Susceptibility to the explosion of commercially pure aluminum powder of different sizes [2]. The explosion parameter indicates the pressure rise per unit time (bar per second) in a dust explosion vessel during a controlled dust explosion experiment. The figure is taken from [2] with the permission from Elsevier.

Table 13.1 Susceptibility of different metal and alloy powders to explosion [3]. The explosion parameter is represented by the pressure rise per unit time (bar per second) in a dust explosion vessel during a controlled dust explosion experiment. The average powder diameter was 75 microns.

Metal or alloy powders	Explosion parameter
Commercially pure aluminum	1379
Aluminum-magnesium alloy	689
Aluminum-nickel alloy	689
Commercially pure magnesium	621
Aluminum-silicon alloy	517
Commercially pure titanium	379
Aluminum-lithium alloy	255
Aluminum-copper alloy	179
Aluminum-iron alloy	124

regulated by the Department of Homeland Security in the United States. Powder storage containers should not be placed near heat sources, hot surfaces, and equipment that can generate sparks. Fires caused by metal powders must be extinguished using an appropriate fire extinguisher. Existing manufacturing facilities may only have a conventional sprinkler that uses water and has the potential to aggravate a fire caused by metal powders. Precautions and corrective measures should be taken for fire safety in AM by following the National Fire Protection Association (NFPA) standards relevant to AM (Table 13.2).

13.2.2 Health hazards

Condensates produced from the evaporation of liquid metal are very dangerous for human health. For the directed energy deposition processes with wire feedstock, the health hazards are similar to those commonly encountered in fusion welding. For example, an apparently harmless alloy such as stainless steel contains nickel and chromium, both of which are known carcinogens. Fumes, vapors, and particulate matter produced during welding are known to cause immune system dysfunction and upper and lower respiratory tract infections among welders.

Fine powders and particulates can cause several health-related risks resulting from inhalation, ingestion, or contact with the skin. Spatters resulting from the use of high-power-density heat sources can also generate fine metal particles within the chamber. Fine powders and particles condensate in the filter making the replacement of filters the most dangerous part of AM operation. Fine metallic powders are also known to affect the eyes and respiratory system and may cause diseases such as asthma. Appropriate precautions should be taken to transport, process, and recycle powders.

Additive manufacturing equipment is often tightly placed to maximize space utilization on the shop floor and enhance production efficiency. However, it can result in an unsafe work environment. Additive manufacturing processes use shielding gases such as argon and nitrogen, which can displace the ambient air in the workspace and reduce the amount of breathable air. A monitoring system for safe oxygen levels should be installed for maintaining worker health and safety. Worked out example 13.1 explains the strategies for increasing safety and reducing health hazards.

Table 13.2 National Fire Protection Association (NFPA) standards relevant to additive manufacturing [4]. NFPA is an international organization for eliminating death, injury, property, and economic loss due to fire, and electrical hazards.

Standards	Descriptions
NFPA 484	Standards for combustible metals
NFPA 654	Standards for the prevention of fire and dust explosions from the manufacturing, processing, and handling of combustible particulate solids
NFPA 77	Recommended practice on static electricity
NFPA 2113	Standards on selection, care, use, and maintenance of flame-resistant garments for the protection of industrial personnel against short-duration thermal exposures from fire
NFPA 68	Standard on explosion protection by rapid venting
NFPA 69	Standard on explosion prevention systems
NFPA 70	The national electric code

Worked out example 13.1

Several factors affect safety and cause potential health hazards during additive manufacturing. What are some strategies for increasing safety and reducing health hazards?

Solution:

Several commonly used strategies are discussed below:

Use of personal protective equipment (PPE): Common PPE used during AM operations include gloves, aprons, safety goggles, protective shields, and gas masks. PPE can shield an operator from possible exposure to hazards.

Follow standard operating procedures: Standard operating procedures are instructions for various activities, handling equipment, and emergency instructions. Generally, these are developed by a shopfloor safety official and should be followed to maintain a safe workplace.

Hazard identification: Identification of hazardous materials or the environment can help to prevent accidents and exposures. Hazard identification is generally done by a facility team by reviewing an entire manufacturing process and identifying specific hazards from the feedstock materials or AM process.

Control of emissions: All AM processes result in some form of emissions such as volatile organic compounds, ultrafine particles, and condensate of metal vapors. Proper shielding, ventilation, and operator training are needed to mitigate the risks.

Prevention of dust exposure, fires, and explosions: Alloy powders can form dust clouds and result in fires and explosions. Safe storage and handling procedure are important.

13.2.3   Laser and electron beam radiation

Commonly used heat sources in additive manufacturing such as the laser, electron beam, or electric arc can be dangerous because of their high energy density. Also, high-power lasers can damage the retina by penetrating a powerful light into the eyes in fractions of a second. The eye's protective blink reflex is not fast enough to prevent the penetration of the laser beam. Symptoms of a laser burn in the eye include excessive watering from the eyes, a headache shortly after exposure, and sudden blurred vision. Laser irradiation can also permanently damage vision. Therefore, proper eye protection should be used against laser radiation during the additive manufacturing operation. Other typical additive manufacturing equipment hazards include electrical energy hazards, irradiation hazards, and thermal hazards. Some of these hazards can have serious consequences if not properly controlled.

13.3   Sustainability

Sustainable manufacturing is defined as the production of components using nonpolluting processes that conserve energy and natural resources, can be applied to a wide range of materials, and are safe for operators, communities, and consumers. This section, covers the energy consumption and carbon footprint of AM processes, the potential for new alloy developments for AM, and several other approaches for making AM sustainable.

13.3.1 Energy consumption

Energy consumption in additive manufacturing not only impacts the sustainability of the process but also affects the microstructure and properties of the manufactured parts. AM is an energy-intensive process because it uses a laser or electron beam or an electric arc with high energy density to melt materials. Energy consumption during AM includes the energy needed to run the machine and other energy demands during the AM process. Energy is consumed by the high-energy beam generator, control system, cooling system, shielding gas flow system, and pumps and vents. The energy consumption varies depending on the type, make, and model of the machine. Energy is also consumed in the production of feedstock which varies widely depending on the type, quality, and materials. For example, for powder production, gas atomization, water atomization, and plasma-based atomization consume different amounts of energy. Part of the energy supplied by the heat source is absorbed by the feedstock to melt the materials and the remaining energy is lost to the environment by conduction, convection, and radiation. For components with simple design (Figure 13.2), it is energy efficient to fabricate them using traditional manufacturing processes such as casting, machining, and forming. However, complex and intricate parts need many steps of operations such as casting, forming, machining, and finishing which will increase the total energy consumption. Therefore, for complex parts, additive manufacturing is more energy efficient (Figure 13.2) because it allows one-step fabrication of the entire part. Worked out examples 13.2 and 13.3 illustrate the energy consumption for additive manufacturing processes.

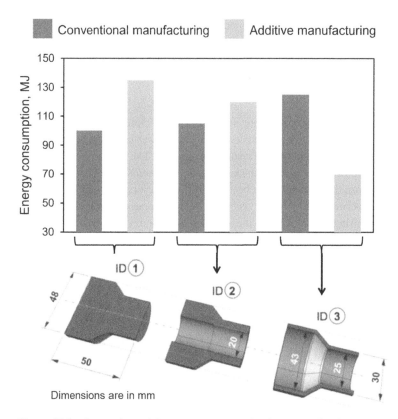

Figure 13.2 Comparison of the energy consumption in conventional manufacturing and additive manufacturing of parts of different complexities. Here, conventional manufacturing includes forming, machining, and finishing. PBF-EB was used to additively manufacture the part. The figure is made by T. Mukherjee and T. DebRoy using the data reported in [5].

Worked out example 13.2

Both powder bed fusion and binder jetting are powder-based additive manufacturing processes for metallic materials. Do a literature review and determine which process consumes more energy and why?

Solution:

In binder jetting, alloy powders are agglomerated using a liquid binder. The binding process does not need supply of energy from a heat source. Some amount of energy is consumed by the furnace where the green parts are cured to achieve the required strength of parts. In contrast, the powder bed fusion process uses a high-energy laser or electron beam to melt the alloy powders. Therefore, the energy consumption of the powder bed fusion process is more. For example, it is reported [6] that about 80 kWh of energy is consumed during the fabrication of a 1 kg titanium alloy part using laser powder bed fusion. However, to make the same part, binder jetting requires only about 3 kWh of energy.

Worked out example 13.3

A laser powder bed fusion machine uses a maximum power level of 8.5 kW during a build. What are the total energy consumption and the cost of energy to make a part that takes 8 hours to print? Useful data: the cost of the energy per kWh is 10.5 cents.

Solution:

The total energy consumption = (maximum power level × the number of hours needed) = (8.5 kW × 8 h) = 68 kWh of energy.

The total cost for the energy consumption = (total energy consumed × cost per kWh of energy) = (68 kWh × 10.5 cents/kWh) = 714 cents = \$7.14.

The cost of the energy is much less compared to the machine and feedstock costs discussed in Section 13.4.

13.3.2 Carbon footprint of additive manufacturing processes

The carbon footprint is indicated by the total amount of greenhouse gases such as carbon monoxide, carbon dioxide, and methane generated by processes. The average carbon footprint for a person in the United States is 16 tons per year [6]. Both the additive manufacturing processes and the feedstock fabrication produces a significant amount of greenhouse gases. For example, Figure 13.3 provides the amount of CO_2 emission during the manufacturing of a 1 kg titanium alloy part using different additive manufacturing processes [6]. The carbon footprint per kg of material in additive manufacturing can be significantly higher than that in traditional manufacturing processes when both the production of the feedstock and post-processing are considered. However, AM can save a large mass of material and often can have lower manufacturing impacts than machining. The Additive Manufacturing Green Trade Association, launched in November 2019 to promote the environmental benefits of AM, provides recommendations to reduce the carbon footprint of additive manufacturing [6]. They encourage optimizing the usage of AM machines, either by sharing machines or reducing unused build plate space. In addition, they recommend selecting metals and alloys that need less energy to melt.

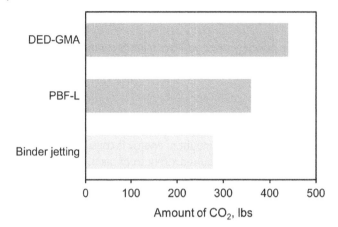

Figure 13.3 Comparison of the amount of CO_2 emission during the manufacturing of a 1 kg titanium alloy part using different additive manufacturing processes. The figure is made by T. Mukherjee and T. DebRoy using the data reported in [6].

13.3.3 New alloy development

Currently, only a handful of the over 5500 commercial alloys can be easily printed because of the difficulties of making defect-free parts with desirable microstructures and properties. In addition, some alloys are not commercially available in the forms of powders and wires typically used in AM. Opportunities exist to develop new alloys for use in AM. These new materials need to be made available in a form or shape optimized for printing. Since the heating and cooling rates in AM differ from traditional processing methods, special consideration should be given to the evolution of microstructure and properties of these new materials. In addition, an adaptation of new materials for AM may also require a strong business case to pay for testing and qualifying the product. Table 13.3 summarizes a few new materials that are being developed for use in AM.

13.3.4 Several approaches to making additive manufacturing sustainable

AM has many positive sustainability advantages over traditional manufacturing processes. For example, AM can fabricate components from recycled feedstock and reduce material waste, energy usage, and emissions. However, there is a need to evaluate the sustainability of AM processes. Efforts are being made to reduce materials waste and energy consumption and find innovative ways to use new materials in AM. Several commonly practiced methods to make additive manufacturing sustainable are summarized in Table 13.4.

13.4 Economic issues

In the 13 trillion-dollar global manufacturing industry, the market value of all AM products was only 7.3 billion dollar in the year 2019 which was just about 0.06% of the global manufacturing economy [1]. Facility upgrades, high costs of machine and feedstock, the safety measures beyond those commonly practiced in conventional manufacturing, often limit the adoption of AM by small to medium businesses. These economic issues hinder the market penetration of AM products. In this section, we discuss several costs associated with AM and the cost-competitiveness issues.

Table 13.3 New materials development for additive manufacturing.

Materials	Descriptions
Composites made by powder addition	Small particles of metals or ceramics are added during the process to perform in-situ composite fabrication.
Compositionally graded materials	The chemical composition varies gradually between two dissimilar materials. They are printed by varying the composition layerwise. Such materials are immensely useful to improve high-temperature creep and fatigue properties.
Functional materials	Specially designed material with a determined function such as semiconductors, polymers, molecular crystals, and nanoparticles. These materials are printed to make parts to support emerging technologies.
Metamaterials	These materials are engineered to have a property that is not found in naturally occurring materials. They are printed in the form of assemblies of multiple elements of composite materials.
Shape memory alloys	These alloys are deformed below a given temperature and they return to their original undeformed shape by being heated. Parts printed using this alloy can be used as temperature sensors, actuators, heart stents, and clamping fixtures.
High entropy alloys	These alloys are formed by mixing equal or relatively large proportions of five or more elements. Printed parts of these alloys have a high strength-to-weight ratio, fracture resistance, tensile strength, and corrosion and oxidation resistance.

Table 13.4 Several approaches to making additive manufacturing sustainable. Adapted from [7].

Approach	Descriptions
Innovative sustainable design	Sustainable design focuses on the reduction in weight, improving the strength-to-weight ratio, and reducing materials wastage. AM supports freeform fabrication allowing freedom in design.
Minimize materials wastage	One-step near-net shape fabrication using AM significantly reduces materials wastage compared to the traditional subtractive processes.
Recycling of material	In powder-based AM processes, unused powders can be collected from the powder bed and recycled.
Less energy consumption	AM is more energy-efficient than most traditional manufacturing processes for the fabrication of complex and intricate parts. An efficient process design can save more energy and make AM sustainable.
Green manufacturing	AM can be a green manufacturing process because it can make parts using fewer natural resources, reducing waste and emissions, and recycling and reusing materials.

13.4.1 Cost of additively manufactured parts

AM machine, processing time, inspection, set up, feedstock materials, and post-processing are major contributors to the cost of additively manufactured parts which can be relatively high compared to cast or wrought products. The two most important costs are the costs of feedstock material and manufacturing costs. The material and manufacturing costs vary widely depending on the AM process and materials used. In some cases, the complexity of a part can increase both the material and manufacturing costs as illustrated in Figure 13.4 for the printing of a stainless steel 316 bottle opener by PBF-L. The complex design of the bottle opener requires a support structure for overhangs that need to be removed by hammering and grinding the chipped surface after manufacturing. Both the

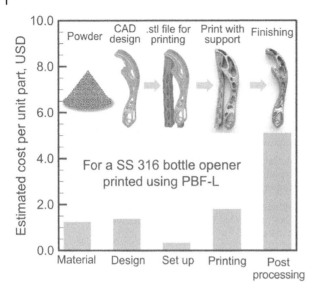

Figure 13.4 Cost of different stages of additive manufacturing. The data are for laser powder bed fusion of stainless steel 316 bottle openers. The figure is adapted from [1] with permission from Springer Nature.

extra material needed for the support structure and the time required for the post-processing add to the cost. Worked out example 13.4 explains how the feedstock cost in additive manufacturing can be minimized. The cost of complexity in AM must be viewed considering the alternatives. When such alternatives are available, the manufacturing of complex parts by traditional routes will require special machines, tooling, assembly, and other expenses, often making AM cost competitive.

Worked out example 13.4

Additive manufacturing feedstocks are very expensive and they significantly contribute to the total cost of the product. How can the feedstock cost be minimized?

Solution:

Below are some of the ways to reduce the cost of the feedstock:
- Feedstock can be produced by cheaper production methods. For example, powder feedstocks can be produced using gas or water atomization processes that are cheaper than plasma-based atomization processes.
- Unused powders can be collected from the machine and reused to save money.
- Feedstocks can be produced from recycled materials. For example, wire feedstocks can be produced from billets made by melting scraps and used materials.
- The parts can be designed by allowing the use of cheaper materials where possible.

13.4.2 Cost-competitiveness

In traditional manufacturing such as casting or machining, the costs are incurred for raw material, equipment, tooling, set-up, part complexity, and the volume of the order. Cost per part is high if only a few parts are made because of the initial cost of the equipment and set-up, but reduces as the production volume increases and the high set-up cost is distributed among many products. This trend is observed in Figure 13.5 where the cost of manufacturing a landing gear part made of

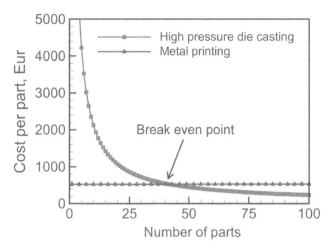

Figure 13.5 The cost of a landing gear part made of AlSi10Mg alloy by PBF-L is compared with that made by high-pressure die casting. The figure is adapted from [1] with permission from Springer Nature.

AlSi10Mg alloy by PBF-L is compared with that of a part made by high-pressure die casting. Since AM does not require any set-up and tooling costs for new parts, the cost per part does not change significantly with product volume. As a result, the economic viability of AM does not depend on minimum order size and there is no cost imperative to sell a target minimum number of components. The data show the significant competitive advantage of AM for manufacturing up to a certain number of parts over high-pressure die casting.

AM is beneficial for fabricating complex parts because AM does not require new equipment or special tooling. In contrast, such complex parts made by conventional manufacturing always require special equipment and in many cases assembly of smaller parts. Figure 13.6 shows a trend of the variation of cost with complexity. AM is cost-effective in manufacturing complex and intricate parts. Several unique aspects of AM make the AM processes cost-competitive. First, the capability of AM to produce a complex component in one step often avoids the cost of joining or assembling multiple small parts. Second, the ability to produce components on demand reduces inventory costs over conventional manufacturing. Third, the same machine can be used to produce a wide variety of materials saving equipment cost and space. Finally, the production of parts for which the supply chain does not exist allows the repair of expensive equipment avoiding rebuilding an expensive plant. Worked out example 13.5 illustrates the method of calculating the cost of additively manufactured parts.

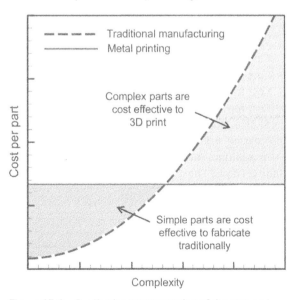

Figure 13.6 Qualitative representation of the cost and part complexity relation for metal printing and traditional manufacturing. The figure is adapted from [1] with permission from Springer Nature.

13.5 Summary

In the last few decades, significant research and development have made the AM processes safer, more sustainable, and cost-competitive in many cases. Maintaining a safe working environment requires the use of personal protective equipment, adherence to standard operating procedures, early detection of dangers, control of emissions, and prevention of dust exposure, fires, and explosions. By reducing material waste, recycling resources, and using less energy, additive manufacturing (AM) is evolving into a sustainable production technique. There is also a need for a reduction in AM machine and feedstock costs. Research collaboration among universities, national labs, and companies can develop promising ideas, standards, and protocols for enhancing safety, maintaining sustainability, and reducing costs. Collaboration among different organizations may also help to train a large workforce to address the issues related to safety, sustainability, and the economy of additive manufacturing.

Worked out example 13.5

Estimate the cost of printing stainless steel 316 parts of 5 cm^3 volume by powder bed fusion. In a batch, 30 specimens are fabricated together. No support structure is needed. Data: The equipment cost, lifetime, idle time, and build time: $1 million, 10 years, 20%, and 20 hours, respectively. The cost of metal powder: $100/kg, density of powder feedstock: 7.6 gm/cm^3. Usage of argon shielding gas: 15 liters per minute. Cost of argon: $0.01 per liter. Power consumption for 20 hours: 30 kWh. Cost of electricity: $0.10/kWh. Cost for post-processing: $500

Solution:

The lifetime usage hour of the machine = (10 × 365 × 24 × 0.8) = 70,080 hours
   The hourly direct cost of machine usage: $1,000,000/70,080 = $14.27/hour
   Machine cost for printing = $14.27/hour × 20 hours = $285.4
   Mass of powder needed to print a batch of 30 specimens = (30 × 5 × 7.6) = 1140 grams
   Cost of powder feedstock for the batch = 1.140 × 100 = $114
   Volume of argon for shielding = (20 × 60 × 15) = 18,000 liters
   Cost of argon for shielding = 18,000 × 0.01 = $180
   Cost of electricity = 30 kWh × $0.10/kWh = $3
   Cost for post-processing: $500
   Cost for 30 specimens printed in a batch =
   $285.4 (machine) + $114.0 (powder) + $180.0 (argon) + $3.0 (electricity) + $500.0 (post-processing) = $1082.40
   In addition, there are costs for the labor, maintenance of the equipment, and a variety of indirect costs. Thus, the true cost is much higher.

Takeaways

Safety

- Metal or alloy powders used as feedstock in additive manufacturing can be a potential source of explosion or fire.
- There are several strict procedures for handling pure aluminum and aluminum alloy powders that are highly susceptible to fire and explosion.

Takeaways (Continued)

- Condensate produced from the evaporation of liquid metal is dangerous for human health.
- Fine powders and particulates can cause several health-related risks resulting from inhalation, ingestion, or contact with the skin.
- High-power lasers can damage the retina by penetrating a powerful light into the eyes and can permanently damage vision.
- Use of personal protective equipment, following standard operating procedures, early identification of hazards, control of emissions, and prevention of dust exposure, fires and explosions are essential for maintaining a safe working environment.

Sustainability

- Machine-level energy consumption in additive manufacturing includes the high-energy beam generator, control system, cooling system, shielding gas flow system, pumps and vents, and the production of feedstocks.
- Process level energy consumption in additive manufacturing includes the energy absorption by the feedstock to melt the materials and the energy transfer out of the system.
- The carbon footprint per kg of material in additive manufacturing can be significantly higher compared to traditional manufacturing processes.
- Carbon footprint can be reduced by optimizing the usage of machines, either by sharing machines or reducing unused build plate space and by selecting metals and alloys that need less energy to melt.
- Printing of new materials such as composites, compositionally graded materials, functional materials, metamaterials, shape memory alloys, and high entropy alloys can make additive manufacturing sustainable.
- Approaches to make additive manufacturing sustainable include innovative sustainable design, minimizing materials wastage, recycling materials, and reducing energy consumption.

Economic issues

- The cost of printed parts includes machine cost, feedstock cost, cost of electricity and shielding, machine setup and operation costs, and cost of post-processing. In addition, the costs of labor, machine maintenance, and other items make the true cost significantly higher.
- Complex and intricate parts are cost-effective to manufacture using additive manufacturing.
- The cost of printed parts does not change significantly with batch size. Therefore, in many cases, it is cost-effective to manufacture parts in a large batch size using traditional processes.

Appendix – Meanings of a selection of technical terms

Atomization: A process of disintegrating solid or liquid into fine particles. For example, molten metals are atomized to make powder feedstock for additive manufacturing.

CAD: The acronym of CAD stands for computer-aided design. CAD represents a three-dimensional design of a part.

Ceramic: Hard, brittle, heat and corrosion-resistant inorganic, nonmetallic material.

Composite: A composite material is produced from two or more constituent materials. Nonmetallic materials are added with alloys to print metal matrix composites.

Cooling rate: The rate with which the deposited material cools down to room temperature. The cooling rate varies significantly depending on the temperature at which the cooling rate is estimated. Cooling rates affect the microstructure and properties of parts.

<u>Electron beam</u>: A focused beam of electrons used as a heat source in additive manufacturing.

<u>Fusion welding</u>: A manufacturing process that uses a heat source to melt two parts to form a sound joint after solidification.

<u>Greenhouse gas</u>: Greenhouse gases absorb and emit radiant energy within the thermal infrared range. The common greenhouse gases are water vapor, carbon monoxide, carbon dioxide, methane, nitrous oxide, and ozone.

<u>Laser beam</u>: The acronym laser stands for "light amplification by stimulated emission of radiation." A focused beam of laser is often used in additive manufacturing as a heat source.

<u>Plasma</u>: An electrically neutral gas consisting of electrons, excited atoms, and positively charged ions. A plasma arc is often used as a heat source in additive manufacturing.

Practice problems

1) How does additive manufacturing contribute to sustainable practices in the manufacturing industry?

2) A DED-L machine uses a maximum power level of 10.0 kW during a build. What are the total energy consumption and the cost associated with it to make a part that takes 8 hours to print? Assume that the cost of the energy per kWh is 10.8 cents. Comment on the impact of the energy consumption on the product cost.

3) Which of the commonly used alloy powders are most susceptible to fire and explosion?

4) Compare DED-L and DED-GMA in terms of the potential health hazards they may cause.

5) A company manufactures a complex metallic part that needs the fabrication of 15 small parts by machining followed by joining them by brazing. Should they replace their fabrication process with additive manufacturing if they want to make a batch of 10 parts?

6) What are the main challenges in developing new alloys for additive manufacturing? What are some of the probable solutions to address those challenges?

7) Recycling powders in powder-based additive manufacturing processes may significantly reduce the feedstock cost. What are the main drawbacks of powder recycling?

8) Why is additive manufacturing a feasible fabrication process to produce compositionally graded materials?

9) The cost of a powder bed fusion machine is $1 million. The machine has a lifetime of 12 years. If the machine is used for 16 hours a day, what is the hourly direct cost of machine usage?

10) Estimate the direct cost of printing Inconel 718 parts of 4 cm^3 volume by powder bed fusion. In a batch, 25 specimens are fabricated together. No support structure is needed. Data: The equipment cost, lifetime, idle time, and build time: $1.2 million, 10 years, 25%, and 18 hours, respectively. The cost of metal powder: $150/kg, and the density of powder feedstock: 8.2 gm/cm^3. Shielding gas usage: 10 liters per minute. Cost of shielding gas: $0.01 per liter. Power consumption for 18 hours: 32 kWh. Cost of electricity: $0.10/kWh. Cost for post-processing: $400.

References

1 DebRoy, T., Mukherjee, T., Milewski, J.O., Elmer, J.W., Ribic, B., Blecher, J.J. and Zhang, W., 2019. Scientific, technological and economic issues in metal printing and their solutions. *Nature Materials*, 18(10), pp.1026–1032.

2 Tascón, A., 2018. Influence of particle size distribution skewness on dust explosibility. *Powder Technology*, 338, pp.438–445.

3 Jacobson, M., Cooper, A.R. and Nagy, J., 1964. *Explosibility of Metal Powders*, Vol. 6516. US Department of the Interior, Bureau of Mines.

4 Lunetto, V., Catalano, A.R., Priarone, P.C., and Settineri, L., 2018. Comments about the human health risks related to additive manufacturing. In *International Conference on Sustainable Design and Manufacturing*, Springer, Cham, pp. 95–104.

5 Liu, Z.Y., Li, C., Fang, X.Y. and Guo, Y.B., 2018. Energy consumption in additive manufacturing of metal parts. *Procedia Manufacturing*, 26, pp.834–845.

6 Huckstepp, A., 2019. Energy consumption in metal additive manufacturing. *Digital Alloys' Guide to Metal Additive Manufacturing - Part 7*. Available at https://www.digitalalloys.com/blog/energy-consumption-metal-additive-manufacturing. (accessed on: 21 February 2023)

7 Javaid, M., Haleem, A., Singh, R.P., Suman, R. and Rab, S., 2021. Role of additive manufacturing applications towards environmental sustainability. *Advanced Industrial and Engineering Polymer Research*, 4(4), pp.312–322.

14

Current Status, Trends, and Prospects

Learning objectives

After reading this chapter the reader should be able to do the following:

1) Appreciate the recent growth of additive manufacturing technology.
2) Understand the diversity of applications of additive manufacturing and the ongoing developments in several critical applications.
3) Recognize the current trends in additive manufacturing.
4) Critically assess additive manufacturing's prospects and appreciate why the outlook for the future is very positive.

CONTENTS

Theory and Practice of Additive Manufacturing, First Edition. Tuhin Mukherjee and Tarasankar DebRoy.
© 2024 John Wiley & Sons, Inc. Published 2024 by John Wiley & Sons, Inc.

14.1 Introduction

Because of the unique capabilities of additive manufacturing, it is now used in aerospace, medical, automotive, consumer products, and other industries as evidenced from Figure 1.2 in Chapter 1. The technology is being widely used in making complex parts that cannot be easily made using conventional manufacturing. Many large corporations such as GE and Siemens are using additive manufacturing to produce high-performance components with excellent properties, good functionality, and durability. In addition, it is used to produce highly customized products such as custom prosthetics for the healthcare industry and consumer goods such as jewelry. It can also quickly and cost-effectively make prototypes of alternate designs of new products to improve them. It is also a rapidly growing technology as evidenced by the expansion of the global additive manufacturing market. Chapter 13 discusses how additive manufacturing can offer significant environmental benefits in the areas of energy consumption, waste generation, and sustainability of the materials used. Making the best use of the technology requires an assessment of the current status, trends, and prospects of additive manufacturing. This chapter provides an outline of the activities and trends that are useful for a critical assessment of the technology and its potential for future growth.

Like all new technologies, additive manufacturing faces significant scientific, technological, and economic challenges and regulatory hurdles. The level of effort that is being made to address these challenges can be appreciated by the growing volume of peer-reviewed literature on additive manufacturing presented in this chapter. The growth of the number of patents in recent years presented here shows the continuing innovations for the refinement of technology. The data on the growth of peer-reviewed literature and patents also show the interest and vitality of the field and its prospects. Additive manufacturing technology is continuously improving in the areas of the use of different types of materials, equipment refinement, and integration of emerging digital technology that are providing benefits of higher quality products with improved productivity, and cost-competitiveness. These data are also consistent with the recent market growth data presented here and provide a window of opportunities in the near future. Also, several impactful case studies that show the growing applications of additive manufacturing are presented. They range from the developments toward the printing of organs to affordable and sustainable housing that can be rapidly built on demand to the printing of food to several other unique manufacturing examples. Finally, the results presented throughout the book serve as a basis for an outlook based on evidence.

14.2 Current status

14.2.1 Publications and patents

Additive manufacturing is a rapidly advancing field with many countries making significant scientific and technological contributions through publications. Web of Science data (Figure 14.1 (a)) shows that the number of publications per year on additive manufacturing has increased by about 25 times from 2013 to 2022. These numbers indicate the rapidly growing interest in additive manufacturing worldwide. Publications originating from different countries are shown in Figure 14.1 (b). Numerous papers on diverse topics in additive manufacturing of metals, polymers, and ceramics have been published in scientific journals, and the work is expected to have a significant impact on the field.

3D printing has viability in diverse industries, the ability to print unique components, and the potential to overcome deficiencies in current manufacturing processes. It is also clear from the

(a)

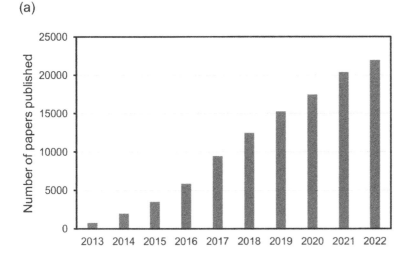

(b)

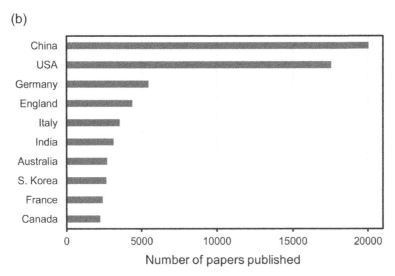

Figure 14.1 (a) Number of papers published on additive manufacturing from 2013 to 2022 (both years inclusive). (b) The number of papers published by the top ten countries in this period. For both figures, data are collected from the Web of Science (all databases) on 02 February 2023 using keywords "additive manufacturing" or "3D printing" as "Topic." *Source:* T. Mukherjee and T. DebRoy.

data on patents published from various countries in the last ten years in Figure 14.2. The total number of global patents for additive manufacturing is comparable to that of other widely used manufacturing processes. The patents are critical to the continuing improvements in additive manufacturing, in protecting innovation and investment in this field, and ensuring that companies that develop new technologies and applications reap the benefits of their efforts.

14.2.2 Growth of additive manufacturing industries

The additive manufacturing industry has grown significantly in recent years, and this trend is expected to continue. For example, the data in Figure 14.3 (a) indicates that the global market value [1] of additively manufactured products has increased more than four times between 2014

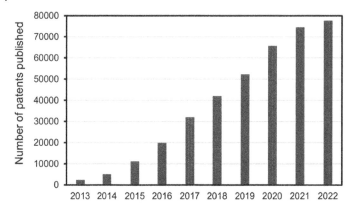

Figure 14.2 Number of patents published on additive manufacturing from 2013 to 2022 (both years inclusive). Data are collected from Google Patents on 02 February 2023 using keywords "additive manufacturing" or "3D printing." *Source:* T. Mukherjee and T. DebRoy.

(a)

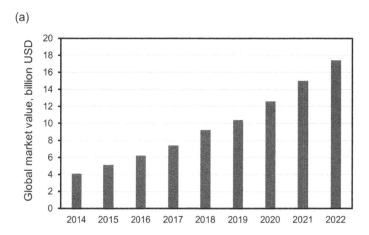

(b)

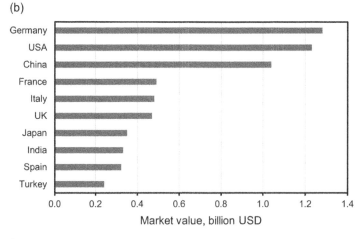

Figure 14.3 (a) Global market value (in billion USD) of additive manufacturing in the last several years. The plot is made based on data reported in [1] where the data are taken from Wohlers Reports. (b) Market value (in billion USD) of additive manufacturing in the top ten countries in 2020. The plot is made based on data reported in [2] where the data are taken from SmarTech Analysis reports. *Source:* T. Mukherjee and T. DebRoy.

and 2022. The industry has already had a significant impact on the global economy, and its continued growth is expected to create new opportunities for businesses and increase the competitiveness of the world manufacturing industry. The increasing availability of 3D printing technology, as well as the development of new applications for the technology, is expected to propel the industry forward in the coming years. Germany has the largest market, with nearly $1.3 billion in annual revenues [2], followed by the United States and China (Figure 14.3 (b)). However, the market value of all 3D printed products in 2019 amounts to only about 0.08% of the 13 trillion US Dollars global manufacturing industry [3]. Facility upgrades, high capital and operational costs of the equipment, expensive feedstocks, safety practices, and training, all of which are more expensive than in traditional manufacturing. These costs frequently limit the adoption of 3D printing by small and medium-sized businesses. The current state of market penetration and growth of additive manufacturing follows the normal course of the advancement of developing technologies. Worked out example 14.1 illustrates the method to forecast the market value.

Worked out example 14.1

Global market value (in billion USD) of additive manufacturing from 2014 to 2021 is shown in Figure 14.3 (a) and is provided in Table E14.1 below. Based on these data forecast the market value for 2022 using linear regression and exponential smoothing algorithms of MS Excel. Estimate the error in forecasting based on the data reported in Figure 14.3 (a).

Table E14.1 Data on the global market value (in billion USD) of additive manufacturing.

Year	2014	2015	2016	2017	2018	2019	2020	2021
Market value (billion USD)	4.1	5.1	6.2	7.4	9.2	10.4	12.6	15.0

Solution:

Forecast using linear regression and exponential smoothing algorithms can be done using the functions FORECAST. LINEAR and FORECAST.ETS, respectively in MS Excel. For the excel sheet shown in Figure E14.1 below, the commands can be written as:

   Linear regression: FORECAST.LINEAR(A10,$B$2:$B$9,$A$2: $A$9)

   Exponential smoothing: FORECAST.ETS(A10,$B$2:$B$9, $A$2:$A$9)

   The estimated market values for 2022 using linear regression and exponential smoothing algorithms are 15.62 and 17.26 billion USD, respectively. The actual value for 2022 as reported in Figure 14.3 (a) is 17.4 billion USD. Therefore, the error in estimation for linear regression and exponential smoothing is 1.78 and 0.14 billion USD, respectively assuming the data in Figure 14.3 (a) to be accurate. In this case, exponential smoothing algorithm is a better forecasting tool because of the non-linearity in market growth.

	A	B
1	Year	Billion USD
2	2014	4.1
3	2015	5.1
4	2016	6.2
5	2017	7.4
6	2018	9.2
7	2019	10.4
8	2020	12.6
9	2021	15
10	2022	15.62

Figure E14.1 Forecast calculations in MS Excel using linear regression.

14.3 Case studies and ongoing efforts

14.3.1 Printing of organs

The need for printing artificial organs of different functionality is driven by donor scarcity and organ shortages. The organ transplant waiting list in the US had over 100,000 people in 2022 according to the US Health Resources & Services Administration and this number far exceeds the 14,000 available donors at the end of 2022 [4]. On average, 17 people died every day in the US while waiting for an organ transplant in 2022 [4]. In addition, organ transplants require immunosuppression medications to avoid rejection by the recipient's body. If 3D-printed organs grown from the recipient's cells become a viable option for transplant, it would satisfy a critical need and potentially avoid the risk of organ rejection. However, the process of printing human organs layer upon layer assembling multiple cell types, growth factors, and functionality is highly complex [5]. Although work has begun and steady progress is being made toward this goal, significant hurdles need to be overcome before the printing of functional organs becomes a viable option.

It is noteworthy that 3D-printed prosthetics have been available for several years and even a 3D-printed eye has been fitted to a patient in Moorfields Eye Hospital, UK in 2021 [6]. The patient has been using prosthetic peepers for over two decades. The 3D-printed prosthetic has a clear definition and real depth to the pupil and mimics a real eye in appearance. Although the prosthetic eye was not meant to restore his vision, it helped him because the 3D-printed eye had the real eye's appearance [6]. In contrast with the 3D-printed eye, a 3D-printed cornea did restore the vision of a patient in 2022 [7, 8]. The cornea was 50 micrometers thin, comparable in thickness to human hair, and made of sterile biocompatible acrylic material to avoid rejection by the host's body.

The 3D-printing technology for printing most human organs is in its early stages and faces many significant technical and scientific challenges. The replacement of the complex functions of human organs remains a major issue. Researchers are examining the use of printed tissue patches for possible repair of damaged organs [9]. Some of the basic tasks for 3D printing of human organs would include printing a scaffold that mimics the structure and function of the target organ and can sustain the survival of the living cells. The scaffold needs to be biocompatible and would be able to support the growth of cells and tissues.

A sufficient volume of living cells must be sourced, preferably from the patient to minimize the chance of the body rejecting the implant or from a donor to populate the 3D-printed scaffold and form functional tissue. A small piece of tissue is taken from the body of a patient and cells are grown in a pressurized sterile stainless steel reactor where the nutrients necessary for cell growth are supplied periodically. The temperature and oxygen concentration in the reactor are also adjusted to promote growth. The bio-ink from which organs are to be printed may consist of a mixture of cells, water-rich molecules called hydrogels, and other chemicals that help the cells continue to grow and differentiate. The cells may be added either during or after the scaffold is printed [10]. For the organ or a part of an organ to survive and function, it must receive adequate oxygen and nutrient supply, which will require the development of a functional vascular network within the printed tissue. The steps necessary for printing organs are an active area of research. Researchers are making progress in printing simple tissues and structures. While we all hope for breakthroughs, the serious challenge of creating functional organs with all of their intricate blood vessels and tissues, and the ethical, and regulatory considerations do not permit any rigorous prediction of a clear timeline for when this will become a reality.

14.3.2 Affordable and sustainable housing on-demand

The 3D printing of a house involves using a large-scale 3D printer to lay down layers of building materials (Figure 14.4 (a)) such as concrete and plastic to create the house. Houses can be printed with exterior designs (Figure 14.4 (b)). Often, the houses are painted or plastered to improve

Figure 14.4 (a) 3D-printing process of a house showing the deposition of concrete materials from a 3D printer nozzle. 3D-printed house (b) during construction and (c) after finishing [11]. The figure is taken from an open-access article [11] under the terms and conditions of the Creative Commons Attribution (CC BY) license.

durability (Figure 14.4 (c)). This technology has the potential to revolutionize the way houses are built, making the construction process faster, more affordable, and more sustainable.

There are four unmistakable benefits of 3D-printed houses. First, 3D printing can reduce the time required to build a house, making it an ideal solution for emergency housing to cope with natural disasters or for rapid urbanization. This speed is attributed to the automation of the building process, which significantly minimizes the need for manual labor, reducing the time it takes to build a house. Additionally, 3D printing technology can print multiple sections of a house simultaneously, further reducing the time it takes to build a complete structure. For example, a company in China 3D-printed the structure of 10 single-room houses in just 24 hours [12]. Apart from the construction of walls and other parts of the house by additive manufacturing, it would be necessary to install the flooring, roofing, utility lines such as water and power, and other necessary facilities which will require additional time. Second, 3D printing allows for greater flexibility in materials and design, making it possible to create unique and customized structures that can be adapted to a range of different needs. For example, a special type of mortar and concrete can be added selectively to improve the mechanical strength of the structure to protect the house from natural disasters such as hurricanes [13]. Worked out example 14.2 illustrates the calculation of the maximum load that a 3D printed slab can withstand without failure. Third, 3D printing reduces the need for manual labor, which significantly minimizes the labor costs associated with traditional building methods. A 1100 square-meter five-story apartment block in Jianghu, China, was printed for $161,000 (US dollars) [12]. Worked out example 14.3 illustrates an approximate calculation of costs of 3D printed structures. Finally, 3D printing uses less material, energy, and time than traditional building methods, reducing the waste and carbon footprint of construction. While there are still challenges to be overcome, such as ensuring the strength and durability of 3D-printed structures, the potential benefits of this technology make it a promising future for the building industry. Worked out example 14.4 explains how the mechanical strength and durability of 3D printed houses can be improved.

Worked out example 14.2

Construction materials are often susceptible to failure due to stresses in bending. Therefore, flexural strength [14] is an important mechanical property of construction materials. A 3D-printed slab of 4 m length with a rectangular cross-section of 0.5 m × 0.2 m is made of a concrete material with a flexural strength of 6 MPa. What is the maximum load that the slab can withstand without a bending failure? Assume that the load is concentrated at the mid-length and mid-width of the slab on the top surface.

Solution:

For a slab supported at both ends and with a load concentrated at the mid-length and mid-width on the top surface, the flexural strength (σ_f) is calculated as [14]:

(Continued)

Worked out example 14.2 (Continued)

$$\sigma_f = \frac{3PL}{2wd^2} \tag{E14.1}$$

where P is the maximum load, L is the length of the slab, w and d are the width and depth of the rectangular section, respectively. Putting the values in the equation, we get,

$$6 \times 10^6 = (3 \times P \times 4)/(2 \times 0.5 \times 0.2^2)$$

After simplification, $P = (6 \times 10^6 \times 2 \times 0.5 \times 0.2^2) / (3 \times 4) = 20,000$ N = 20 kN.

Therefore, the maximum load (P) that the slab can withstand without a bending failure is 20 kN, sufficient to support 20 people each weighing 100 kgs.

Worked out example 14.3

The four walls of a 4 m × 3 m room are made by 3D printing. The thickness and height of the wall are 100 mm and 4 m, respectively. On an average, the 3D printing nozzle can deposit 20 mm thick layer of concrete in one pass. The printing head can move at a speed of 100 mm/s. How long time should it take to print the four walls. If the cost of the concrete material is $150 per cubic meter, what is the cost of material to print the walls?

Solution:

Number of passes needed to print the walls = height of the wall/layer thickness = 4 m/20 mm = 200 passes.

Time needed to finish one pass = perimeter of the room/speed of the printing head = 2 × (4 m + 3 m)/100 mm/s = 140 s.

Therefore, time needed to print the walls using 200 passes = 200 × 140 = 28,000 s = 7 hours 47 minutes (approximately).

Volume of concrete needed = perimeter of the room × thickness of wall × height of the wall = [2 × (4 m + 3 m)] × 0.1 m × 4 m = 5.6 m³.

Therefore, the total material cost = 5.6 m³ × $150 per m³ = $840. The cost of material is only a small portion of the total cost.

Worked out example 14.4

How the mechanical strength and durability of 3D-printed houses can be improved?

Solution:

Materials Selection: Choose a strong and durable material such as reinforced concrete for the structure of the house. These materials offer better strength and durability compared to traditional concrete.

Reinforcement techniques: Adding reinforcement techniques like adding steel bars or other reinforcements to the structure can improve the mechanical properties of the 3D-printed house.

Printing orientation: Printing in an orientation that maximizes the strength of the material will help improve the mechanical properties of the house.

Layer thickness: The layer thickness of the 3D-printed structure has a significant impact on its strength. Thinner layers can increase the strength of the structure, but also increase printing time and cost.

Post-processing: Sanding, polishing, and painting can improve the surface finish and durability of the 3D-printed house.

14.3.3 Food printing

3D printing is often used to create intricate shapes and designs of foods, such as pureed fruits, chocolate, candies, and even meat. For example, Figure 14.5 shows 3D-printed candies of different sizes made by selective laser sintering of sugar powders. 3D printing can be used to create meals for people with dietary restrictions. For example, people with food allergies can have their meals created with ingredients that are safe for them. Furthermore, people with special diets, such as those who are gluten-free or vegan, can have their meals customized to their specific needs. 3D printing can help to reduce waste in the food industry. Traditional food production methods often result in large amounts of food being thrown away, due to overproduction, spoilage, or because the food does not meet the quality standards required for sale. However, with 3D printing, the amount of food waste produced is greatly reduced, as the printer only creates the exact amount of food required. In addition, 3D-printed foods are also highly customizable, allowing for the creation of unique and creative meals that would not be possible with traditional cooking methods. For example, 3D printing can be used to create intricate shapes, such as letters, animals, or even logos, out of food. This opens up a whole new world of possibilities for chefs to create visually stunning meals that not only taste great but are also works of art. However, 3D-printed foods are often very expensive and not affordable for mass consumption.

14.3.4 Alloy design and discovery

Additive manufacturing offers many unique capabilities over conventional manufacturing discussed in the Chapter 1. However, there are many variants of additive manufacturing and the process parameter window is very large in comparison with conventional manufacturing processes such as welding and casting. For example, the heat source power, scanning speed, cooling rates, and other variables vary by several orders of magnitude. Since the microstructures and properties of metallic materials are highly sensitive to external impulses such as the cooling rate, printing defect-free components with good microstructure and properties remains a difficult task. Of over 5500 commercial alloys, only a handful of alloys can now be printed and additive manufacturing contributes only a negligible portion of the manufacturing economy.

The main difficulties of using the current commercial alloys in additive manufacturing are attributed to large columnar grain structures, the presence of brittle phases in the microstructure,

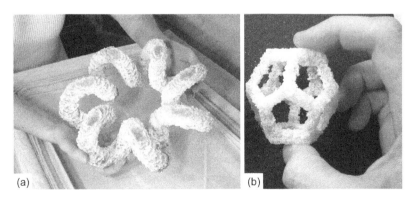

(a)   (b)

Figure 14.5 3D-printed (a) large [15] and (b) small [16] candies using selective laser sintering of sugar powders. Both figures are taken from open-access articles under the terms and conditions of the Creative Commons Attribution (CC BY) license.

and other common defects resulting in poor printability. The processing conditions and alloy composition are the two important factors responsible for the difficulty. For example, the selection of process variables affects solidification growth rate and temperature gradient and they influence the morphology of grains. So, in some systems, a parameter space may exist where equiaxed grains may be obtained. However, this approach requires access to a well-tested model to calculate grain morphology and the ability to undertake a large volume of calculations. Moreover, the computed process variables may be impossible if not difficult to implement in the machine available for the job. A more practical approach is to redesign an existing alloy to obtain a favorable grain morphology and prevent cracking and other difficulties. Alloys with favorable grain structures, phases, and minimum defects need to be designed and tested for printability.

In conventional alloy design, additions of the nature and amounts of solute atoms were made to achieve the desired solidification behavior, grain morphology, and phases to attain the target mechanical properties. However, the microstructures of metallic materials are highly sensitive to variables such as cooling rates and additive manufacturing involves multiple thermal cycles that affect the microstructure and properties of metallic parts. The existing alloys of iron, aluminum, nickel, titanium, and copper were not designed to provide a largely defect-free microstructure with favorable grain structure and phases. One approach of alloy design to prevent the common occurrence of long columnar grains, time tested in casting, is the addition of inoculants and grain refiners to avoid solidification cracking and achieve small equiaxed grains and good mechanical properties of parts. Similarly, some alloying elements may be added to meet other specific goals. Table 14.1 shows several examples of alteration of the composition of commercial alloys to improve microstructures. In addition to new alloys, it is important to mention a few examples of existing established alloys that are also additively manufactured to achieve specific functions. For example, compositions of silicon-containing steels that have been printed to achieve a desirable crystallographic texture for use in electric motors and transformers [17] may even be further refined to optimize microstructure in additively made parts. Similarly, low-cost iron alloys containing nickel and cobalt [18] for applications requiring low thermal expansion coefficients, are other potentially appealing applications for new alloys with desirable microstructures.

Table 14.1 Examples of the redesigned alloys [after reference [19]].

Alloy	Design	Purpose	References
Al 7075	Silicon addition	Reduce crack sensitivity by forming low melting eutectic and grain refinement	[20]
Ti6Al4V	Molybdenum addition	Stabilize beta phase	[21]
Ti	Copper addition	Expand constitutional supercooling zone and promote heterogeneous nucleation	[22]
Ti	Lanthanum addition	Reduce texture and promote equiaxed grains	[23]
Co-Cr-Fe-Ni	Aluminum addition	Thin layers tend to have epitaxy with the grains on which they grow	[24]
ABD-850AM	New age-hardenable nickel-based superalloy	Reduce freezing range and improve cracking resistance	[25]
ABD-900AM	New age-hardenable nickel-based superalloy	The alloy is designed to resist solidification cracking and corrosion	[25]

14.3.5 Printing of solid-state batteries

Solid-state batteries replace the liquid or gel electrolyte with a solid electrolyte which is safer than traditional batteries and can achieve a high energy density. In addition, these batteries are smaller and lighter than traditional batteries, which makes them ideal for mobile devices and electric vehicles. However, manufacturing large solid-state batteries that are often used in electric vehicles needs the time-consuming assembly of many small cells. Small solid-state batteries such as the cell-phone batteries need manufacturing and assembly of many small and intricate parts. To overcome these difficulties, additive manufacturing of the entire battery structure is gaining attention [26].

Figure 14.6 schematically shows the 3D-printing process to make solid-state batteries. The ink of electrolyte material is printed on a substrate. Multiple layers are printed to achieve the desired height of the features. The geometry of the printed parts, as well as how they bond with each other and the substrate, is affected by the properties of the ink. Different designs and patterns can be printed on either side of the substrate. After printing, the printed parts and substrate are placed in a furnace for binder burnout and sintering. In the end, electrodes are placed inside the printed scaffolds to complete the battery assembly.

3D printing enables the fabrication of custom-designed batteries with complex geometries that would be difficult to achieve using traditional manufacturing methods. It can also create intricate shapes and structures, which can be used to optimize the performance of the battery. Furthermore, the 3D printing process is much faster than traditional manufacturing methods, making it possible to produce large numbers of batteries in a relatively short amount of time.

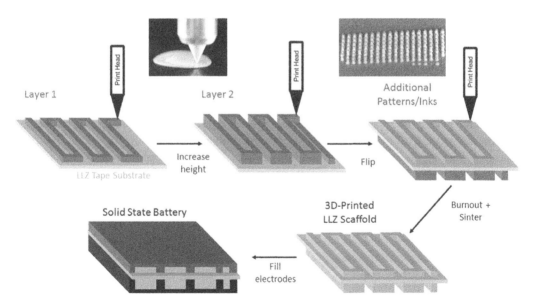

Figure 14.6 Schematic of 3D printing of solid electrolyte batteries. LLZ (garnet-type solid-state lithium conductors $Li_7La_3Zr_2O_{12}$) ink is printed on an LLZ substrate. The height is gradually increased by depositing more layers. Different designs and patterns can be printed on either side of the substrate. After printing, the printed parts and substrate are placed in a furnace for binder burnout and sintering. In the end, electrodes are placed inside the printed scaffolds to complete the battery assembly [27]. The figure is taken from an open-access article [27] under the terms and conditions of the Creative Commons Attribution (CC BY) license.

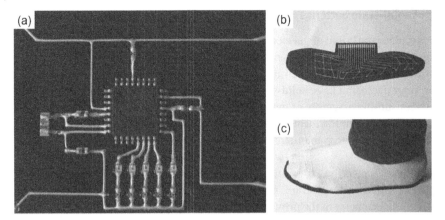

Figure 14.7 (a) 3D-printed microcontroller circuit. (b) 3D-printed wearable soft electronics for socks to monitor the pressure of a human foot. (c) The socks with wearable soft electronics embedded inside [28]. The figure is taken from an open-access article [28] under the terms and conditions of the Creative Commons Attribution (CC BY) license.

14.3.6 Printing of electronic devices

3D printing is also gaining attention to fabricate complex and intricate electronic devices. It can print complex circuits much more rapidly than conventional techniques. Figure 14.7 (a) shows a 3D-printed microcontroller circuit. 3D printing can be used to create sensors integrated into customized wearable electronic devices such as smartwatches and fitness trackers that can monitor vital signs, heart rate, and other health parameters. For example, Figure 14.7 (b) shows 3D-printed wearable flexible electronics for socks. The socks with wearable soft electronics embedded inside can track the pressure of a human foot (Figure 14.7 (c)). 3D-printed customized wearable devices can be designed to fit snugly and comfortably on the body, providing a more natural and seamless experience for the wearer.

One of the biggest advantages of this technology is the ability to quickly produce prototypes of devices without the need for expensive manufacturing equipment. This allows designers and engineers to iterate on designs and test out different ideas without the need for significant capital investment. Another advantage of 3D printing electronics is the ability to produce custom devices for specific applications. For example, an engineer might need a specialized part for a particular device that is not readily available on the market. Using 3D printing, they can design and print that part exactly to their specifications, making it possible to complete the project in a fraction of the time it would take to order and receive the part from a manufacturer. In addition to the benefits of speed and customization, 3D printing of electronic devices is also environmentally friendly. Unlike traditional manufacturing methods, which often involve waste and hazardous chemicals, 3D printing uses relatively little waste material and is much less harmful to the environment.

14.4 Trends

Recent trends in additive manufacturing indicate progress in multiple areas. For example, additive manufacturing allows mass customization of medical implants specific to patients. It realizes unique low-weight designs for load-bearing aerospace and automotive components. It allows the making of compositionally graded materials with site-specific properties. The manufacture of large structural and functional components using wire arc additive manufacturing is gaining wider applications. Improved online monitoring and process control are helping to improve part quality and cost competitiveness. These recent trends are an outcome of the integration of emerging digital technology with manufacturing like never before and the ongoing research and development activities indicated in Section 14.2. Below are several recent trends.

14.4.1 Manufacturing aided by emerging digital technology

The growth of additive manufacturing will not follow the path adapted by the established manufacturing processes such as casting and welding. In the past, the knowledge base of welding and casting was developed from rigorous experimental work that involved decades of painstaking work. The expenses for undertaking such comprehensive experimental work in additive manufacturing using expensive feedstock and machines will not be affordable to most small and medium businesses. However, additive manufacturing has its origin deeply rooted in digital technology. From the Computer-Aided Design (CAD) of products to the stereolithography (STL) file, to the process control using an integrated sensing system that can gather temperature fields and other data during the printing of each layer to reduce defects, represents an unmistakable radical change in manufacturing technology. Learning from the empirical testing of the past is replaced by models based on scientific principles often aided by machine learning where every new data is a foundation for a better product.

Offline verifiable mechanistic models bring out the interdependence of process microstructure property performance relationships without any time-consuming and expensive empirical testing. Knowledge of such foundational interdependence can aid manufacturing in an unprecedented manner. For example, many important parameters which affect product quality such as the temperature and velocity fields, cooling rates, solidification morphology, and the scale of microstructure can be computed using mechanistic models before building parts and testing them. The small size of the fusion zone and the spatially variable, transient temperatures make measurement of temperature and other important parameters difficult if not impractical. Apart from a deep scientific understanding of the process, the mechanistic models enable the selection of parameters for the printing of sound parts guided by scientific principles avoiding time-consuming and expensive trial-and-error testing. Mechanistic models and machine learning are increasingly being used to choose process parameters, which will enhance component quality, lower costs, and cut down on the amount of trial-and-error testing for qualifying parts.

14.4.2 The unique combination of desirable properties

Although the mechanical properties of parts produced by AM are in many cases comparable with their conventionally processed counterparts, properties can vary with process parameters and locally within a part. However, for some alloys, combinations of superior properties of parts, not attainable by conventional manufacturing, have been reported [29]. For instance, contrary to what is typically expected in conventional manufacturing, the strength and ductility of stainless steel parts were concurrently improved, thus defying the strength-ductility trade-off. Hierarchal microstructure, dislocation networks that retard but do not prevent dislocation movements, nano-cellular structure in fine grains, and extremely fine solidification cells have all been attributed to this unusual behavior. The presence of twinning in stainless steel 316 has been suggested as a contributing factor to its good ductility. Strength improvements without a loss of ductility have also been reported in titanium alloys, SS 316L, 12CrNi2, and Al-12Si alloys.

Rapid cooling of Ti-22Al-25Nb during PBF-L, for example, increases dislocation density due to stress accumulation and forms a nanoscale hexagonal omega precipitate. High strength without degradation of ductility may be attributed to both high dislocation density and tiny precipitates that hinder dislocation movement. During the solidification of Ti-6Al-4V, columnar grains of BCC beta phase form, and subsequently, the hexagonal-close-packed (HCP) alpha phase grows inside the beta phase. The strength is attributed to the small sizes of both alpha and beta phases during DED. In addition, the toughness of the part is enhanced by the globular shape of alpha during heat treatment. During powder bed fusion, rapid cooling of Ti-6Al-4V results in HCP martensite which increases strength. Similarly, in a Ti-Al-V-Fe alloy, Ti-185, fine-grain microstructure with nanoscale precipitates enhances the mechanical properties of parts. Although improved properties are encouraging, they are not realized under all processing conditions. This is consistent with the large process parameter window where the cooling rate, temperature gradient, and solidification growth rate vary considerably for different AM processes and variables. As a result, the parts produced may have a wide variety of microstructures and properties, and controlling the mechanical properties of parts will require a greater understanding of both the evolution of microstructure and properties under complex thermal cycles. Since attaining repeatable production of parts to achieve the desired combination of properties is an important milestone, understanding the underlying principles of the improved combination of properties will improve cost competitiveness and enrich metals science and engineering.

14.4.3   4D printing

An exciting extension of 3D printing static objects is 4D printing which can print parts that change attributes in response to external impulses or with time. So, a 4D-printed structure can change its form or function with time or environmental stimulus. For example, a composite polymeric structure composed of a rigid part and a hydrophilic part, when immersed in water, changes shape [30]. This is because the hydrophilic portion swells considerably when in contact with water.

There are many similarities and differences between 3D and 4D printing. A 3D-printed structure is a static structure and can be made of single or multiple materials. In 4D-printed materials, the differences in material properties are the sources of shape changes and so 4D printing often uses multiple materials. It is necessary to consider the stimulus that is to be used to bring about the change in shape or other attributes. Both the material and the design are important for 4D printing. For example, polymers can modify their structure in response to temperature, moisture, or radiation of a particular wavelength range. The modification of the shape of the objects may take place by various mechanisms. Shape-memory polymers can retrieve their original shape from a deformed shape when heated. Some polymers display several configurations depending on temperature and time. 3D printing can use multiple polymers having different glass transition or crystal-melt transition temperatures so that appropriate shape memory effect applications may be created. Similarly, a combination of materials may be initially stressed and the stress may be released resulting in a change in shape with time.

14.5   Outlook

The recent increase in commercial AM equipment sales, the number of international patents awarded, and market revenue all indicate that AM will continue to grow in the future. Additive manufacturing has many advantages over traditional manufacturing and many large corporations will undoubtedly help the expansion of AM in niche applications. Additive manufacturing has advanced significantly in recent years, and its applications have become more diverse. The variety of important products ranges from customized prosthetics to jewelry to rapid-build sustainable housing, food printing, and many other important applications. Applications in a variety of industries for critical applications, processing of a wide range of materials, customization of products in a cost-competitive manner, and integration of emerging digital technologies make the outlook for additive manufacturing very positive. However, the economic impact of additive manufacturing is small in comparison with the impact of the global manufacturing industry. Significant market penetration of additive manufacturing will depend on overcoming many of the scientific, technological, and economic problems faced by additive manufacturing discussed in Chapter 1.

14.6   Summary

Additive manufacturing has experienced rapid growth in recent years and is now being used in aerospace, healthcare, automotive, consumer goods, and other industries. There are considerable ongoing activities in additive manufacturing to solve scientific, technological, and economic issues as evidenced by the growing number of papers and patents. Significant ongoing research in the fields of printing biological materials, affordable housing, foods, solid-state batteries, new alloys, and multiple materials all point to expanding the diversity of printed products and innovative applications for 3D printing in a variety of industries. Integration of emerging digital technology, achieving unique combinations of desirable properties, and 4D printing of parts are also examples of the diversity of activities and trends in the field. All the ongoing work points to achieving higher productivity, better product quality, and improved cost competitiveness. Applications in diverse industries, processing of a variety of materials, customization of products in a cost-competitive manner, and integration of emerging digital technologies make the future of additive manufacturing promising.

Takeaways

Current status

- The number of publications per year on additive manufacturing has increased by about 25 times from 2013 to 2022. These numbers indicate the rapidly growing interest in additive manufacturing worldwide.
- The global market value of additively manufactured products has increased more than four times between 2014 and 2022.
- The increasing number of patents is critical to the continuing improvements in additive manufacturing, in protecting innovation and investment in this field, and ensuring that companies that develop new technologies and applications.

Ongoing efforts

- 3D-printed organs are in the process of development but the process of printing artificial organs assembling multiple cell types, growth factors, and functionality of living organs is highly complex, and significant work lies ahead for them to be viable.
- Houses are printed by depositing layers of building materials such as concrete and plastic to create the house. This technology can revolutionize the way houses are built, making the construction process faster, more affordable, and more sustainable.

(Continued)

Takeaways (Continued)

- 3D printing is often used to create intricate shapes and designs of foods, such as pureed fruits, chocolate, candies, and meat. However, 3D-printed foods are often expensive and not affordable for mass consumption.
- New alloys are being designed to produce the desired solidification behavior, grain morphology, and phases to achieve the target mechanical properties of printed parts.
- 3D printing of solid-state batteries, flexible and customized wearable electronics, and electronic circuits are gaining attention.

Trends

- Manufacturing aided by emerging digital technology, achieving unique combination of desirable properties, and 4D printing are examples of important recent trends in additive manufacturing.

Outlook

- Applications in a variety of industries for critical applications, processing of a wide range of materials, customization of products in a cost-competitive manner, and integration of emerging digital technologies make the outlook for additive manufacturing very positive.
- Significant market penetration of additive manufacturing will depend on overcoming many of the scientific, technological, and economic problems faced by additive manufacturing.

Appendix – Meanings of a selection of technical terms

Arterial tree: The branching system of arteries that terminates in short, narrow, muscular vessels.

Cartilage: Firm, whitish, and flexible connective tissue found in the respiratory tract, external ear, and the articulating surfaces of joints.

Cornea: The transparent part of the eye that covers the iris and the pupil and allows light to enter the inside. Its main job is to help the eyes to focus.

Dislocation: In materials science, a dislocation is a crystallographic defect or irregularity within a crystal structure that contains an abrupt change in the arrangement of atoms.

Grain structures: The microstructure of metallic parts is made up of individual crystalline areas known as grains. The structure of these grains is affected by the alloy composition and the manufacturing process.

Microcontroller: A microcontroller is a small computing device on a single integrated circuit chip. A microcontroller contains one or more CPUs along with memory and programmable input/output peripherals.

Mortar: Mortar is a paste that is used to bind stones, bricks, and concrete masonry units, to fill and seal the irregular gaps between them. It is a mixture of cement, lime, and sand.

Prosthetics: A prosthesis is an artificial device that replaces a missing body part, which may be lost through trauma, disease, or a condition present at birth.

Renal pelvis: The renal pelvis is a hollow part in the middle of each kidney.

Solidification morphology: Different types of grain structures form during the solidification process such as planar, cellular, and dendritic.

Practice problems

1) Discuss briefly several important indicators for the growth of the additive manufacturing industry.
2) Discuss the trends in the number of scientific publications and patents related to additive manufacturing.
3) Discuss the ongoing efforts in additive manufacturing that are likely to have major impacts on the lives of people.
4) What is the current market size of the 3D printing industry and what is the main lesson from the recent trends for the last several years?
5) What role do government policies play in the industrial growth of additive manufacturing?
6) How is the additive manufacturing industry working to reduce its environmental impact?
7) What are the current trends and future outlook for the use of 3D printing in the production of medical devices, and aerospace components?
8) Discuss the important factors that affect the construction time and cost of 3D-printed houses.
9) What are some of the challenges associated with 3D printing foods, and how to overcome them?
10) Based on a literature search discuss how did 3D printing technology help in the production of essential medical supplies during the COVID-19 pandemic?
11) How is the increasing use of artificial intelligence and machine learning affecting the 3D printing industry?
12) Discuss a few indicators that point to a positive outlook for the additive manufacturing industry.

References

1 HUBS, 2021. A Protolabs Company. *Additive Manufacturing Trend Report.* Available at https://f. hubspotusercontent10.net/hubfs/4075618/Additive%20manufacturing%20trend%20report%20 2021.pdf?utm_campaign=Gated%20Content%20Downloads&utm_medium=email&_ hsmi=82605589&_hsenc=p2ANqtz-9pM5X9Hu5jL_49jLKLdBARaGb5PVPhM3kYwzk2c36z2bmd pMZRLJiSd4n5ZKSZOJoCbPnSwFXfi64pLJvjvFjceDPf3w&utm_content=82605589&utm_ source=hs_automation (accessed on 06 February 2023).

2 Sher, D., 15 January 2020. *This is how much additive manufacturing is worth in the top 20 global AM markets.* Available at https://www.3dprintingmedia.network/the-top-20-global-am-markets (accessed on 06 February 2023).

3 Mukherjee, T. and DebRoy, T., 2019. A digital twin for rapid qualification of 3D printed metallic components. *Applied Materials Today*, 14, pp.59–65.

4 Rogers, K., 15 July 2022. *When we'll be able to 3D-print organs and who will be able to afford them.* Available at https://www.cnn.com/2022/06/10/health/3d-printed-organs-bioprinting-life-itself-wellness-scn/index.html (accessed on 10 February 2023).

5 Yi, H.G., Lee, H. and Cho, D.W., 2017. 3D printing of organs-on-chips. *Bioengineering*, 4(1), article no.10.

6 David, S., 27 November 2021. *UK patient receives world's first 3D printed eye prosthetic.* Available at https://www.3dprintingmedia.network/uk-patient-receives-worlds-first-3d-printed-eye (accessed on 10 February 2023).

7 Hutton, D., 18 January 2021. *Artificial cornea restores patient's vision.* Available at https://www. ophthalmologytimes.com/view/artificial-cornea-restores-patient-s-vision (accessed on 10 February 2023).

8 Hanaphy, P., 23 March 2022. Available at https://3dprintingindustry.com/news/israeli-surgeons-carry-out-worlds-thinnest-3d-printed-cornea-transplant-206587 (accessed on 10 February 2023).

9 Ledford, H., 2015. Printed body parts come alive. *Nature*, 520(7547), pp.273-273.

10 Collins, F., 03 November 2015. *Building a better scaffold for 3D bioprinting*. Available at https://directorsblog.nih.gov/2015/11/03/building-a-better-scaffold-for-3d-bioprinting (accessed on 12 February 2023).

11 Xu, W., Huang, S., Han, D., Zhang, Z., Gao, Y., Feng, P. and Zhang, D., 2022. Toward automated construction: the design-to-printing workflow for a robotic in-situ 3D printed house. *Case Studies in Construction Materials*, 17, article no.e01442.

12 Massie, C., 28 January 2015. *China's WinSun Unveils two new 3D printed buildings*. Available at: https://www.architectmagazine.com/technology/chinas-winsun-unveils-two-new-3d-printed-buildings_o (accessed on 10 February 2023).

13 Tomlinson, C., 21 March 2022. *First complete 3D-printed home can withstand hurricanes*. Available at https://www.houstonchronicle.com/business/columnists/tomlinson/article/3-D-printed-house-offers-resiliency-and-beauty-17007516.php (accessed on 10 February 2023).

14 Callister, W.D. and David, G.R., 2010. *Materials Science and Engineering*, 8th Edition. John Wiley & sons, New York.

15 Ligon, S.C., Liska, R., Stampfl, J., Gurr, M. and Mülhaupt, R., 2017. Polymers for 3D printing and customized additive manufacturing. *Chemical Reviews*, 117(15), pp.10212–10290.

16 Tejada-Ortigoza, V. and Cuan-Urquizo, E., 2022. Towards the development of 3D-printed food: a rheological and mechanical approach. *Foods*, 11(9), article no.1191.

17 Haghdadi, N., Laleh, M., Moyle, M. and Primig, S., 2021. Additive manufacturing of steels: a review of achievements and challenges. *Journal of Materials Science*, 56, pp.64–107.

18 DebRoy, T., Wei, H., Zuback, J., Mukherjee, T., Elmer, J., Milewski, J. et al., 2018. Additive manufacturing of metallic components–process, structure and properties. *Progress in Materials Science*, 92, pp.112–224.

19 Liu, Z., Zhao, D., Wang, P., Yan, M., Yang, C., Chen, Z., Lu, J. and Lu, Z., 2022. Additive manufacturing of metals: microstructure evolution and multistage control. *Journal of Materials Science & Technology*, 100, pp.224–236.

20 Montero-Sistiaga, M.L., Mertens, R., Vrancken, B., Wang, X., Van Hooreweder, B., Kruth, J.P. and Van Humbeeck, J., 2016. Changing the alloy composition of Al7075 for better processability by selective laser melting. *Journal of Materials Processing Technology*, 238, pp.437–445.

21 Vrancken, B., Thijs, L., Kruth, J.P. and Van Humbeeck, J., 2014. Microstructure and mechanical properties of a novel β titanium metallic composite by selective laser melting. *Acta Materialia*, 68, pp.150–158.

22 Zhang, D., Qiu, D., Gibson, M.A., Zheng, Y., Fraser, H.L., StJohn, D.H. and Easton, M.A., 2019. Additive manufacturing of ultrafine-grained high-strength titanium alloys. *Nature*, 576(7785), pp.91–95.

23 Barriobero-Vila, P., Gussone, J., Stark, A., Schell, N., Haubrich, J. and Requena, G., 2018. Peritectic titanium alloys for 3D printing. *Nature Communications*, 9(1), article no.3426.

24 Sun, Z., Tan, X., Wang, C., Descoins, M., Mangelinck, D., Tor, S.B., Jägle, E.A., Zaefferer, S. and Raabe, D., 2021. Reducing hot tearing by grain boundary segregation engineering in additive manufacturing: Example of an AlxCoCrFeNi high-entropy alloy. *Acta Materialia*, 204, article no.116505.

25 Tang, Y.T., Panwisawas, C., Ghoussoub, J.N., Gong, Y., Clark, J.W., Németh, A.A., McCartney, D.G. and Reed, R.C., 2021. Alloys-by-design: Application to new superalloys for additive manufacturing. *Acta Materialia*, 202, pp.417–436.

26 Pang, Y., Cao, Y., Chu, Y., Liu, M., Snyder, K., MacKenzie, D. and Cao, C., 2020. Additive manufacturing of batteries. *Advanced Functional Materials*, 30(1), article no.1906244.

27 Quartarone, E. and Mustarelli, P., 2020. Emerging trends in the design of electrolytes for lithium and post-lithium batteries. *Journal of the Electrochemical Society*, 167(5), article no.050508.

28 Valentine, A.D., Busbee, T.A., Boley, J.W., Raney, J.R., Chortos, A., Kotikian, A., Berrigan, J.D., Durstock, M.F. and Lewis, J.A., 2017. Hybrid 3D printing of soft electronics. *Advanced Materials*, 29(40), article no.1703817.

29 DebRoy, T., Mukherjee, T., Wei, H.L., Elmer, J.W. and Milewski, J.O., 2021. Metallurgy, mechanistic models and machine learning in metal printing. *Nature Reviews Materials*, 6(1), pp.48–68.

30 Tibbits, S., 2014. 4D printing: multi-material shape change. *Architectural Design*, 84(1), pp.116–121.

Index

Theory and Practice of Additive Manufacturing, First Edition. Tuhin Mukherjee and Tarasankar DebRoy.
© 2024 John Wiley & Sons, Inc. Published 2024 by John Wiley & Sons, Inc.

Printed and bound by CPI Group (UK) Ltd, Croydon, CR0 4YY

16/04/2025

14658589-0004